Second Edition of

RODD'S CHEMISTRY OF CARBON COMPOUNDS

ELSEVIER SCIENCE B.V.
Sara Burgerhartstraat 25
P.O. Box 211, 1000 AE Amsterdam, The Netherlands

First edition 2001

Library of Congress Cataloging in Publication Data
A catalog record from the Library of Congress has been applied for.

ISBN: 0 444 50901 1

The paper used in this publication meets the requirements of ANSI/NISO Z39.48-1992 (Permanence of Paper).
Printed in The Netherlands.

Second Edition of

RODD'S CHEMISTRY OF CARBON COMPOUNDS

VOLUME I

ALIPHATIC COMPOUNDS

*

VOLUME II

ALICYCLIC COMPOUNDS

*

VOLUME III

AROMATIC COMPOUNDS

*

VOLUME IV

HETEROCYCLIC COMPOUNDS

*

VOLUME V

TOPICAL VOLUMES
MISCELLANEOUS
GENERAL INDEX

*

Related Titles of Interest

Books

CARRUTHERS: Cycloaddition Reactions in Organic Synthesis
CLARIDGE: High-Resolution NMR Techniques in Organic Chemistry
FINET: Ligand Coupling Reactions with Heteroatomic Compounds
GAWLEY & AUBÉ: Principles of Asymmetric Synthesis
GRIBBLE & GILCHRIST: Progress in Heterocyclic Chemistry
HASSNER & STUMER: Organic Syntheses Based on Name Reactions and Unnamed Reactions
KATRITZKY & POZHARSKII: Handbook of Heterocyclic Chemistry, 2nd Edition
LEVY & TANG: The Chemistry of *C*-Glycosides
LI & GRIBBLE: Palladium in Heterocyclic Chemistry
McKILLOP: Advanced Problems in Organic Reaction Mechanisms
OBRECHT: Solid Supported Combinatorial and Parallel Synthesis of Small-Molecular-Weight Compound Libraries
PELLETIER: Alkaloids: Chemical and Biological Perspectives
SESSLER & WEGHORN: Expanded, Contracted and Isomeric Porphyrins
WONG & WHITESIDES: Enzymes in Synthetic Organic Chemistry

Major Reference Works

BARTON, NAKANISHI, METH-COHN: Comprehensive Natural Products Chemistry
BARTON & OLLIS: Comprehensive Organic Chemistry
KATRITZKY & REES: Comprehensive Heterocyclic Chemistry I CD-Rom
KATRITZKY, REES & SCRIVEN: Comprehensive Heterocyclic Chemistry II
KATRITZKY, METH-COHN & REES: Comprehensive Organic Functional Group Transformations
SAINSBURY: Rodd's Chemistry of Carbon Compounds
TROST & FLEMING: Comprehensive Organic Synthesis

Journals

BIOORGANIC & MEDICINAL CHEMISTRY
BIOORGANIC & MEDICINAL CHEMISTRY LETTERS
CARBOHYDRATE RESEARCH
HETEROCYCLES (distributed by Elsevier)
JOURNAL OF SUPRAMOLECULAR CHEMISTRY
PHYTOCHEMISTRY
TETRAHEDRON
TETRAHEDRON: ASYMMETRY
TETRAHEDRON LETTERS

Full details of all Elsevier Science publications, and a free specimen copy of any Elsevier Science journal, are available on request from your nearest Elsevier Science office.

Second Edition of

RODD'S CHEMISTRY OF CARBON COMPOUNDS

A modern comprehensive treatise

Edited by
MALCOLM SAINSBURY
School of Chemistry, The University of Bath,
Claverton Down, Bath BA2 7AY, England

VOLUME V

TOPICAL VOLUME

ASYMMETRIC CATALYSIS

2001

ELSEVIER
Amsterdam-London-New York-Oxford-Paris-Shannon-Tokyo

Contributors to this Volume

J. EAMES

Dept of Chemistry, Queen Mary & Westfield College, University of London, Mile End Road, London, E1 4NS

D.C. FORBES

Department of Chemistry, University of South Alabama, Mobile, Alabama 36688-0002, USA

A.S. FRANKLIN

Department of Chemistry, University of Exeter, Stocker Road, Exeter, EX4 4QD

C. FROST

Dept of Chemistry, University of Bath, Claverton Down, Bath, BA2 7AY

D.M. HODGSON

Dyson Perrins Laboratory, Department of Chemistry, University of Oxford, South Parks Road, Oxford, OX1 3QY

M. LIU

Maybridge plc, Trevillett, Tintagel, Cornwall, PL35 OAL

B. LYGO

Department of Organic Chemistry, University of Nottingham, Nottingham, NG7 2RD

A. MERRITT

Group Leader, Core Combinatorial Group, Medicinal Sciences, Glaxo-Wellcome Research & Development Ltd, Gunnells Wood Road, Stevenage, Herts SG1 2NY

P.A. STUPPLE

Discovery Chemistry, Pfizer Global Research and Development,
Ramsgate Road, Sandwich, Kent, CT13 9NJ

N.C.O. TOMKINSON

Dept of Chemistry, Cardiff University, P O Box 912, Cardiff, CF10 3TB

M. WATKINSON

Dept of Chemistry, Queen Mary & Westfield College, University of London,
Mile End Road, London, E1 4NS

M.C. WILLIS

Dept of Chemistry, University of Bath, Claverton Down, Bath, BA2 7AY

Preface

In the 2nd edition of Rodd's Chemistry of the Carbon Compounds and its Supplements some topics did not receive special attention since the foot print of the series was laid down at a time when many areas were, at best, in their infancy. It is for this reason that reviews of asymmetric catalysis and similar more specialised subjects are to form the basis for a limited number of Topical Volumes.

Asymmetric synthesis initiated by metal catalysts is, of course, not a new topic and asymmetric reduction and epoxidation of alkenes, in particular, are major subject areas covered widely in the Supplements to the 2nd Edition of Rodd's Chemistry of the Carbon Compounds. However, there has been a huge increase in the use of asymmetrically catalysed reactions in other areas in the last five years or so, and this is why this subject has been specially selected to form the content of a Topical Volume.

The book does not dwell at length upon topics of great importance such as Sharpless epoxidation, or upon the control of stereochemistry through catalytic reduction of double bonds. These subjects have been exhaustively surveyed elsewhere. Instead it concentrates upon those topics that appear to be 'up and coming'. Only time can tell if all will be adopted as widely and successfully as their predecessors, but it now seems clear that chemists have at their disposal an ever increasing and formidable array of methods to effect total topological control of even the most complex target molecule.

This book written by some of the best young chemists, concentrating upon work in their own subject areas, provides a snap shot of the current 'state of play' in this a key test of their scientific ingenuity.

Malcolm Sainsbury

Bath
May 2001

Contents

Contents

List of Common Abbreviations and Symbols Used

$[\alpha]$	specific optical rotation
δ	NMR chemical shift (in ppm)
λ	wavelength
ν	frequency; wave number
(+)	dextrorotatory
(−)	laevorotatory
a	axial
Å	Ångström units
aq	aqueous
Ar	aryl
as, asymm.	asymmetrical
B	base
BOC	^{t}butoxycarbonyl (or carbo-*t*-butoxy)
binap	2,2′-bis(diphenylphosphino)-1,1′-binaphthyl
binol	1,1′-bi-2-naphthol
Bipy	2,2′-bipyridyl
Bn	benzyl
BSA	*N*,*O*-bis(trimethylsilyl)acetamide
BSTFA	*N*,*O*-bis(trimethylsilyl)trifluoroacetamide
BTAF	benzyltrimethylammonium fluoride
Bu	butyl
Bu	*n*-butyl
^{i}Bu	iso-butyl
^{s}Bu	*sec*-butyl
^{t}Bu	tert-butyl
Bz	benzoyl
CIDNP	chemically induced dynamic nuclear polarization
con	conrotatory
conc.	concentrated
Cp	η^5-cyclopentadiene
Cp*	pentamethylcyclopentadiene
CSA	camphorsulfonic acid
D	Debye unit, 1×10^{-18} e.s.u.
D	dextro-rotatory; dextro configuration
DL	optically inactive (externally compensated)
DABCO	1,4-diazabicyclo[2.2.2]octane
DBA	dibenz[*a*,*h*]anthracene
DBN	1,5-diazabicyclo[4.3.0]non-5-ene

DBU	1,8-diazabicyclo[5.4.0]undec-7-ene
DCC	dicyclohexylcarbodiimide
DDQ	2,3-dichloro-5,6-dicyano-1,4-benzoquinone
de	diastereomeric excess
DEAD	diethyl azodicarboxylate
dec. or decomp	with decomposition
DET	diethyl tartrate (+ or −)
DHP	dihydropyran
DIBAH	diisobutylaluminum hydride
Diglyme	diethylene glycol dimethyl ether [or bis(2-methoxyethyl)ether]
DiHPhe	2,5-dihydroxyphenylalanine
Dimsyl Na	sodium methylsulfinylmethide
DIOP	2,3-*O*-isopropylidene-2,3-dihydroxy-1,4-bis(diphenylphosphino)butane
dipt	diisopropyl tartrate (+ or −)
dis	disrotatory
DMA	dimethylacetamide
DMAD	dimethyl acetylenedicarboxylate
DMAP	4-dimethylaminopyridine
DME	1,2-dimethoxyethane (glyme)
DMF	dimethylformamide
DMF-DMA	dimethylformamide dimethyl acetal
DMSO	dimethyl sulfoxide
DMTSF	dimethyl(methylthio)sulfonium fluoroborate
E^+	electrophile
*E*1, *E*2	uni- and bi-molecular elimination mechanisms
ElcB	unimolecular elimination in conjugate base
EDTA	ethylenediaminetetraacetic acid
EDG	electron donating group
Ee	enantiomeric excess
ent	reversal of stereocentres
epi	inversion of normal configuration
ESR	electron spin resonance
Et	ethyl
EWG	electron withdrawing group
GABA	4-aminobutyric acid
GLC	gas liquid chromatography
h	hour
HMPA	hexamethylphosphoramide (or hexamethylphosphorous triamide)
HOMO	highest occupied molecular orbital
HPLC	high performance liquid chromatography, high pressure liquid chromatography
Hz	hertz
IR	infrared

J	coupling constant in NMR spectra
K	dissociation constant
kcal	kilocalories
L	laevo-rotatory; laevo configuration
LAH	lithium aluminum hydride
LDA	lithium diisopropylamide
LLC	liquid liquid chromatography
MCPBA	*m*-chloroperoxybenzoic acid
Me	methyl
MEM	methoxyethoxymethyl
MEM-Cl	ß-methoxyethoxymethyl chloride
MOM	methoxymethyl
m.p.	melting point
MS	mass spectrometry
Ms	mesyl (methanesulphonyl)
MSA	methanesulfonic acid
MsCl	methanesulfonyl chloride
MVK	methyl vinyl ketone
NBS	*N*-bromosuccinimide
NCS	*N*-chlorosuccinimide
NMR	nuclear magnetic resonance
NOE	Nuclear Overhauser Effect
Nu^-	nucleophile
o	*ortho-*
op	optical purity
ORD	optical rotatory dispersion
oxidn	oxidation
p	*para-*; orbital
PCC	pyridinium chlorochromate
PDC	pyridinium dichromate
PG	prostaglandin
Ph	phenyl
PPA	polyphosphoric acid
PPE	polyphosphate ester (or ethyl *m*-phosphate)
ppt	precipitate
PPTS	pyridinium *p*-toluenesulfonate
Pr	propyl
^{i}Pr	isopropyl
Py	pyridine
rac	racemic
Red-Al	sodium bis(methoxyethoxy)aluminium dihydride

redn	reduction
S_N1, S_N2	uni- and bi-molecular nucleophilic substitution mechanisms
S_Ni	internal nucleophilic substitution mechanisms
K-Selectride	potassium tri-*s*-butylborohydride
L-Selectride	lithium tri-*s*-butylborohydride
SEM	β-trimethylsilylethoxymethyl
symm.	symmetrical
TBDMS	^{t}butyldimethylsilyl
TBDMS-Cl	^{t}butyldimethylsilyl chloride
TBDPS	^{t}butyldiphenylsilyl
Tf	trifyl (trifluoromethanesulfonyl)
TFA	trifluoracetic acid
TFAA	trifluoroacetic anhydride
THF	tetrahydrofuran
THP	tetrahydropyran (or tetrahydropyranyl)
TIPBS-Cl	2,4,6-triisopropylbenzenesulfonyl chloride
TIPS-Cl	1,3-dichloro-1,1,3,3-tetraisopropyldisiloxane
TLC	thin-layer chromatography
TMEDA	*N,N,N',N'*-tetramethylethylenediamine[1,2-bis(dimethylamino)ethane]
TMS	trimethylsilyl
TMS-Cl	trimethylsilyl chloride
TMS-CN	trimethylsilyl cyanide
Tol	toluene
Tos/Tosyl	*p*-toluenesulphonyl
TosMIC	tosylmethyl isocyanide
Tr/Trityl	triphenylmethyl
Trp	tryptophan
Ts	tosyl (or *p*-toluenesulfonyl)
UV	ultraviolet
vic	vicinal

Second Edition of Rodd's Chemistry of Carbon Compounds,
Volume V, Topical Volumes
Asymmetric Catalysis, edited by M.Sainsbury

Chapter 1

EPOXIDATION OF ALKENES

M. LIU

Introduction

The single most important method for the asymmetric epoxidation of allylic alcohols is that due to Sharpless and since its introduction many diverse applications have been reported. The success of the Sharpless approach has initiated much research into other techniques of asymmetric oxidation of alkenes in general and this review considers progress in the area in the last six years.

1. Recent applications of Sharpless asymmetric epoxidation

Immobilisation of the Sharpless Ti-tartrate ester based asymmetric alkene epoxidation catalyst has been reported. Linear poly(tartrate esters) (2a)-(f) have been synthesised from L-(+)-tartaric acid and diols (1) or αα'-dichloroxylene and used in the epoxidation of *trans*-hex-2-en-1-ol with $Ti(O^iPr)_4$-*tert*-butyl hydroperoxide (TBHP). Isolated yields of the corresponding epoxide are in the range 42-80% and e.e. generally fall between 41-79% (L. Canali *et al.*, J. Chem. Soc., Chem. Commun., 1997, 123).

CO_2H / H—OH / HO—H / CO_2H + HO—$(CH_2)_n$—OH →(pTSA, 120°C, 3 days) HO, CO_2; HO, $CO_2(CH_2)_n$]$_x$

(1a), n = 2
(1b), n = 6
(1c), n = 8
(1d), n = 12

(2a)-(d)

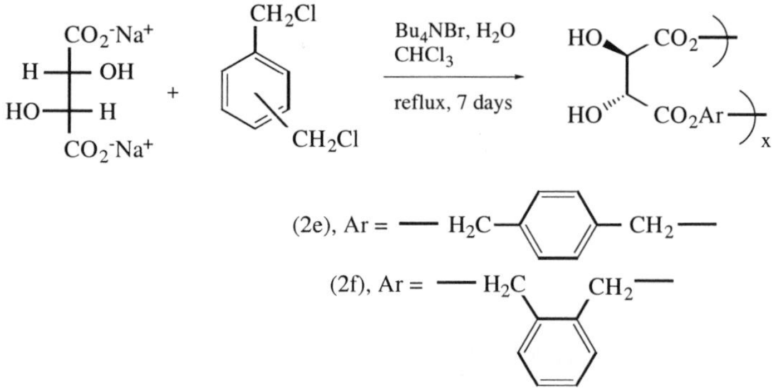

Total synthesis of the anti-tumour styryl lactones (+)-goniotriol (3) and (+)-goniofufurone (4) have been achieved from cinnamyl alcohol *via* the Sharpless protocol (Z-C. Yang and W-S. Zhou, Heterocycles, 1997, **45**, 367).

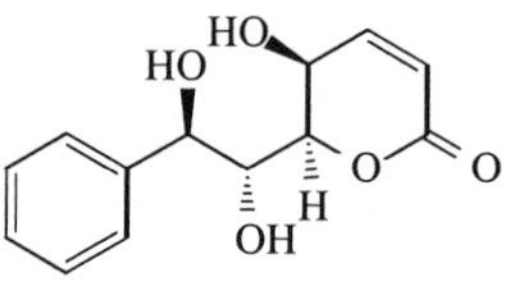

(3)

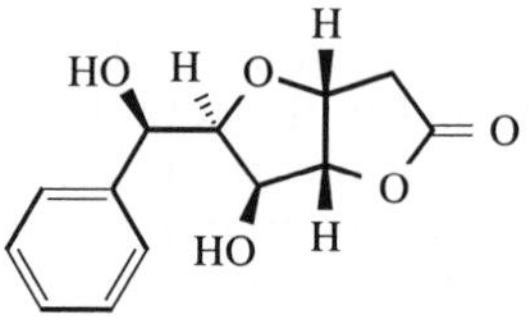

(4)

Application of the Sharpless kinetic resolution to (±)-(*E*)-1-(2-furyl)but-2-en-1-ol affords the (S)-epoxide in 42% yield and high optical purity. This epoxide has been further converted in six steps to the antitumour antibiotic asperlin (5) (T. Honda *et al.*, Heterocycles, 1995, **41**, 425).

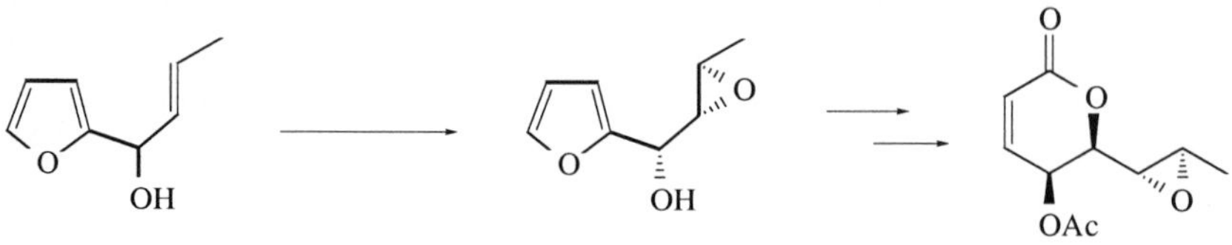

(5)

Similar kinetic resolution and subsequent epoxidation of racemic unsymmetrical divinylmethanols having different substituents at the β-position of both double bonds of the starting material proceeds in a regio- and diastereo-selective manner to give the corresponding epoxy alcohols with high e.e. (T. Honda *et al.*, J. Chem. Soc., Perkin Trans. 1, 1996,

1729). The selectivity is believed to be controlled by the relative reactivity and steric circumstances of both double bonds in the starting materials.

D-(-)-DIPT, $Ti(O^iPr)_4$, TBHP

MS 3Å, CaH_2, CH_2Cl_2, -20°C

racemic

D-(or L)-DIPT, $Ti(O^iPr)_4$, TBHP

MS 3Å, CaH_2, CH_2Cl_2, -20°C

The methodology has been utilised in the enantiocontrolled synthesis of naturally occurring octadecadienoic acid derivatives (6)-(9) isolated from rice plants and used as self-defensive agents against rice blast disease (T. Honda *et al.*, J. Chem. Soc., Perkin Trans. 1, 1999, 23).

(6), R^1 = OH, R^2 = H

(7), R^1 = H, R^2 = OH

(8), R^1 = OH, R^2 = H

(9), R^1 = H, R^2 = OH

2. Jacobsen's catalyst and related reagents

The preparation of a chiral (salen)Mn complex, analogous to the Jacobsen catalyst, inside the supercages of large pore Y zeolite has been accomplished. The performance has been analysed *via* the epoxidation of four prochiral unfunctionalised alkenes and the results compared with those obtained using (salen)Mn(III)Cl under homogeneous conditions. Heterogeneous oxidations carried out at 5°C using NaOCl in CH_2Cl_2 show good enantioselectivity for *trans*-β-methylstyrene and indene, although the e.e.s obtained using the zeolite-bound catalyst are lower than

those reported for the (salen)Mn(III)Cl complex. This has been attributed to two unfavourable factors (i) the occurrence of a non-catalysed, unselective epoxidation route in the liquid phase and/or (ii) the existence of residual amounts of uncomplexed Mn^{2+} acting as catalytic sites. Lower reaction rates are also observed for the heterogeneous (salen)Mn(III)Y catalyst compared to its homogeneous counterpart. This is believed to be due to the restrictions imposed on the diffusion of the substrate and the product through the micropores of the solid, especially when the reaction is run at low temperatures (M. J. Sabater *et al.*, J. Chem. Soc., Chem. Commun., 1997, 1285).
Manganese complexes of C_2-symmetric salen ligands (10) and (11), which are selectively soluble in perfluorocarbons, have been synthesised from 4,6-perfluorodioctyl salicylaldehyde and the appropriate chiral 1,2-diamine and used in the epoxidation of alkenes under fluorous biphasic system (FBS) conditions. Although both Mn complexes are catalytically active and require lower amounts of catalyst compared to homogeneous conditions, enantioselectivity is generally low for styrene and its derivatives, stilbene and 1,2-dihydronaphthalene. Indene, however undergoes epoxidation in 92% e.e. and 90% e.e. with Mn-(10) and Mn-(11) complexes respectively. A similar trend is observed when reactions are carried out in the presence of other oxygen donors commonly used in combination with chiral (salen)Mn complexes (G. Pozzi *et al.*, J. Chem. Soc., Chem. Commun., 1998, 877).

R R H H H_2N NH_2 + C_8F_{17} CHO OH C_8F_{17} → EtOH reflux R R H H N N C_8F_{17} OH HO C C_8F_{17} C_8F_{17}

(10), R-R = $-(CH_2)_4-$
(11), R = Ph

Polymer supported analogues of Jacobsen's catalyst, P(12)-(16) have been synthesised. They are based on the following design criteria (i) the local molecular structure of the Mn complex should mimic precisely the optimum structure of Jacobsen's catalyst; (ii) the complex should be attached by a single flexible linkage to the polymer support to minimise local steric restriction; (iii) the catalyst should be attached to the polymer support with sufficiently low loading to maximise site isolation of catalytic centres, hence minimising possibility of oxo-bridged dimer formation; and (iv) the morphology of the support should be such that no mass transfer limitation arises, with all active sites freely accessible (L. Canali *et al.*, J. Chem. Soc., Chem. Commun., 1998, 2561).

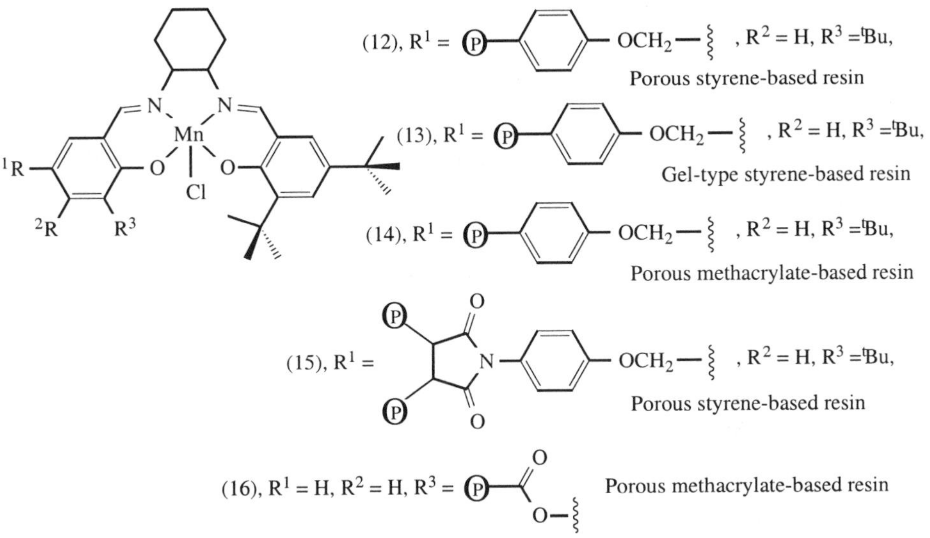

(12), R^1 = Ⓟ–C_6H_4–OCH_2–, R^2 = H, R^3 = ^tBu, Porous styrene-based resin

(13), R^1 = Ⓟ–C_6H_4–OCH_2–, R^2 = H, R^3 = ^tBu, Gel-type styrene-based resin

(14), R^1 = Ⓟ–C_6H_4–OCH_2–, R^2 = H, R^3 = ^tBu, Porous methacrylate-based resin

(15), R^1 = (Ⓟ)₂-succinimido–C_6H_4–OCH_2–, R^2 = H, R^3 = ^tBu, Porous styrene-based resin

(16), R^1 = H, R^2 = H, R^3 = Ⓟ–C(=O)O–, Porous methacrylate-based resin

With polymer catalysts P(15) and P(16), both yield and enantiomeric purity are low in the asymmetric epoxidation of 1-phenylcyclohex-1-ene, carried out in CH_2Cl_2 using MCPBA as oxidant and NMO as the activator. The catalysts P(12) and P(13) performed better with yields of 30% and 47% and e.e.s of 61% and 66% respectively. Catalyst P(14) shows similar activity to P(12) and P(13) but the e.e. is significantly improved (>90%). In this case the catalyst is derived from a methacrylate-based resin rather than a styrene, although the immediate link to the polymer is *via* a styryl residue. High enantioselectivity is believed to be due to this matrix, which although macroscopically rigid, has improved local mobility relative to the styrene-based analogue and this allows higher polarity along with a lower loading of catalyst sites.
The epoxidation of 6-acetamido-7-nitro-2,2-dimethylchromene, has been achieved in 60% e.e. and 11% yield under optimised conditions employing an achiral (salen)Mn(III) complex in the presence of chiral amine (-)-sparteine. The asymmetric induction can be increased to 73% e.e. on the addition of water to the reaction medium (T. Hashihayata *et al.*, Tetrahedron, 1997, **53**, 9541).

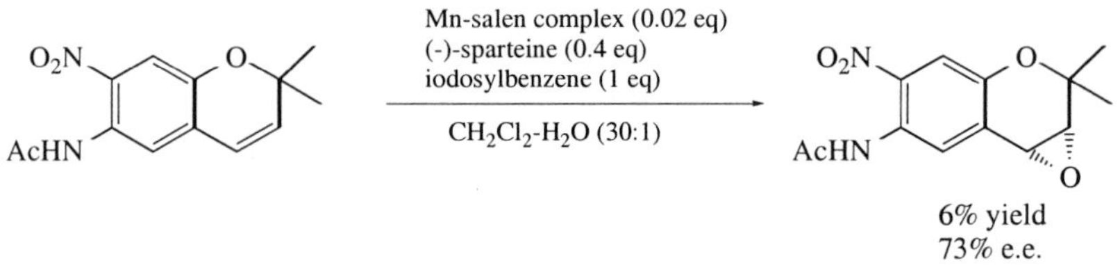

Mn-salen complex (0.02 eq)
(-)-sparteine (0.4 eq)
iodosylbenzene (1 eq)

CH_2Cl_2-H_2O (30:1)

6% yield
73% e.e.

The same substrate has been shown to exhibit unusual solvent dependent stereochemistry when the asymmetric epoxidation is catalysed with a (salen)Cr-complex. In general, reaction in a less polar solvent *e.g.* toluene and CH_2Cl_2 gives the (3S,4S)-epoxide, while reaction in a polar solvent *e.g.* CH_3CN affords the (3R,4R)-epoxide preferentially. A similar trend is observed in the epoxidation of 6-cyano-2,2-dimethylchromene (H. Imanishi and T. Katsuki, Tetrahedron Lett., 1997, **38**, 251).

The mechanism of the asymmetric epoxidation of indene using NaOCl, catalysed by Jacobsen's chiral Mn-salen complex and accelerated by the axial ligand 4-(3-phenylpropyl)pyridine *N*-oxide (P_3NO) has been studied. In a two-phase system of aqueous NaOCl and chlorobenzene, the *N*-oxide transports HOCl, the active oxidant, into the organic phase where HOCl oxidises Mn(III) to the catalytically active Mn^{V}=O in the rate determining step. The *N*-oxide also acts as a stabilising agent for the catalyst as well as a surfactant (D. L. Hughes *et al.*, J. Org. Chem., 1997, **62**, 2222).

A dimeric form of Jacobsen's catalyst has been shown to provide comparable activity and enantioselectivity as Jacobsen's catalyst itself under homogeneous conditions in the epoxidations of 1-phenyl-1-cyclohexene, 1,3-cyclooctadiene and *trans*-β-methylstyrene (74.9%, 28.5%, 27% e.e. *vs.* 69.1%, 24.5%, 28.8% e.e. respectively). Under heterogeneous conditions, accomplished by inclusion of the dimer in a polydimethylsiloxane (PDMS) membrane, enantioselectivity in the epoxidation of *trans*-β-methylstyrene is low (14.4% e.e.), although this excess is comparable to that obtained by the use of the homogeneous monomer (K. B. M. Janssen *et al.*, Tetrahedron: Asymmetry, 1997, **8**, 3481).

Phosphodiesterase (PDE) IV inhibitor CDP840 (21) and its enantiomer have been prepared *via* a Jacobsen epoxidation of the *Z*-triaryl olefin (17) and *E*-triaryl olefin (18). Enantiomers (19) and (20) are obtained in 89% and 48% e.e. respectively. The disparate results in asymmetric induction have been explained in terms of the 'skewed side-on approach' model proposed by Jacobsen (J. E. Lynch *et al.*, J. Org. Chem., 1997, **62**, 9225).

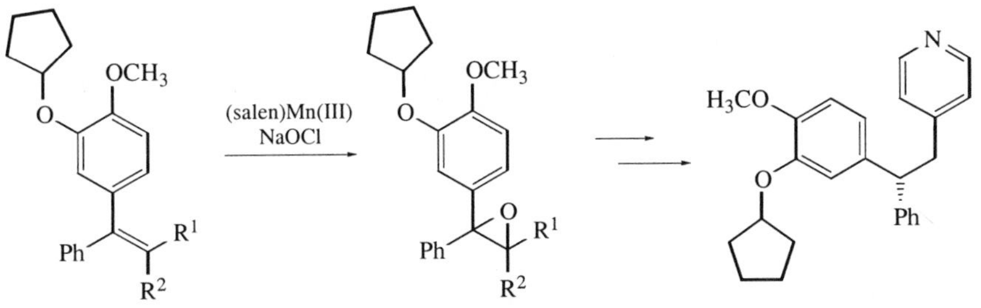

(salen)Mn(III)
NaOCl

(17), R^1 = pyr, R^2 = H
(18), R^1 = H, R^2 = pyr

(19), R^1 = pyr, R^2 = H
(20), R^1 = H, R^2 = pyr

(21)

Enantioenriched 3-chlorostyrene oxide of >99% e.e. has successfully been obtained from the (salen)cobalt catalysed hydrolytic resolution of the product previously obtained in lower e.e. *via* the Mn(salen) catalysed asymmetric epoxidation of 3-chlorostyrene (B. D. Brandes and E. N. Jacobsen, Tetrahedron: Asymmetry, 1997, **8**, 3927).

A polymeric chiral Mn(III)-salen complex, obtained from chiral metal complex monomer co-polymerised with ethyleneglycol dimethylacrylate as the cross linking monomer, has been used in the asymmetric epoxidation of a range of unfunctionalised alkenes. Inferior enantioselectivity is however observed when the heterogeneous polymeric system is compared with the homogeneous system. The difference between the polymeric and low molecular weight chiral (salen)Mn(III) catalyst has been attributed to steric and/or certain microenvironmental effects associated with the macro molecular system (B. B. De *et al.*, Tetrahedron: Asymmetry, 1995, **6**, 2105).

1-Phenylsulfonylcyclohexa-1,3-diene (22) has been shown to undergo epoxidation with high enantioselectivity when catalysed by symchiral (salen)Mn(III)Cl complexes. Epoxidation of (22) and the 7-membered ring analogue (25) affords the corresponding epoxides (23 and 27) in 65% and ≥ 99% e.e. respectively. For (22), diphenyl sulfone is a co-product. The octadiene (26) is also oxidised to the epoxide (28) in high e.e. (90%) although the reaction is sluggish. In all cases the double bond ‘remote’ from the sulfonyl group is selected for attack but problems over the rates of epoxidation are noticed when acyclic dienyl sulfones are the substrates (M. F. Hentemann and P. L. Fuchs, Tetrahedron Lett., 1997, **38**, 5615).

(salen)Mn(III)
NaOCl
CH_2Cl_2

(22) (23) (24)

(25), n = 1
(26), n = 2

(27), n = 1
(28), n = 2

3. Polyamino acid mediated catalysts

Synthetic applications of polymeric α-amino acids have been reviewed (S. Ebrahim and M. Wills, Tetrahedron: Asymmetry, 1997, **8**, 3163; L. Pu, *ibid.*, 1998, **9**, 1457).
The polyamino acid-catalysed asymmetric epoxidation of enones was first reported in 1980. The transformation, now commonly referred to as the Julia-Colanna epoxidation, originally utilised a triphasic system comprising an insoluble polyamino acid (PAA) catalyst such as poly-L-leucine (PLL), an aqueous solution of NaOH and H_2O_2 and an organic solvent.
Polyoxygenated (*E*)-chalcones (29)-(31) have been epoxidised with H_2O_2 in a triphasic system consisting of aqueous NaOH-poly L- or D-alanine-CCl_4 to afford the (-)-*trans*-epoxides (32a), (33a), (34a) and (+)-*trans*-epoxides (32b), (33b), (34b) in high yields (97-99%). The (-)-chalcone oxiranes exhibit higher optical purities (70-84% e.e.) than the (+)-isomers (53-64% e.e.) due to the considerably higher purity of natural L-alanine (versus synthetic D-alanine) used to form the catalyst (H. van Rensburg *et al.*, J. Chem. Soc., Chem. Commun., 1996, 2747).

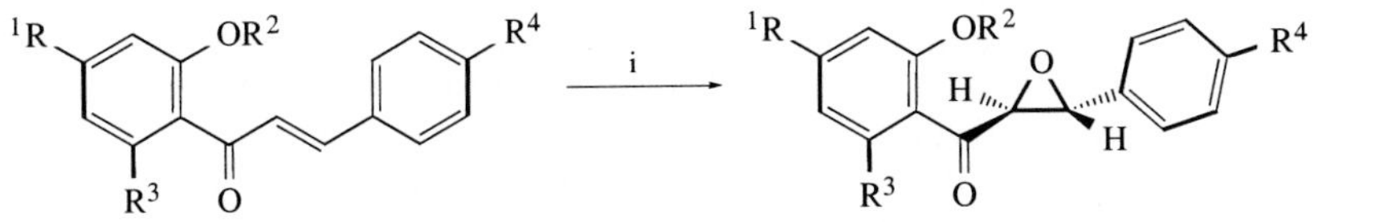

(29), $^1R = R^3 = H$, $R^2 = MOM$, $R^4 = OMe$
(30), $^1R = R^4 = OMe$, $R^2 = MOM$, $R^3 = H$
(31), $^1R = R^3 = R^4 = OMe$, $R^2 = MOM$

(32), $^1R = R^3 = H$, $R^2 = MOM$, $R^4 = OMe$
(33), $^1R = R^4 = OMe$, $R^2 = MOM$, $R^3 = H$
(34), $^1R = R^3 = R^4 = OMe$, $R^2 = MOM$

i, 30% H_2O_2 : 6M NaOH 1:0.32 (v/v), poly-D- or poly-L-alanine : chalcone 1:1 (m/m), CCl_4, RT, 24h.

The synthetic usefulness of chiral polyamino acids in the Julia-Colanna asymmetric epoxidation has been extended to include the oxidation of enones other than chalcones (M. E. Lasterra-Sanchez and S. M. Roberts, J. Chem. Soc., Perkin Trans. 1, 1995, 1467). For example, the enones (35)-(40) all undergo epoxidation in 72->96% e.e. catalysed by poly-(L)- or poly-(D)-leucine in a three phase system in yields of 67-98%.

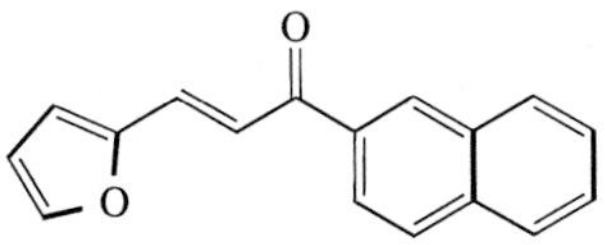

(35) 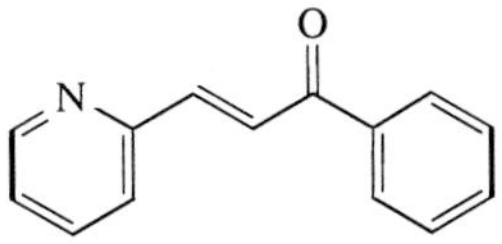 (36)

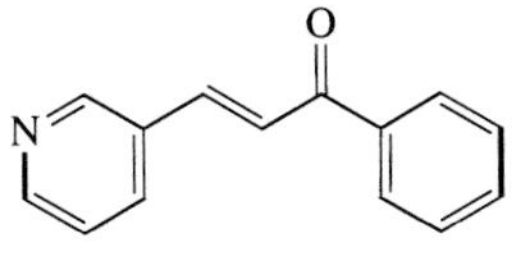

(37) 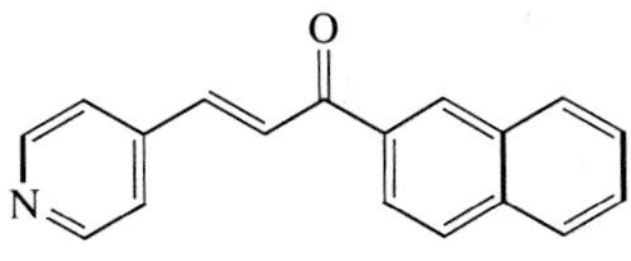 (38)

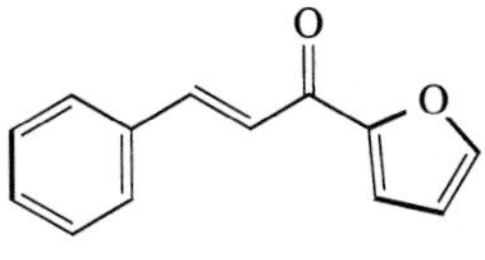

(39) 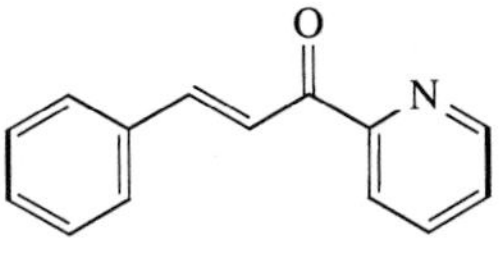 (40)

The scope of the asymmetric epoxidation has been broadened further by replacement of the ring structures attached to C-1 and/or C-3 of the enone unit with a *tert*-butyl group. Enones (41), (42) and (43) are oxidised to their corresponding epoxides with good to excellent stereocontrol (76-

>98% e.e.) and yields in the range 85-92%. Satisfactory enantioselectivities are also obtained with substrates containing more than one double bond. Thus, (bis)epoxides (44) and (45) are obtained in 90% and 80% e.e. respectively from the corresponding diene and tetralene. The PAA catalyst (*e.g.* poly-(L)-leucine, PLL) can be made from the chosen *N*-carboxyanhydride by a heterogeneous method (humidity cabinet) or a homogeneous method (1,3-diaminopropane, DAP). Other oxidants such as sodium perborate, sodium percarbonate and *tert*-butylhydroperoxide are reported to be effective at giving the desired oxiranes in good enantiomeric excess (M. E. Lasterra-Sanchez *et al.*, J. Chem. Soc., Perkin Trans. 1, 1996, 343).

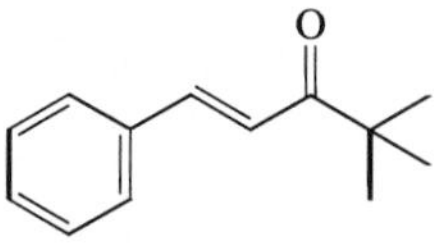

(41)

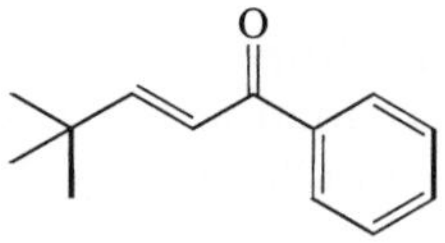

(42)

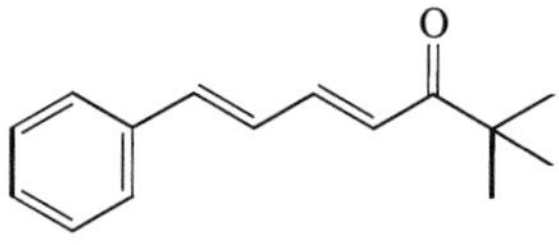

(43)

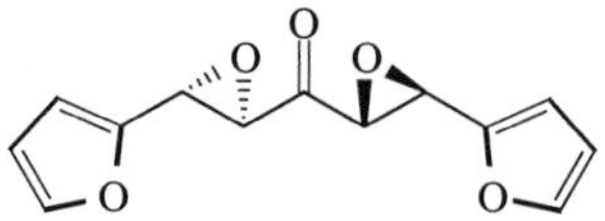

(44)

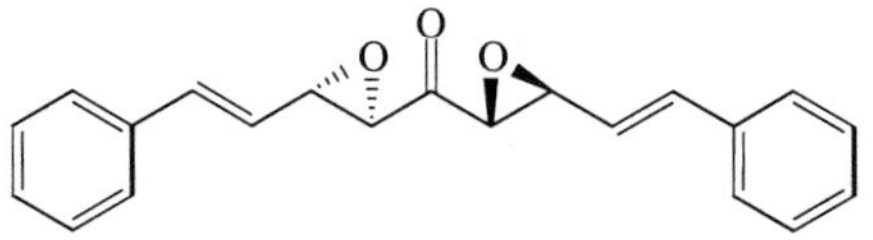

(45)

If one of the methyl groups in the *tert*-butyl unit at C-1 is replaced by hydrogen (46) or a methoxy group (47), prolonged reaction times are required and optical purities are only modest (63% and 62% e.e., respectively). Substrates (48)-(51), with a cyclopropyl group adjacent to the carbonyl group or at the β-position give good to excellent e.e.s (74-≥98%) and yields in the range 52-85%. Enone (52), with an alkyne group next to the carbonyl, is also oxidised to the corresponding epoxide in 90% e.e. and 57% yield (W. Kroutil *et al.*, J. Chem. Soc., Chem. Commun., 1996, 845).

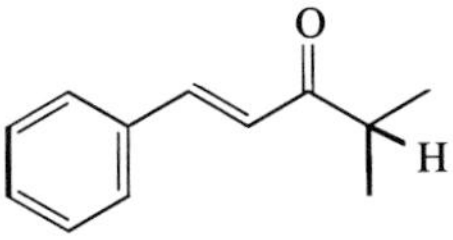

(46) 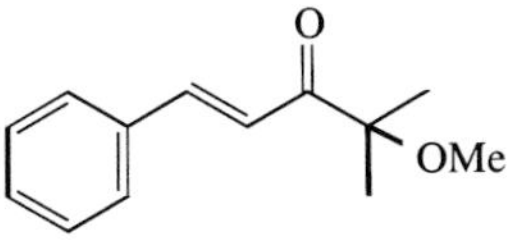 (47)

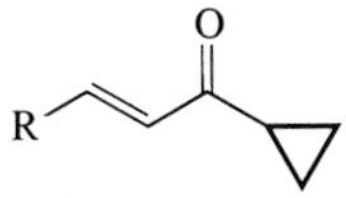

(48), R =Ph
(49), R = 2-naphthyl
(50), R = PhCH=CH

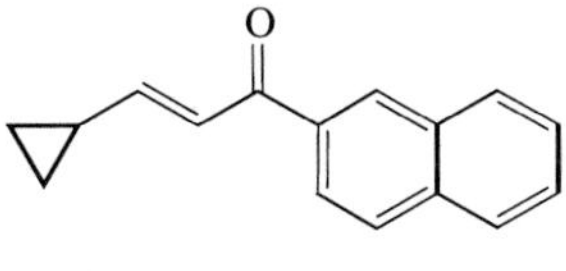

(51) 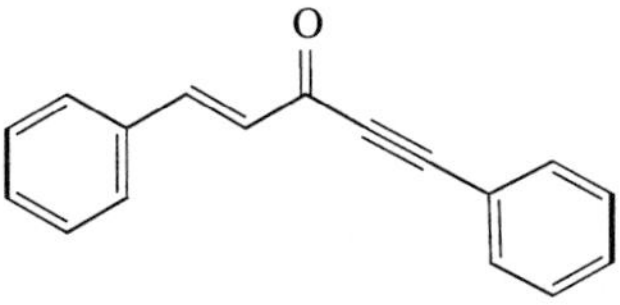 (52)

Enone (53) has been oxidised with moderate stereoselectivity using DAP-PLL and is the first example of an α-substituted, β-unsubstituted enone undergoing asymmetric epoxidation with any degree of efficiency using this methodology. The absolute configuration of the resultant epoxide was not established. Enediketones (54)-(57) and unsaturated ketoester (58) are oxidised efficiently using PLL in yields of 60->95% and 76->95% e.e. (W. Kroutil *et al.*, J. Chem. Soc., Perkin Trans. 1, 1996, 2837).

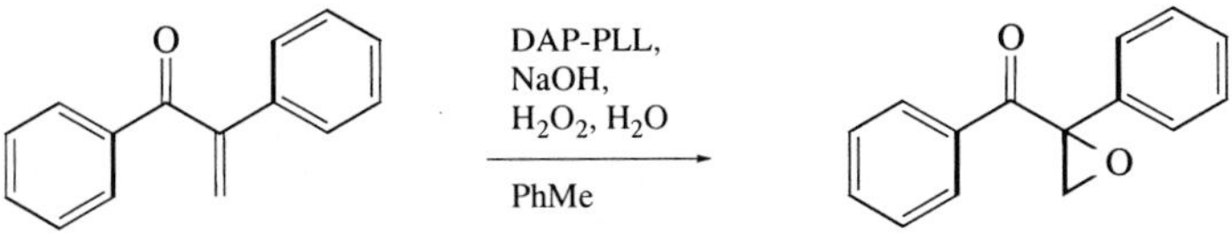

DAP-PLL, NaOH, H_2O_2, H_2O / PhMe

(53)

78% yield
59% e.e.

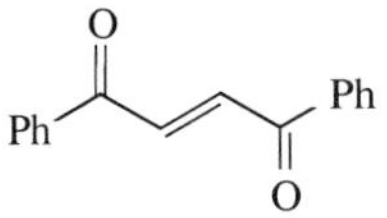

(54) 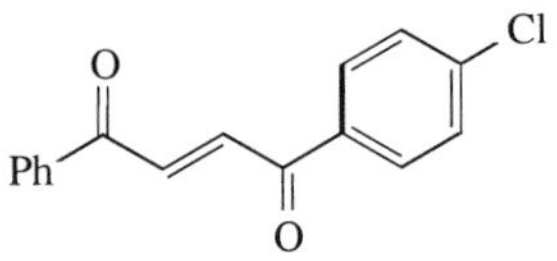 (55) 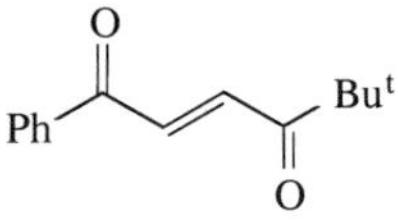 (56)

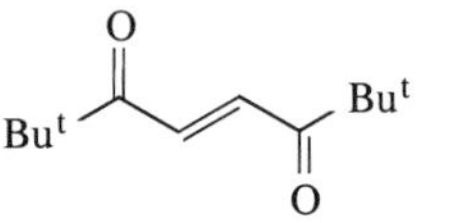

(57) 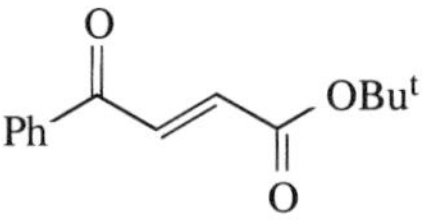 (58)

Reaction times for the epoxidation of certain substituted chalcones in a three-phase system, catalysed by polyleucine, can be reduced when sodium perborate tetrahydrate and ultrasound are utilised. Yields are comparable with similar reactions using hydrogen peroxide with polyalanine or with polymer supported poly-L-leucine, although e.e.s are slightly lower (R. M. Savizky *et al.*, Tetrahedron Asymmetry, 1998, **9**, 3967).

Several limitations to the triphasic protocol including long reaction times, slow degradation of PAA by NaOH, incompatibility to epoxidation of compounds with alkali sensitive functional groups, and required pre-treatment of PAA to form a gel, as well as slow oxidation of compounds with active hydrogen atoms can be overcome by conducting the epoxidation in a two phase non-aqueous system made up of oxidant, a non-nucleophilic base, immobilised poly amino acid and an organic solvent (P. A. Bentley *et al*, J. Chem. Soc., Chem. Commun., 1997, 739). Using this improved procedure for the Julia-Colanna asymmetric epoxidation of α,β-unsaturated ketones, the epoxy ketones (59)-(64) have been obtained in 70-91% yield and 80->95% e.e. using urea hydrogen peroxide (UHP) in THF or *tert*-butyl methyl ether containing 1,8-diazabicyclo[5.4.0]undec-7-ene (DBU) with immobilised poly-L-leucine (I-PLL) as the insoluble catalyst (B. M. Adger *et al.,* J. Chem. Soc., Perkin Trans. 1, 1997, 3501). Epoxy ketones (63) and (64) were used as precursors in the syntheses of diltiazem (65), a blood pressure lowering agent, and the Taxol™ side chain (66) respectively both giving single enantiomer forms.

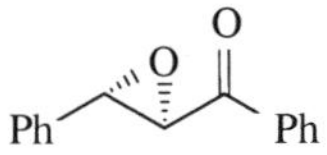

(59) 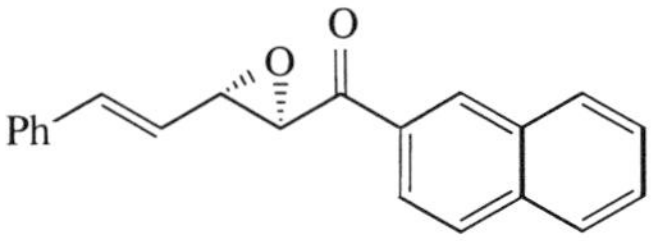 (60)

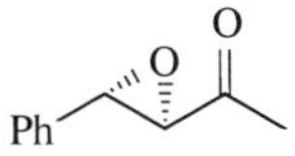

(61) 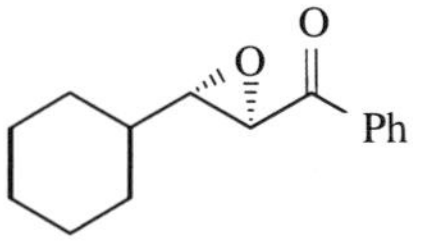 (62)

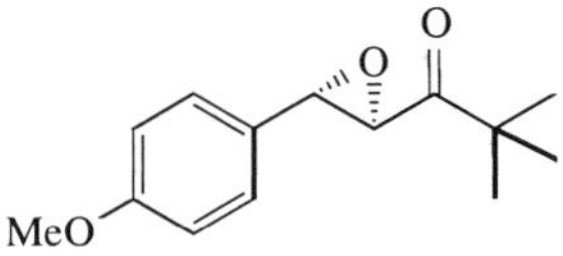

(63) 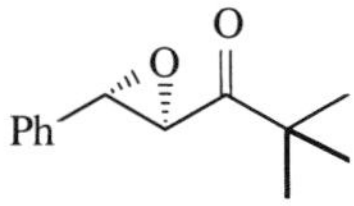 (64)

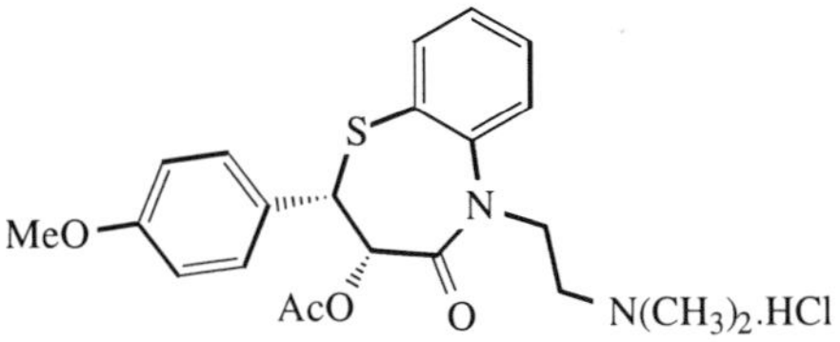

(65) 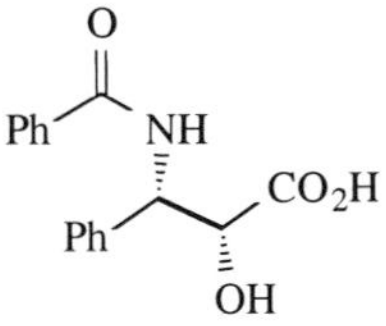 (66)

5-Benzoyl-2,4-pentadienoate esters (67) and (68) are readily oxidised at the 4,5-double bond to give the epoxides (69) and (70) respectively showing the marked difference in the reactivity of different alkene groups in the same substrate (J. V. Allen *et al.,* J. Chem. Soc., Perkin Trans. 1, 1997, 3297). Chlorodiene (71), trienone (72) and alkylated dienone (73) also undergo oxidation regioselectively with UHP under catalysis by PLL to furnish epoxides (74) (57% yield, 86% e.e.), (75) (43% yield, 90% e.e.) and (76) (70% yield, 92% e.e.), respectively.

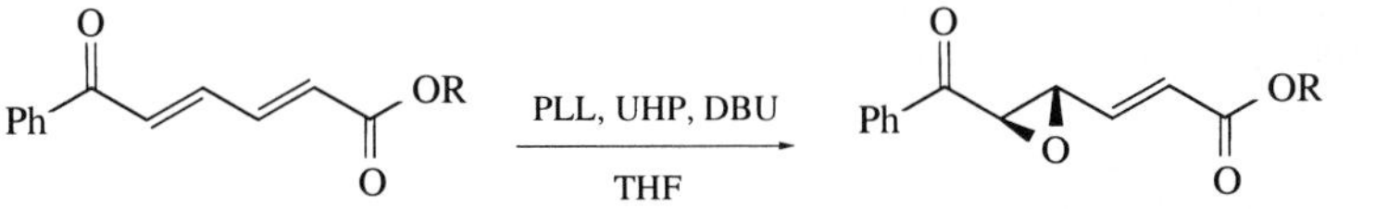

PLL, UHP, DBU / THF

(67), R = Bu^t
(68), R = Me

(69), R = Bu^t, 95% yield, 90% e.e.
(70), R = Me, 90% yield, 90% e.e.

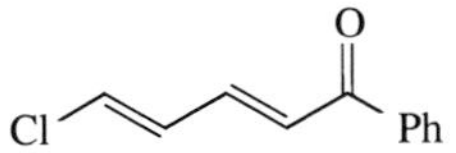

Cl, Ph, O (71) 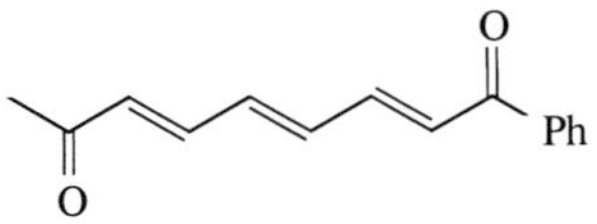 (72) 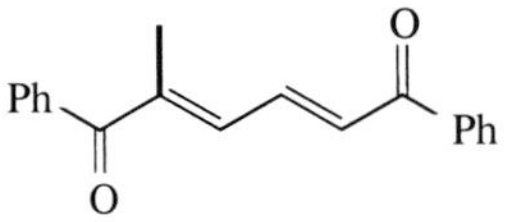 (73) 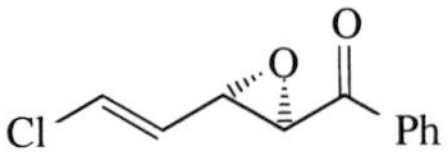 (74) 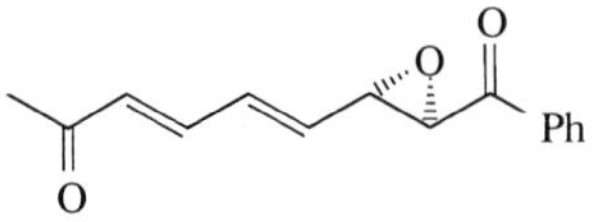 (75) 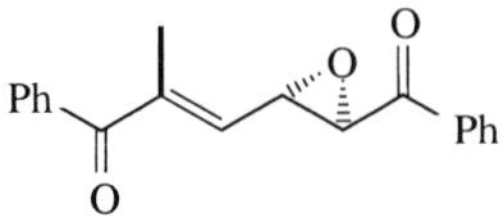 (76)

In the two-phase non-aqueous system employing immobilised PLL as catalyst, chalcone methyl ethers (29) and (30) (*vide supra*) undergo epoxidation to afford the corresponding (-)-*trans*-epoxychalcones in 64-80% yield and improved optical purities of 85-95% e.e. compared to the three-phase protocol (67-86%). The enantiomeric (+)-*trans*-epoxychalcones were similarly obtained (61-76% yield, 81-90% e.e.) using immobilised poly-D-leucine (PDL) in the same two-phase system (R. J. J. Nel *et al.*, Tetrahedron Lett., 1998, **39**, 5623).
The poly-L-leucine catalyst can be recycled under the non-aqueous protocol (J. V. Allen, J. Chem. Soc., Perkin Trans. 1, 1998, 3171). The catalyst has been shown to suffer little damage during the transformation of chalcone into its corresponding epoxide. The yield and e.e. remain high, although reaction times need to be lengthened. The catalyst becomes progressively less efficient with substrates which require longer reaction times such as *tert*-butyl styryl ketone. However, reactivation of the catalyst may be achieved by simply washing with 4M aqueous sodium hydroxide.
More recently, a new procedure for polyleucine catalysed asymmetric enone epoxidation is reported in which sodium percarbonate serves a dual role of oxidant and base (J. V. Allen *et. al.*, Tetrahedron Lett., 1999, **40**, 5417). Epoxidation of selected substrates using a DME/water solvent system affords the corresponding epoxides in high e.e., results, which are comparable with the biphasic conditions. For example, enone (41) is epoxidised in 94% isolated yield and 94% e.e. (biphasic method: 76%

yield, 94% e.e.). The conditions also allow epoxidation to proceed with lower catalyst to substrate ratios and with less effective catalysts, but without a diminished enantioselectivity normally caused by rapid non-polyleucine-catalysed epoxidation.

A modified procedure for the Julia-Colanna stereoselective epoxidation of α,β-enones has been developed, which utilises polyleucine absorbed onto silica as catalyst (T. Geller and S. M. Roberts, J. Chem. Soc., Perkin Trans. 1, 1999, 1397). The polyamino acid silica-based catalyst (PaaSiCat) is easily prepared and has a significantly higher activity than non-adsorbed PLL. For example, in the standard Weitz-Scheffer test reaction, *i.e.* epoxidation of chalcone to its corresponding epoxide, only 23% of the PLL normally employed may be used, without significant influence on the rate of the reaction and enantiomeric excess. The work-up procedure is also easier, due to rapid phase-separation and filtration. Overall this makes catalyst recycling easier and minimises losses. The catalyst is robust and can be heated at 150°C for 12h under vacuum and still retain full catalytic activity. Two substrates, (77) and (78), tested with the new procedure both gave 93% e.e. and conversions of 85% and 78%, respectively.

PLL/silica
UHP, DBU
THF

(77), R = *o*-aminophenol
(78), R = $CH(CH_3)_2$

(79), R = *o*-aminophenol
(80), R = $CH(CH_3)_2$

The PaaSiCat methodology has been employed in the preparation of 2-arylpropanoic acids in high enantiomeric purity. For example, the initial step in the preparation of the optically active non-steroidal anti-inflammatory agent (+)-(S)-fenoprofen (83) involves epoxidation of enone (81) to epoxide (82) (L. Carde *et al.*, Tetrahedron Lett., 1999, **40**, 5421).

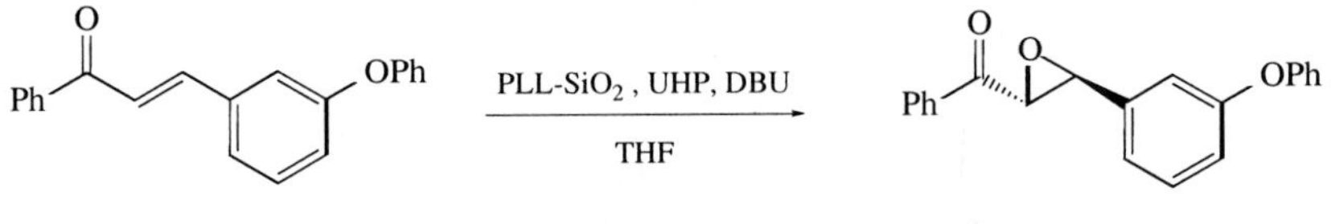

Ph, O, OPh — PLL-SiO_2, UHP, DBU / THF → Ph, O, O, OPh

(81) (82)

98% yield
94% e.e.

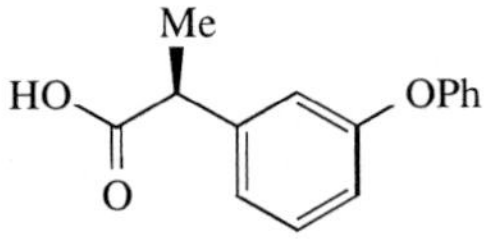

Me, HO, O, OPh

(83)

An optically active chalcone epoxide, a precursor of (+)-clausenamide (84), has been obtained using a modified Julia-Colanna system using a 15-mer or 20-mer of L-leucine bound to a polyethylene glycol (PEG) based support (M. W. Cappi *et al*, J. Chem. Soc., Chem. Commun., 1998, 1159).

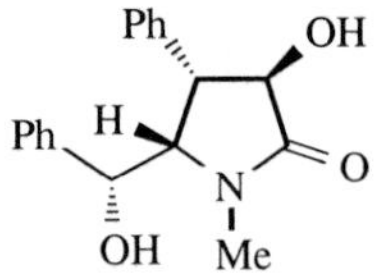

Ph, OH, Ph, H, O, N, OH, Me

(84)

4. Porphyrin catalysts

Up to 88% e.e. has been obtained in the asymmetric epoxidation of *cis*-disubstituted olefins, such as 1,2-dihydronaphthalene, catalysed with threoitol strapped manganese porphyrin complexes in the presence of 1,5-dicyclohexylimidazole (J. P. Collman *et al.*, J. Am. Chem. Soc., 1995, **117**, 692). Styrene derivatives undergo similar epoxidation in e.e.s of up to 79% .
Enantioselectivities of 82-88% are obtained for styrene and its halogenated derivatives with a C_2-symmetric iron porphyrin catalyst and iodosylbenzene as oxidant (J. P. Collman *et al.*, J. Am. Chem. Soc., 1999, **121**, 460). Simple, non-conjugated terminal alkenes such as 3,3-dimethylbutene and vinyltrimethyl silane are also epoxidised in high e.e. (87->90%).

The effect of metal, solvent and oxidant on metalloporphyrin-catalysed enantioselective epoxidation of unfunctionalised olefins has been explored (Z. Gross and S. Ini, J. Org. Chem., 1997, **62**, 5514). Of the three metal complexes of one particular porphyrin, much better results are obtained in the epoxidation of styrene with iron and ruthenium than with manganese catalysts. Indeed, under optimised conditions an iron-porphyrin complex allowed the epoxidation of styrene or 4-chlorostyrene to proceed at an efficiency equal to that of the best previously reported.
The enantiomerically pure ruthenium porphyrin (85) has been used in conjunction with 2,6-dichloropyridine *N*-oxide as the terminal oxidant to obtain enantioselectivities of 77% for (1R,2S)-1,2-epoxy-1,2,3,4-tetrahydronaphthalene, whilst a 70% e.e. is obtained for styrene (A. Berkessel and M. Frauenkron, J. Chem. Soc., Perkin Trans. 1, 1997, 2265).

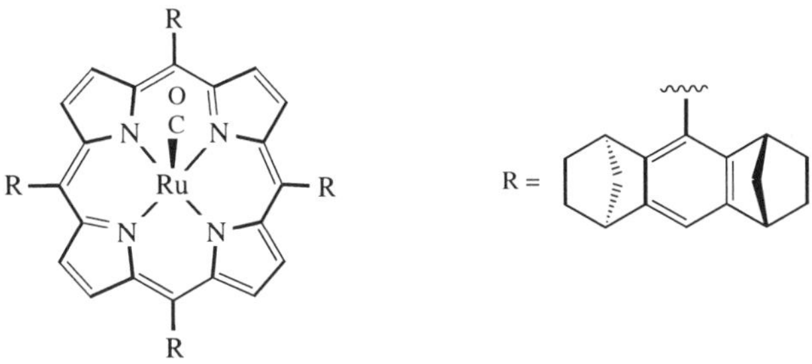

R =

(85)

A new D_2-symmetric ruthenium porphyrin complex has been shown to be a selective catalyst for the asymmetric epoxidation of terminal and *trans*-disubstituted olefins (Z. Gross and S. Ini, Org. Lett., 1999, **1**, 2077). Styrene and its 3- and 4-chloro substituted derivatives undergo epoxidation in 79-83% e.e. while the highest e.e. for β-substituted styrenes using chiral metalloporphyrins is reported for *trans*-β-methylstyrene (69% e.e.). *trans*-Stilbene, a notoriously poor substrate for metalloporphyrin catalysis, is epoxidised to the *trans*-epoxide in 38% enantiomeric excess. Chiral induction, however, is significantly poorer for *cis*-olefins.
A chiral D_4-symmetric porphyrin, derived from the commercially available terpenic ketone R-(+)-nopinone, has been synthesised and its chloromanganese derivative evaluated as a catalyst for the epoxidation of several aromatic alkenes using LiOCl as terminal oxidant and 1,5-dicyclohexylimidazole as the axial ligand in a phase-transfer system. Terminal alkenes such as styrene, 1-vinylnaphthalene and α-methyl

styrene are the best substrates for this catalyst with e.e.s in the range 65-70%. Aliphatic alkenes are generally poor substrates giving low e.e.s, although certain 1,1-disubstituted aliphatic alkenes provide e.e.s in excess of 80% (J. F. Barry *et al.*, Tetrahedron, 1997, **53**, 7753).

The D_4-symmetric porphyrin (86), derived from the C_2-symmetric benzaldehyde containing two norbornane groups fused to the central benzene ring has also been used as a catalyst. Upto 7200 turnovers and upto 76% e.e. (*cis*-β-methylstyrene) and >90% yield can be achieved in the epoxidation of aromatic substituted alkenes in the presence of excess sodium hypochlorate (R.L. Halterman *et al.*, Tetrahedron, 1997, **53**, 11257). Moderate reactivity changes are observed when (86) is modified to (87) and (88), Y=Br and Y=OCH_3, respectively. The methoxy derivative gives slightly improved selectivity, 83% e.e. with *cis*-β-methylstyrene (R.L. Halterman *et al.*, Tetrahedron, 1997, **53**, 11277).

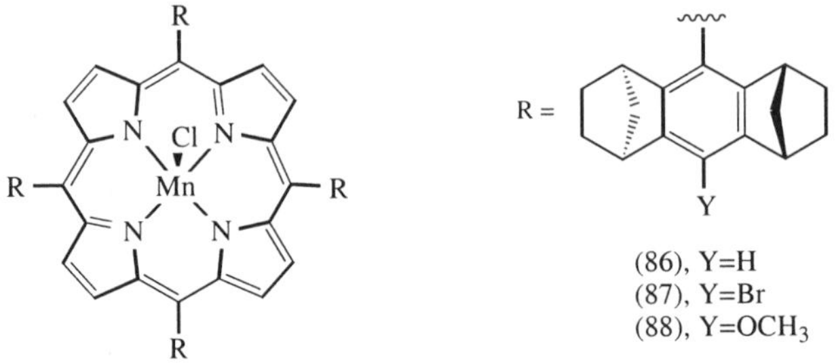

(86), Y=H
(87), Y=Br
(88), Y=OCH_3

Styrene undergoes epoxidation catalysed by the *trans*-dioxoruthenium porphyrin complex (89)-$Ru(O)_2$ in CH_2Cl_2 with iodosylbenzene as oxidant, but the e.e. is only 4% and the chemical yield is also low (11-12%); however, a change of solvent to benzene significantly increases the selectivity to 42% e.e. and the yield to 47%. Enantioselectivity is further enhanced to 57% e.e. by changing the oxygen source from iodosylbenzene to 2,6-dichloropyridine *N*-oxide. The sensitivity to the solvent indicates that several high valent intermediates with different selectivities participate in the oxygen atom transfer from catalyst to substrate (Z. Gross *et al.*, Tetrahedron Lett., 1996, **37**, 7325).

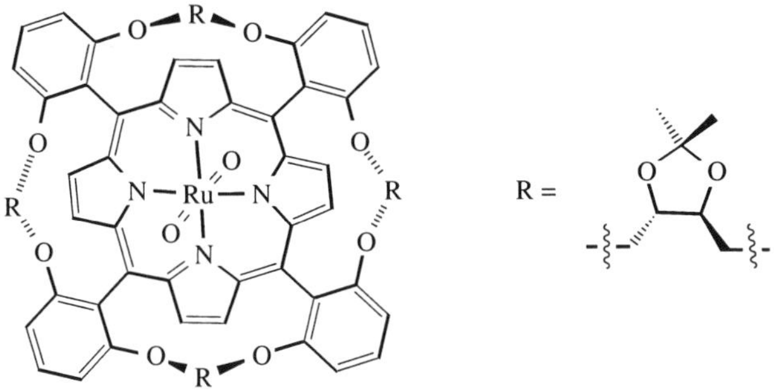

(89)

The most reactive and selective intermediates in the ruthenium porphyrin-catalysed epoxidation of olefins are oxoruthenium(IV) complexes in which the pyridine *N*-oxide is co-ordinated in such a manner that the chiral environment of the oxoruthenium bond is modified (Z. Gross and S. Ini, Inorg. Chem. 1999, **38**, 1446).

5. Oxone® mediated epoxidation

Chiral dioxiranes generated *in situ* by the reaction of Oxone® (potassium monoperoxysulfate, $KHSO_5$) with a suitable chiral ketone precursor continues to attract attention as a method for the asymmetric epoxidation of alkenes. The ketone is regenerated after epoxidation and therefore, in principle, the process can be regarded as catalytic in nature.

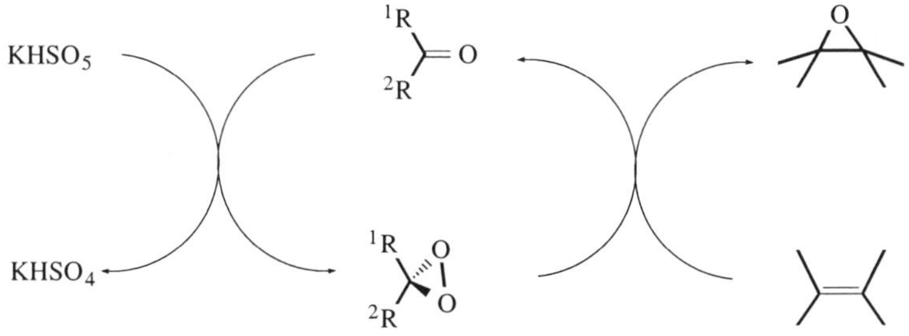

(+)-Isopinocamphone (90) and (S)-(+)-3-phenylbutan-2-one (91) were screened as suitable chiral ketones for Oxone® mediated epoxidations of alkenes (R. Curci *et al.*, J. Chem. Soc., Chem. Commun., 1984, 155). Enantiomeric excesses for the epoxidation of 1-methyl-cyclohexene and *E*-β-methylstyrene were low (9-12.5%) and substrate conversion rates slow in a two-phase system consisting of CH_2Cl_2-buffered H_2O (pH 7-8) with $Bu_4N^+HSO_4^-$ as the phase transfer agent. Other ketones, R-(+)- and S-(-)-3-methoxy-3-phenyl-4,4,4-trifluoro-butan-2-one (92) and (+)-3-(trifluoroacetyl)camphor (93) were also assessed as they contain electron withdrawing groups that enhanced the electrophilicity of the carbonyl carbon atom (R. Curci *et al.*, Tetrahedron Lett., 1995, **36**, 5831). Although conversion rates were increased, the selectivity remained low (13-20% e.e.) for the epoxidation of *trans*-β-methylstyrene, *trans*-2-octene and *cis*-2-methyl-7-octadecene.

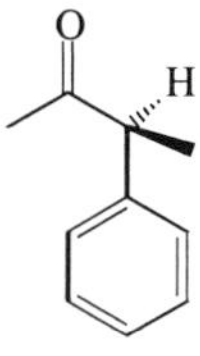

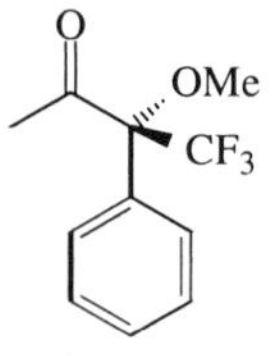

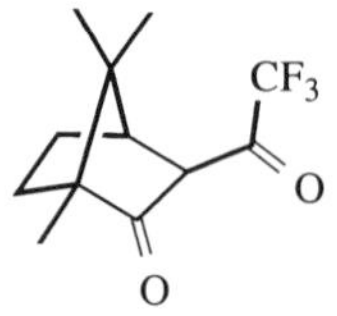

(90) (91) (92) (93)

The concept seems to be a good one, however, and α–fluoro ketones are more efficient catalysts than the parent compounds (S.E. Denmark *et al.*, J. Org. Chem., 1997, **62**, 8288).

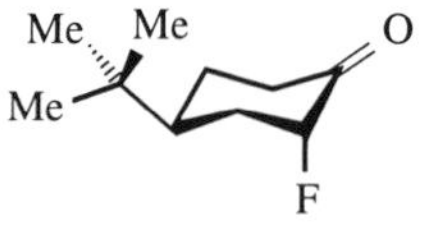

(94) (95) (96)

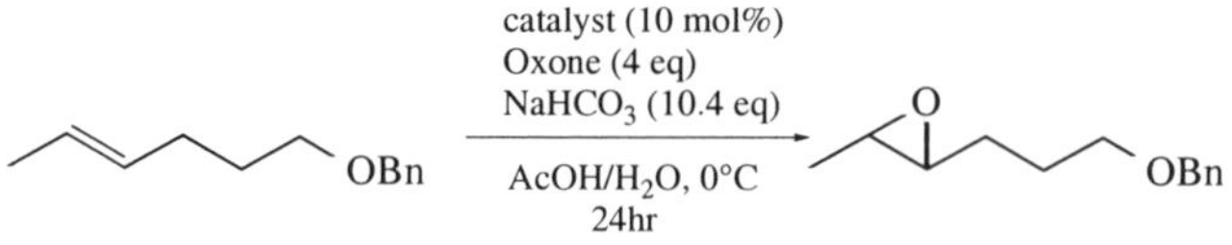

catalyst (10 mol%)
Oxone (4 eq)
$NaHCO_3$ (10.4 eq)
$AcOH/H_2O$, 0°C
24hr

E-(97)

Both monoketones (94) and (95) are superior to 4-*tert*-butylcyclohexanone in the epoxidation of *E*-(97) under homogeneous (CH_3CN/H_2O) conditions with a catalyst loading of 10 mol%. Epoxidation efficiency is highly dependent upon the orientation of the fluorine substituent. Ketone (94) with an equatorial fluorine atom is a much better epoxidation catalyst than (95), which has an axial fluorine atom. A second fluorine substituent also placed in an equatorial position

enhances the activity of the ketone as a epoxidation catalyst, *i.e. cis*-2,6-(96), while further axial fluorine atoms attenuate the activity.
The first C_2-symmetric ketone (98) incorporating a chiral binaphthalene unit for use in the asymmetric epoxidation of unfunctionalised *trans*-olefins and trisubstituted olefins was reported in 1996 (D. Yang *et al.*, J. Am. Chem. Soc., 1996, **118**, 491). Enantiomeric excess of up to 87% was obtained for the epoxidation of *trans*-stilbene derivative (106) using (R)-98 (10 mol%) under monophasic conditions. Selectivity for *cis*-olefins and terminal olefins are however poor.

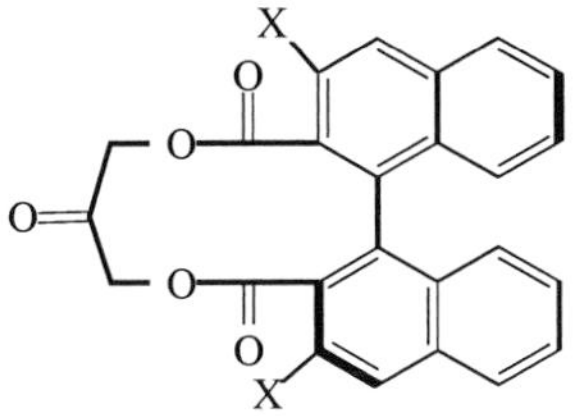

(98), X = H
(99), X = Cl
(100), X = Br
(101), X = I
(102), X = Me
(103), X = CH_2OMe
(104), X = (1,3-dioxan-2-yl)
(105), X = $SiMe_3$

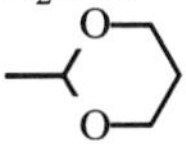

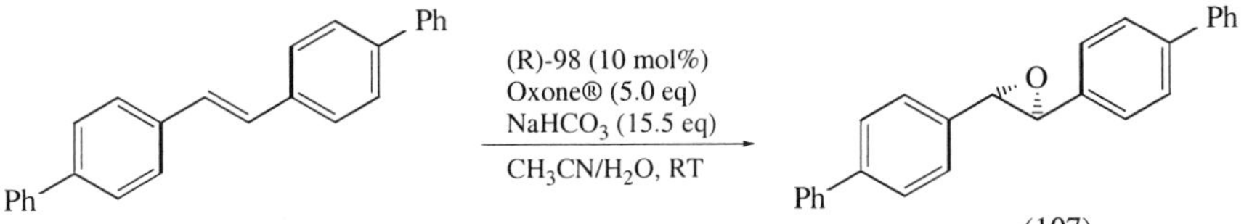

(106) → (R)-98 (10 mol%), Oxone® (5.0 eq), $NaHCO_3$ (15.5 eq), CH_3CN/H_2O, RT → (107) (-)-(S,S) 87% e.e.

Structural modifications to (R)-(98), ketones (99)-(105), which incorporate larger "steric sensors" positioned at H-3 and H-3' allow improved enantioselectivity in the epoxidation of *trans*-stilbene (D. Yang *et al.*, J. Am. Chem. Soc., 1996, **118**, 11311). In general, enantioselectivity first increases and then decreases as the size of the steric sensor rises from H→Cl→Br→I and from H→Me→MOM→acetyl→TMS (Table 1). This indicates that there is an optimum size for a steric sensor and additionally electronegative atoms on a sensor are important. The last point is emphasised by the results obtained for catalysts (99) and (102) where the chlorine atom of (99) is replaced by a methyl group in (102), and the fall off in efficiency when bromine is replaced by iodine. Interestingly, the acetal units in catalyst (104) promote reactivity, perhaps by modifying the properties of the adjacent ester unit in some way.

Table 1: Dependence of catalyst on efficiency of *trans*-stilbene epoxidation

entry	catalyst	reaction time (h)	epoxide yield (%)	epoxide configuration	e.e. (%)
1	(R)-98	1	91	(-)-(S,S)	47
2	(R)-99	2	95	(-)-(S,S)	76
3	(R)-100	3	92	(-)-(S,S)	75
4	(R)-101	22	90	(-)-(S,S)	32
5	(S)-102	1	93	(-)-(R,R)	56
6	(R)-103	1.8	92	(-)-(S,S)	66
7	(R)-104	0.7	95	(-)-(S,S)	71
8	(S)-105	20	nc	(-)-(R,R)	44

nc -not completed

Evidence has been provided for a spiro-transition state of dioxirane epoxidation and through a ^{18}O-labelling experiment, chiral dioxiranes have been found to be the intermediates in chiral ketone catalysed epoxidation reactions under homogeneous conditions (D. Yang *et al.*, J. Am. Chem. Soc., 1998, **120**, 5943). However, in a biphasic medium (CH_2Cl_2-H_2O), an experiment using ^{18}O-labelled 4-*tert*-butylcyclohexanone suggests that dioxirane is not responsible for alkene epoxidation using a ketone-Oxone® system (A. Armstrong *et al.*, J. Chem. Soc., Chem. Commun., 1996, 849).
C_2-Symmetric chiral ketones (108) and (109) have been used as precursors for chiral dioxiranes in the asymmetric epoxidation of *trans*-stilbene to *trans*-stilbene oxide (C.E. Song *et al.*, Tetrahedron: Asymmetry, 1997, **8**, 2921). Moderate e.e. (59%) under homogeneous conditions has been achieved although a ketone:olefin ratio of 1:1 is required.

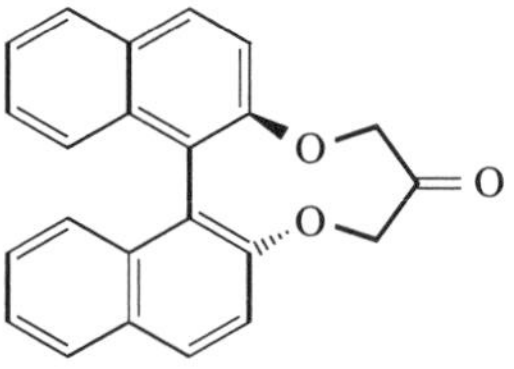

(108) 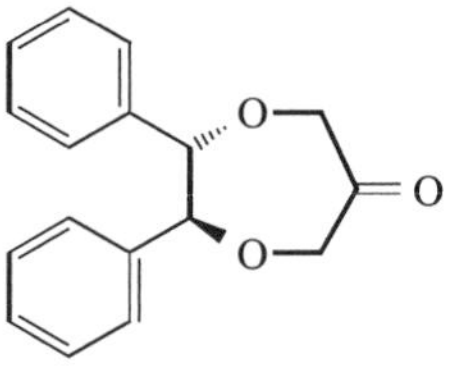 (109)

Also reported are dioxiranes generated *in situ* from chiral ketones (110) and (111), derived from mannitol and (R,R)-(+)-tartaric acid, respectively (W. Adam and C-G. Zhao, Tetrahedron: Asymmetry, 1997, **8**, 3995). Only moderate e.e. (39%) is obtained with ketone (110) in the epoxidation of *trans*-stilbene and two equivalents of the ketone are required. Good to high enantiomeric excesses (up to 79%) can be obtained in the epoxidation of triphenylethylene using ketone (111) with a catalyst loading of 10 mol% and at higher pH (10.5), but the conversion is low. In the epoxidation of silyl ether (112), catalyst (111) (0.5 eq) affords the corresponding (R,R)-(+)-epoxide in 79% e.e. and a conversion of 80%.

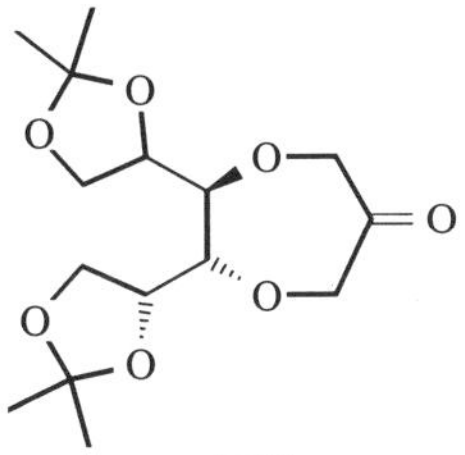

(110) 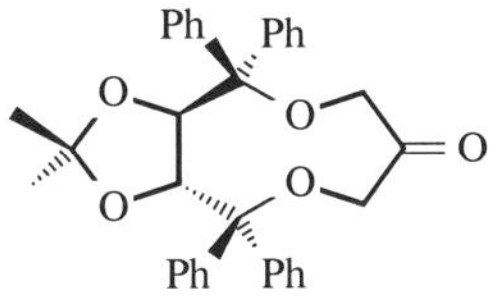 (111) Ph OSiMe$_2$Bu (112)

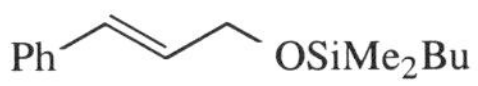

The C$_2$-symmetric ketone (113) has been prepared from fulvene, utilising the Sharpless asymmetric dihydroxylation methodology, as a promoter in the epoxidation of alkenes by Oxone® (A. Armstrong and B.R. Haytor, Tetrahedron: Asymmetry, 1997, **8**, 1677).

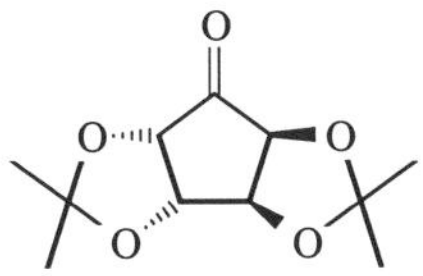

(113)

Electronic effects imparted by non-conjugated remote substituents in asymmetric catalysis has been examined through the synthesis of chiral ketones (114)-(118), derived from (R)-carvone. All the catalysts have a quaternary carbon at C-2 position but differ in the remote substituent at the C-8 position. The significant effect of C-8 substituents on enantioselectivity (42-87% e.e.) demonstrates that electronic tuning is important in the rational approach to catalyst design (D. Yang *et al.*, J. Am. Chem. Soc., 1998, **120**, 7659).

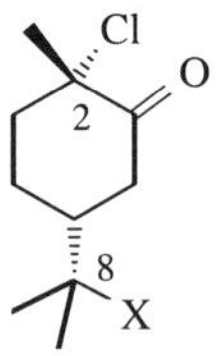

(114), X = F
(115), X = Cl
(116), X = OH
(117), X = OEt
(118), X = H

Enantiomeric excesses of up to 83% have been obtained with tropinone derived chiral ketone (+)-(119) (10 mol%) in the epoxidation of a range of alkenes (A. Armstrong and B. R. Haytor, J. Chem. Soc., Chem. Commun., 1988, 621). α,β-Unsaturated esters such as *E*-methyl cinnamate can also be epoxidised in moderate e.e. (64%), although longer reaction times and a higher catalyst loading (25 mol%) is required.

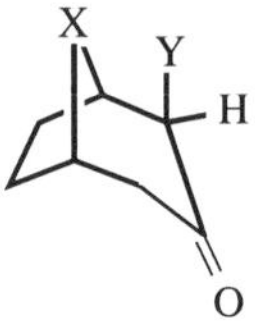

(119), X = N-CO_2Et, Y = F
(120), X = N-CO_2Et, Y = Cl
(121), X = N-CO_2Et, Y = OAc
(122), X = O, Y = F
(123), X = O, Y = OAc

Additional 2-substituted bicyclo[3.2.1]octan-3-ones along with oxabicycles (120)-(123) have been prepared and tested as catalysts for alkene epoxidation by Oxone® (A. Armstrong *et al.*, Tetrahedron: Asymmetry, 2000, **11,** 2057). Although these chiral ketones were not enantiomerically pure (76-80% e.e.), e.e.s of up to 74% have been achieved for the epoxidation of *E*-stilbene.
Chiral iminium salt (S)-(+)-(124), based on the binaphthyl system, are reported to catalyse the oxidation of a range of unfunctionalised alkenes using Oxone® as oxidant (V. K. Aggarwal and M. F. Wang, J. Chem. Soc., Chem. Commun., 1996, 191). With a catalyst loading of 5 mol%, epoxidation of 1-phenylcyclohexene affords the corresponding oxirane in 71% e.e. while a 31% e.e. is obtained for *trans*-stilbene as substrate.
Iminium salts with the general structure (125), derived from 3,4-dihydroisoquinoline with a chiral exocyclic *N*-substituent can be used to achieve catalytic asymmetric epoxidation by Oxone® (P.C. Bulman-Page *et al.*, J. Org. Chem., 1998, **63**, 2774). For *trans*-stilbene, the *N*-isopinocamphenyl derivative (10 mol%) affords the (+)-(R,R)-epoxide in 73% e.e. and 78% yield.

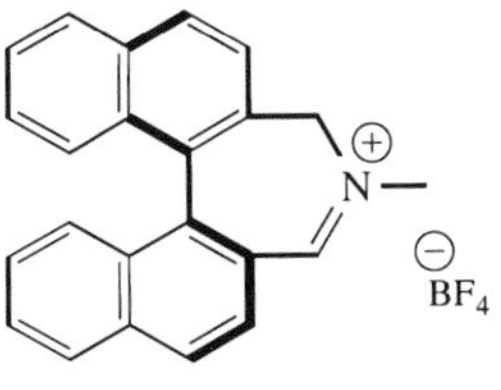

(124)

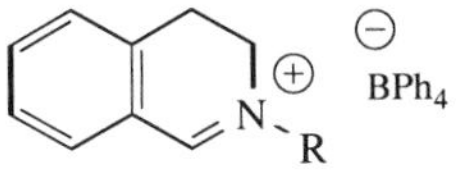

(125)

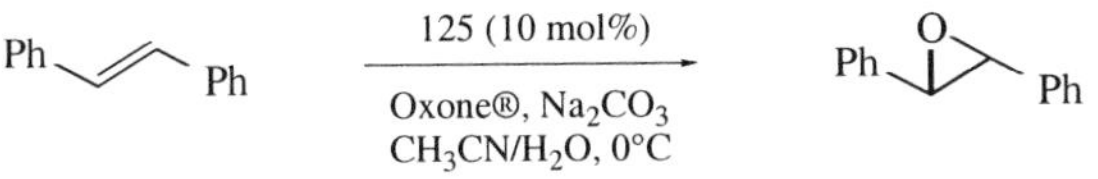

125 (10 mol%)

Oxone®, Na_2CO_3
CH_3CN/H_2O, 0°C

125, R =

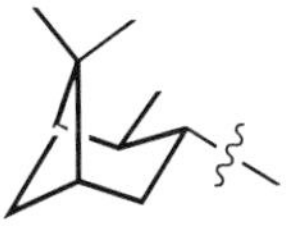

Ketone (126), derived from D-fructose, has been shown to give high enantiomeric excess for the epoxidation of *trans*-disubstituted and trisubstituted prochiral olefins carrying a variety of functional groups in the alkene substrate (Y. Tu *et al.*, J. Am. Chem. Soc., 1996, **118**, 9806). High enantioselectivity (90%) has been obtained in the epoxidation of *trans*-7-tetradecene, a simple unfunctionalised *trans*-olefin. Unfortunately good conversion rates of substrate require the use of excess ketone because rapid decomposition of the catalyst occurs under the neutral (pH 7-8) reaction conditions.
Subsequent studies have shown that the pH has a dramatic effect on the epoxidation efficiency of ketone (126) (Z-X. Wang *et al.*, J. Org. Chem., 1997, **62**, 2328). The Bayer-Villiger reaction is believed to be one of the major pathways for the decomposition of the catalyst. The competing pathway may be suppressed by raising the pH of the reaction, this favours the equilibrium towards intermediate (128) and hence the formation of dioxirane (129). When the epoxidation is carried out at pH 10.5, conveniently achieved by the addition of potassium carbonate, ketone/substrate ratios of 0.2-0.3 can be used to achieve good yields with slightly improved enantioselectivities.

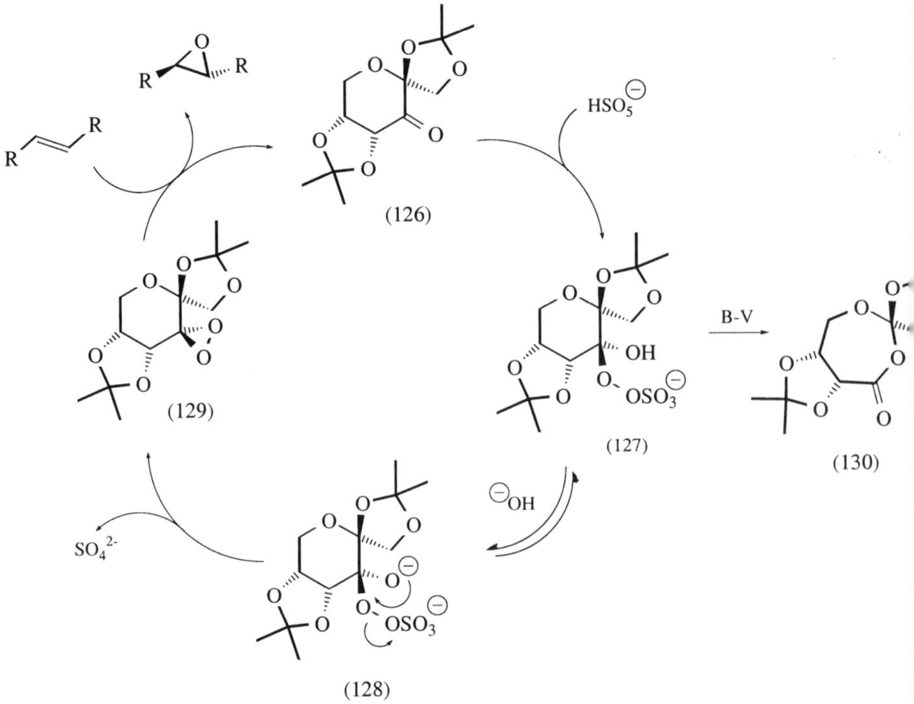
HSO5
B-V
OH
SO4 2-
OH
OSO3
(126)
(127)
(128)
(129)
(130)

Ketone (126) has been used in the regioselective and enantioselective monoepoxidation of conjugated *trans*-disubstituted and trisubstituted dienes (M. Frohn *et al.*, J. Org. Chem., 1998, **63**, 2948). Functional groups such as hydroxyl groups, TBS ethers or esters can be tolerated and the enantiomeric excess for the major monoepoxides range from 89% to 97%. For unsymmetrical dienes, regioselectivity of epoxidation can be regulated by electronic and steric control. For diene (131), regioselectivity is complimentary to the Sharpless asymmetric epoxidation in that the olefin distal to the hydroxy group is selectively epoxidised.

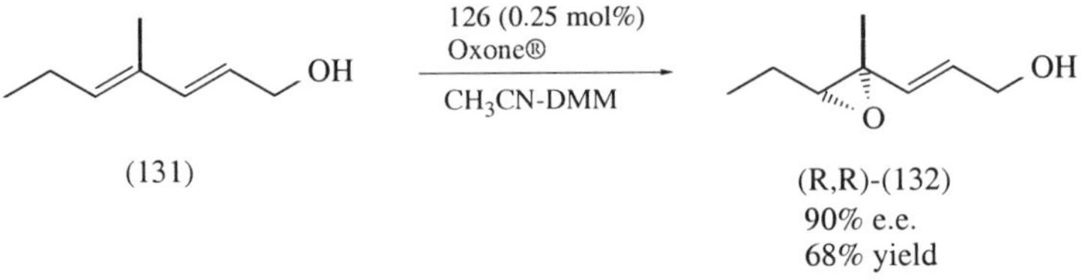
OH
126 (0.25 mol%)
Oxone®
CH3CN-DMM
OH
O
(131)
(R,R)-(132)
90% e.e.
68% yield

Asymmetric epoxidation of hydroxyalkenes catalysed by the chiral ketone (126) is highly pH dependent (Z-X. Wang and Y. Shi, J. Org. Chem., 1998, **63**, 3099). At low pH, direct epoxidation by Oxone® is dominant, which results in racemisation and therefore low enantioselectivity. Epoxidation by Oxone® is believed to be facilitated by the hydroxyl group in the substrate since epoxidation of a mixture of geraniol and its TBS ether results only in the epoxidation of the alcohol. At high pH, increased nucleophilicity of Oxone® towards the chiral ketone increases formation of the corresponding dioxirane. This subsequently out competes direct epoxidation by Oxone® yielding higher enantioselectivity. Under the optimised conditions and a catalyst loading of 30 mol% of (126), asymmetric epoxidation of a series of *trans*-disubstituted and trisubstituted allylic, homoallylic, and bishomoallylic alcohols can be achieved in good to high e.e. (70-94%).
Conjugated enynes (133) undergo catalytic asymmetric epoxidation with the chiral ketone (126) (G. A. Cao *et al.*, Tetrahedron Lett., 1998, **39**, 4425). Epoxidation occurs chemoselectively at the alkene and the acetylene can bear a variety of substituents such as alkyl, TMS, and ester groups. Enantioselectivities are generally high, in the range 89-97%.

R

126 (0.3 eq)
Oxone®
solvent-H_2O

R
O

(133)
R = H, CH_3, TMS, CO_2Et

(R,R)-(134)
yield = 71-88%
e.e. = 89-97%

Enol silyl ethers and esters, such as enol benzoate (135), are epoxidised with high enantioselectivity with fructose derived ketone (126) and Oxone® as oxidant (Y. Zhu *et al.*, Tetrahedron Lett., 1998, **39**, 7819).

OBz

126 (0.3 eq)
Oxone® (1.38 eq)
CH_3CN-DMM

OBz
O

(135)

82% yield
93% e.e.

Further chiral ketone catalysts, (136)-(138), derived from carbohydrates, have been prepared and investigated as epoxidation mediators (Y. Tu *et al.*, J. Org. Chem., 1998, **63**, 8475). Comparison between these three ketones and the structurally related ketone (126) has shown that ketone structure has a profound impact on both the rate of conversion and the enantioselectivity of epoxidation (Table 2).

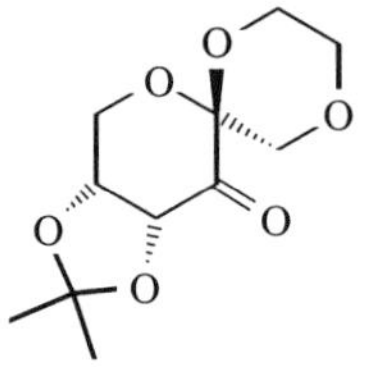

(136) 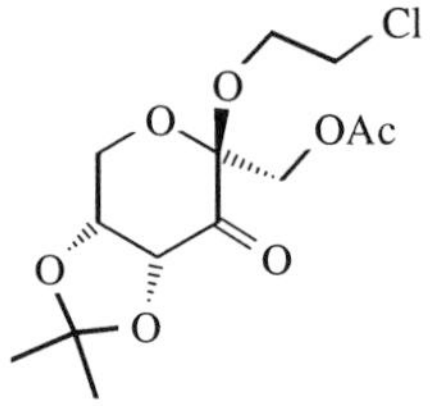 (137) 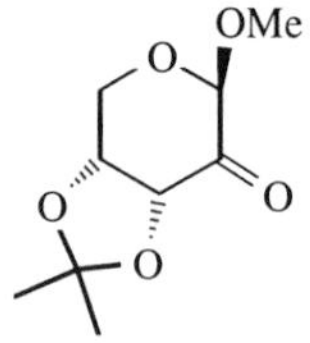 (138)

Table 2: Epoxidation of β-methylstyrene

ketone	conversion (%)	e.e. (%)
126	93	92
136	44	61
137	5	66
138	15	59

Ketones (140a)-(j), prepared from (-)-quinic acid (139), which differ from one another in the nature of the substituents at the β-position to the carbonyl group, have been shown to be effective catalysts for the asymmetric epoxidation of *cis* and terminal alkenes (Z-X. Wang and Y.Shi, J. Org. Chem., 1997, **62**, 8622).

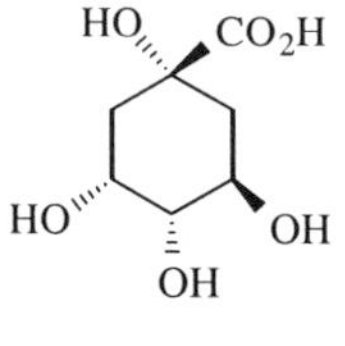

a, R = H
b, R = CO_2Me
c, R = CH_2OAc
d, R = CH_2OBz
e, R = CH_2OTs
f, R = CH_2OTBS
h, R = CMe_2OH
i, R = CPh_2OH
j, R = CPh_2OMe

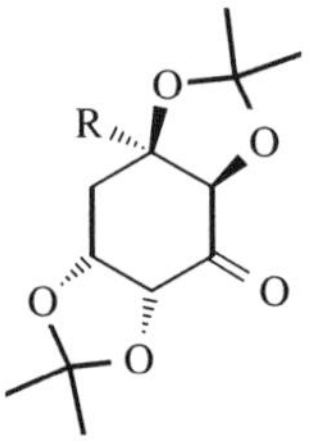

(139) (140)

Chiral ketones (140b)-(f) are more stable and reactive than fructose derived ketone (126) and a smaller catalyst loading (5-10 mol%) is required to achieve good conversion. Enone (141) undergoes epoxidation in high e.e. suggesting that the catalyst can effectively compete with ketones in the substrate and the epoxide product.

140c (10 mol%)
Oxone® (1.38 eq)
DME-DMM-buffer
O°C, 6hr

(141)

(+)-(2S,3R)
80% (94% e.e.)

Substituents at the β-position affect the efficiency and selectivity of epoxidation to varying degrees (Z-X. Wang *et al.*, J. Org. Chem., 1999, **64**, 6443). The unsubstituted C_2-symmetric ketone (140a), and (140i), with the most sterically hindered β-substituent, are the least reactive to *trans*, *cis* and terminal alkenes. However, (140i) gives the highest enantiomeric excess for the *cis*-olefin. Catalysts (140h) and (140j) give the highest conversion and enantiomeric excesses for *trans* and terminal olefins but lowest for the *cis*-olefin.

ketone	mol%	conv. (e.e.%)	mol%	conv. (e.e.%)	mol%	conv. (e.e.%)	mol%	conv. (e.e.%)
140a	5	29 (73)	10	11 (93)	10	13 (67)	10	9 (66)
140h	5	97 (80)	10	91 (96)	5	79 (69)	10	55 (45)
140i	5	7 (50)	-	-	10	7 (59)	10	7 (88)
140j	5	95 (80)	10	94 (96)	5	100 (70)	10	47 (40)

Dioxiranes generated *in situ* from optically active ketones (142) and (143), also derived from D-(-)-quinic acid, serve as effective oxidants for the asymmetric epoxidation of prochiral olefins (W. Adam *et al.*, Tetrahedron: Asymmetry, 1999, **10**, 2749). Up to 87% e.e. is obtained for the corresponding epoxides although excess ketone is required.

(142), R^1, R^2 = Me, Me
(143), R^1, R^2 = $-(CH_2)_5-$

6. Miscellaneous methods of chiral epoxidation

Cinchona alkaloid derived quaternary ammonium phase transfer catalysts bearing an *N*-anthracenylmethyl function have been applied as chiral control elements in the epoxidation of α,β-unsaturated ketones. Epoxidation proceeds with good enantioselectivity for a range of enone substrates (69-87% e.e.). For example, chalcone undergoes epoxidation in 81% e.e. to (-)-chalcone epoxide with catalyst (144) (10 mol%) in toluene when sodium hypochlorite is used as the oxidant. Access to the opposite (+)-enantiomeric epoxide may be obtained with broadly similar selectivities with the *pseudo*-enantiomeric catalyst (145) (B. Lygo and P.G. Wainwright, Tetrahedron Lett., 1998, **39**, 1599).

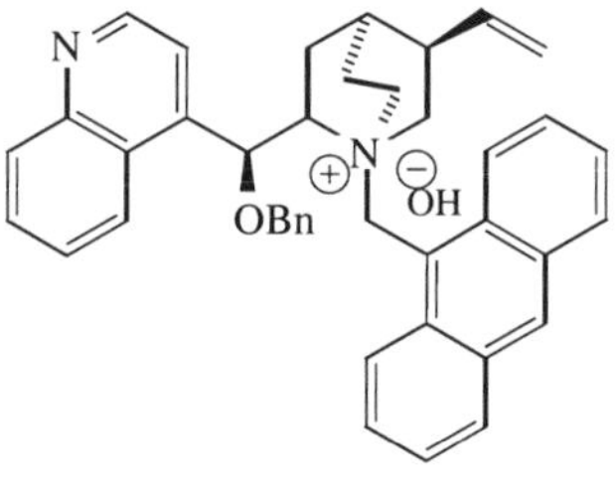

(144) 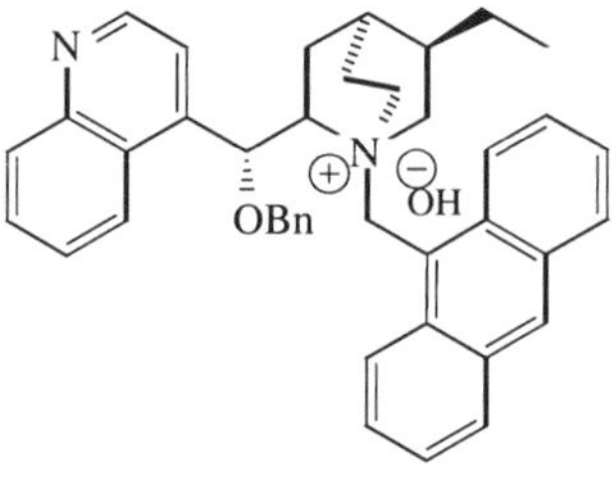   (145)

Ruthenium complexes with chiral bis(dihydrooxazolylphenyl)oxalamide ligands (146) catalyse the epoxidation of *E*-stilbene and *E*-1-phenylpropene in 69% and 58% e.e., respectively, using $NaIO_4$ as oxidant in a two-phase system (N. End and A. Pfaltz, J. Chem. Soc., Chem. Commun., 1998, 589).

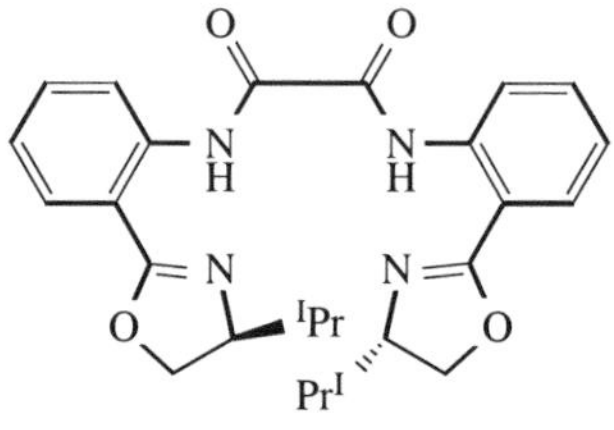

(146)

Catalytic asymmetric epoxidation of chalcone and its derivatives (147)-(151) with *tert*-butylhydroperoxide (TBHP) using dibutylmagnesium (10 mol%) and a slight excess of (+)-diethyl tartrate (11 mol%) gives the corresponding epoxides in 81-94% e.e. and yields in the range 36-61%.

The active catalyst is presumed to be a magnesium bis(alkoxide) derived from (+)-DET. Suprisingly, it was noted that in the case of chalcone, epoxidation with Mg-TBHP gives the opposite absolute configuration when compared to stoichiometric asymmetric epoxidation using lithium alkyl peroxide (C. L. Elstron *et al.*, Angew. Chem. Int. Ed. Eng., 1997, **36**, 410).

^1Ar–CO–CH=CH–Ar^2 → [*t*-BuOOH (1.5 eq), Bu_2Mg (0.1 eq), (+)-DET (0.11 eq); $CH_3C_6H_5$ (+ THF)] → epoxide

	Ar^1	Ar^2
(147)	Ph	Ph
(148)	Ph	*p*-ClPh
(149)	Ph	*p*-MePh
(150)	*p*-MePh	Ph
(151)	2-naphthyl	Ph

The enantioselective epoxidation of functionalised and unfunctionalised alkenes using *m*-chloroperbenzoic acid incorporated into egg phosphatidylcholine liposomes (LIP) has been reported. E.e.s of 92% and 95% are obtained in the epoxidation of methyl *trans*-cinnamate and its 3-methoxy derivative in yields of 70% and 75%, respectively. Chalcone and its chloro-derivatives are epoxidised in moderate to good e.e. (62-70%). *trans*-Methylstyrene epoxide is obtained in 82% e.e. and 76% yield while the terminal alkene (152) undergoes LIP-MCPBA epoxidation in 95% e.e. (A. Kumar and V. Bhakuni, Tetrahedron Lett., 1996, **37**, 4751).

(152) → LIP-MCPBA → epoxide

82% yield
95% e.e.

Chiral lanthanoid-BINOL catalysts, prepared from $La(O^iPr)_3$ and (R)-BINOL in the presence of 4Å molecular sieves have been employed in the catalytic asymmetric epoxidation of enones. Treatment of enones (147), (155) and (156) with cumene hydroperoxide (CMHP) in the presence of La-BINOL complex La-(153) (5 mol%) in THF at room temperature affords the corresponding epoxides with selectivities in the range 83-86% e.e. and 85-93% yield. Selectivity may be increased in the epoxidation of (147) and (155) to 91% and 94% e.e. respectively by using (R)-hydroxymethyl-BINOL (154) instead of (153). However, enones (157)-(160) are best epoxidised using ytterbium complexes (8 mol%), generated from $Yb(O^iPr)_3$ and (R)-(154), where epoxides are generated in selectivities in the range 88-94% e.e. (M. Bougauchi *et al.*, J. Am. Chem. Soc., 1997, **119**, 2329).

$Ln(O^iPr)_3$ + (R)-BINOL derivative, Ln = La or Yb; MS 4Å, THF, RT →

La-BINOL cat. (La-153)
La-3-hydroxymethyl-BINOL cat. (La-154)
Yb-BINOL cat. (Yb-153)
Yb-3-hydroxymethyl-BINOL cat. (Yb-154)

(153), R = H
(154), R = CH_2OH

^{2}R–CO–CH=CH–R^1 → (R)-Ln cat., MS 4Å, RT, THF → epoxy ketone ^{1}R–(epoxide)–CO–R^2

	R^1	R^2
(147)	Ph	Ph
(155)	^{i}Pr	Ph
(156)	Ph	o-$MOMOC_6H_4$
(157)	Ph	CH_3
(158)	Ph	^{i}Pr
(159)	$Ph(CH_2)_2$	CH_3
(160)	$Ph(CH_2)_4$	CH_3

Second Edition of Rodd's Chemistry of Carbon Compounds,
Volume V, Topical Volumes
Asymmetric Catalysis, edited by M.Sainsbury

Chapter 2

OXIDATION NOT INVOLVING EPOXIDATION

J. EAMES and M. WATKINSON

This chapter deals with the catalytic oxidation of substrates excluding the direct oxidation of an alkene and alkyne functional group. A brief account of some asymmetric oxidation reactions involving homogeneous and heterogeneous catalysis has recently appeared (F. Fache, E. Schulz, M.L. Tommasino and M. Lemaire, Chem. Rev. 2000, **100**, 2159-2231).

1. Enantioselective C-H Oxidation.

(a) Enantioselective allylic oxidation of alkenes

The enantioselective transition metal catalysed allylic oxidation of alkenes has recently become a popular transformation. The procedure is relatively straightforward - addition of a substituted perester to an alkene in the presence of a suitable catalyst (typically at loadings of 5 mol %) leads directly to an optically active allylic ester. This reaction is an asymmetric variant of the Kharasch-Sosnovsky reaction (M.S. Kharasch, G. Sosnovsky and N.C. Yang, J. Am. Chem. Soc., 1959, **81**, 5819-5824). Due to the infancy of this procedure, there is currently a limited number of available catalysts, all of which are based on the use of Cu(I) or (II) complexes of amino acids or ligands derived from them.

(i) Use of Copper(I) based oxazoline complexes.

Stereoselective functionalisation of unactivated alkenes has rapidly become a growing and challenging research area. The first synthetically useful reports came independently in 1995 from both Pfaltz (A.S. Gokhale, A.B.E. Minidis and A. Pfaltz, Tetrahedron Lett., 1995, **36**, 1831-1834) and Andrus (M.B. Andrus, A.B. Argade, X. Chen and M.G. Pamment, Tetrahedron Lett., 1995, **36**, 2945-2948).

Pfaltz has shown the use of a series of chiral C_2 symmetric copper(I) bis(oxazoline)s **1a-c** as the asymmetric catalyst for this Kharasch-Sosnovsky reaction. The enantioselectivity was generally good (up to 84% e.e.), whereas the yields were more variable – indicating a modest catalytic turnover. These allylic oxidations were postulated to proceed

1a; R = *i*-Pr
1b; R = *t*-Bu
1c; R = Ph

2 → PhCO$_3$*t*-Bu (**3**), 5 mol% Cu(I)OTf, 6-8 mol% ligand **1a-c**

4 up to 84% e.e. yield = 61% using **1b**

5 up to 67% e.e. yield = 77% using **1c**

6 up to 82% e.e. yield = 44% using **1a**

via a radical chain mechanism. The stereochemical-determining step in this process is presumed to occur in the radical recombination step involving the allylic radical **10** and the copper(II) complex **8**. Addition can occur on either enantiotopic side/face giving the allyl cuprate **11**. This stereochemistry is transferred in the formation of the resulting C-O bond *via* an allylic rearrangement to give the allylic benzoate **12**.

PhCO$_3$*t*-Bu **3** → Cu(I)L **7** → PhC(O)OCu(II)L **8** + *t*-BuO• → **9**

10 + *t*-BuOH → **8** → **11** → **12**

The catalysts **1a-c** were screened against a series of cyclic alkenes. In all cases so far reported the facial preference of the newly formed C-O bond was the same, giving an (*S*)-configuration at the formed allylic stereocentre. It was clear from this initial study there was no single oxazoline **1a-c** which gave good control over a wide range of cycloalkenes and a systematic screening approach appeared to be the best method of choosing a particular catalyst for a given structure of alkene. The better results for the synthesis of allylic benzoates **4-6** are given.

Andrus has further probed the effect of differing substituents within the oxazoline framework of copper(I) complex, **13**, on the stereoselectivity of the reaction, and found that those with the same relative C(3)-configuration gave the same (*S*)-allylic benzoate (M.B. Andrus and X. Chen, Tetrahedron, 1997, **53**, 16229-16240). A smaller ring size in the cycloalkene, such as cyclopentene and cyclohexene gave much better control than their larger more flexible counterparts. The stereocontrol was excellent, but was clearly dependent on the structure of the catalyst and alkene used. It has also been noted that a small change in the substituent pattern R^1 positioned at C(2) (away from the reactive site) did have a dramatic effect on the resulting stereocontrol; for example, changing from R^1 = H to Me, the enantioselectivity dropped from 80% to 67% e.e. when R^2=*t*-Bu.

13 (5 mol%), **3** (100 mol%)

2 → **4**; **9** → **5**; **14** → **6** (O_2CPh)

Substrate → Product	Ligand (R^1,R^2)	% yield	% e.e.
2 → **4**	*t*-Bu, Me	41	42
	Ph, Me	49	81
9 → **5**	*t*-Bu, H	43	80
	t-Bu, Me	49	67
	Ph, Me	44	47
14 → **6**	*t*-Bu, H	44	13
	t-Bu, Me	43	0

Reports into the allylic oxidation of open chain alkenes is far less common than those of cyclic alkenes (M.B. Andrus, A.B. Argade, X. Chen and M.G. Pamment, Tetrahedron Lett., 1995, **36**, 2945-2948). In these cases, the facial preference was no different than those of the cyclic cases, but the enantioselectivity was much lower.

Ph 15 → (13; R^1= t-Bu; R^2=H / $PhCO_3t$-Bu 3) → **16**; 34%; 36% e.e.

C_5H_{11} 17 → (13; R^1= t-Bu; R^2=H / $PhCO_3t$-Bu 3) → **18**; 50%; 50% e.e.

Andrus have also shown that this methodology was sensitive to the nature of the copper(I)-complex (M.B. Andrus and X. Chen, Tetrahedron, 1997, **53**, 16229-16240). The more poorly co-ordinating the counter anion, the better the enantioselectivity [TfO^- (entry 1) > PF_6^- (entry 2) > Br^- (entry 3)]. This presumably indicates that a more Lewis acidic copper complex leads to a higher degree of stereocontrol due to more efficient binding of the intermediate allylic radical in the stereochemical determining step.

9 → CuX/**1b** (15 mol%), O_2N-C_6H_4-CO_3-t-Bu **19** → **20**

Entry	copper(I)-**1b**	% yield	% e.e.
1	CuOTf	43	80
2	$CuPF_6$	60	76
3	CuBr	53	63

Andrus has additionally reported the use of a Cu(I)-SbF_6 complex involving the binaphthyl oxazoline **21a**. Good levels of stereocontrol (up to 73% e.e.) were observed with the 'bench-mark' cyclohexene, **9**. The yields were also particularly high (76%). However, it is worthy of note, that higher yields and evidently better turnover were shown to occur with activated peresters, such as *p*-nitro, *t*-butylperbenzoate **19** (Andrus, D. Asgari and J.A. Sclafani, J. Org. Chem., 1997, **62**, 9365-9368).

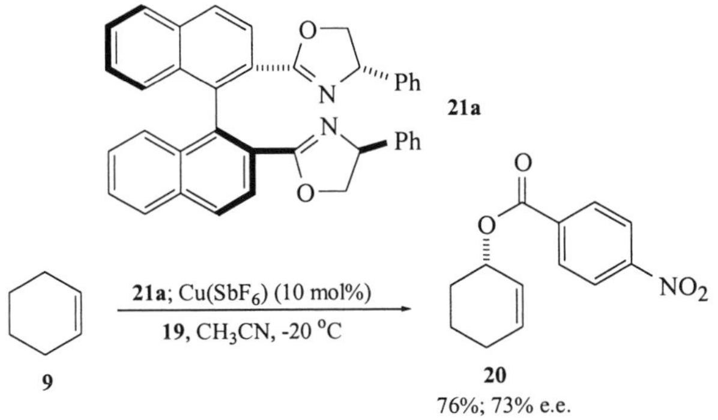

O
N
Ph
N
Ph
O
21a
O
O
NO2
21a; Cu(SbF6) (10 mol%)
19, CH3CN, -20 °C
9
20
76%; 73% e.e.

Andrus has recently probed the effect of the additional stereogenicity in the back-bone of these biaryl atropisomeric oxazolines **21a** and **21b** (M.B. Andrus and D. Asgari, Tetrahedron, 2000, **56**, 5775-5780). The *syn*-stereoisomeric oxazoline **21a**; where R=Ph, appears to give much better stereocontrol than the related *anti*-stereoisomer **21b** (entry 1, 73% e.e. versus entry 2, 0%). This trend is still apparent on changing the R substituent within the oxazoline framework. It appears that a phenyl substituent gave better facial control than either a benzyl or *t*-butyl substituent. The structural nature of the activated perester **22** was also shown to be important; the more sterically demanding the perester the lower the facial control (*e.g.*, **22**; X=2,4,6-Cl_3; 11% e.e. – entry 3). Furthermore, the absolute stereochemistry in the product allylic ester was remarkably predictable; the (*S*)-configuration was exclusively observed.

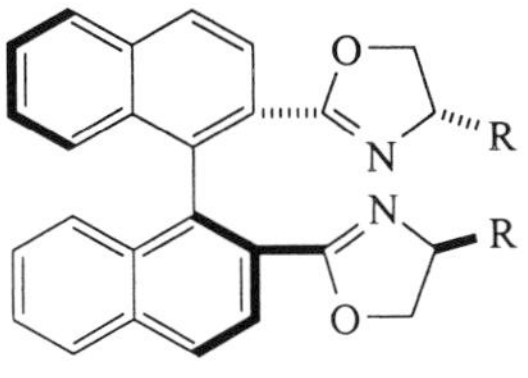

O
N
R
N
R
O

21a

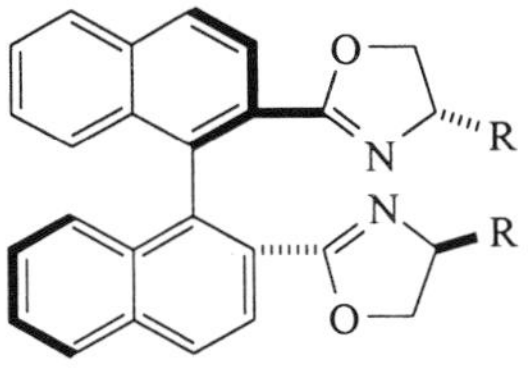

O
N
R
N
R
O

21b

Entry	catalyst	**22**, X	% yield	% e.e.
1	**21a**; R=Ph	*p*-NO_2	78	73
2	**21b**; R=Ph	*p*-NO_2	76	0
3	**21a**; R=Ph	2,4,6-Cl_3	20	11
4	**21a**; R=*t*-Bu	*p*-NO_2	5	18
5	**21b**; R=*t*-Bu	*p*-NO_2	0	0
6	**21a**; R=Bn	*p*-NO_2	48	52
7	**21b**; R=Bn	*p*-NO_2	67	11

(ii) Use of copper(II) oxazoline complexes.

The effect on the oxidation level of the copper complex has been thoroughly investigated. The remainder of this section discusses the use of copper(II) chiral complexes; the most popular being copper(II) acetate. Feringa has used a series of substituted oxazolines to probe the effect of the change of oxidation level in the enantioselective allylic oxidation of cyclohexene, **9** (C. Zondervan and B.L. Feringa, Tetrahedron: Asymmetry, 1996, **7**, 1895-1898). They found that the better the chelating oxazoline (such as **24a**) the higher the facial control which presumably indicates that a tighter transition state leads to higher stereocontrol.

Katsuki has also elegantly shown that this procedure can be extended towards the de-racemisation of racemic substituted cyclopentenes **27** to give the enantiomerically enriched allylic benzoates **29** (80% e.e), **30** (12% e.e.) and **31** (42% e.e.) in an overall 78% yield (Y. Kohmura and T. Katsuki, Synlett, 1999, **8**, 1231-1234). The reaction must proceed via an inter-converting pair of enantiomeric allylic radicals **28a** and **28b**. The C_3-symmetric tris-oxazoline **26** gave the best facial control, whereas, the more common C_2-symmetric based oxazoline ligand **1** gave lower levels of control.

26

(*rac*)-**27**

$Cu(OTf)_2$-**26** (5 mol%)
$PhCO_3t$-Bu **3**
acetone
yield = 78%

28a **28b**

29; R^1 = OCOPh, R^2 = H (80% e.e.)
30; R^1 = H, R^2 = OCOPh (12% e.e.)

31 (42% e.e.)

Katsuki has also further investigation the positional effect associated with his C_3-symmetrical oxazoline **26** in the regioselective oxidation of an unsymmetrical alkene, 1-methylpentene, **32** (K. Kawasaki and T. Katsuki, Tetrahedron Lett., 1997, **53**, 6337-6350). Three positional isomeric allylic benzoates **33-35** were isolated in the following ratio (67:25:8).

The enantioselectivity was evidently dependent on the structural nature of the intermediate allylic radical. It is worth noting that the positional oxidation occurs *endo*- to the cyclopentene framework and thus indicates that the formation of the more substituted allylic radical is clearly favoured.

BzO
OBz
$PhCO_3t$-Bu **3**
$Cu(OTf)_2$-**26** (5 mol%)
acetone, 0 °C
32
33; 41% e.e.
34; 1% e.e.
OBz
35; 93% e.e.
ratio **33**:**34**:**35** = 67:25:8

Singh has further extended this methodology by using a tri-coordinate pyridine based oxazoline **36a** and **36b** (A. DattaGupta and V.K. Singh, Tetrahedron Lett., 1996, **37**, 2633-2636). The stereoselectivity was shown to be good. The more substituted oxazoline **36b** gave the best selectivity for all substrates tested. Performing the reaction in the presence of 4Å molecular sieves was shown to increase the enantioselectivity further; presumably indicating that the reaction is sensitive to the presence of water. The configuration of the newly generated stereocentre was the same for all cycloalkenes studied. Furthermore the facial control appeared to be independent of the ring size of the starting alkene.

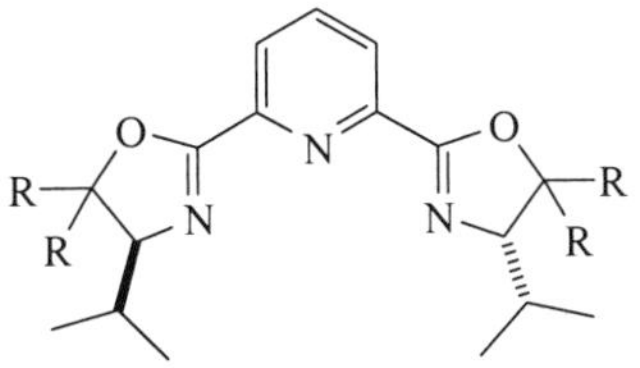

36a; R=H
36b; R=Ph

36a or 36b (3 mol%)
$Cu(OTf)_2$ (2 mol%)

Reaction	Ligand	reagents	% yield	% e.e.
2 → **4**	**36a**	$Cu(OTf)_2$, **3**	48	45
	36a	$Cu(OTf)_2$, **3**, 4Å mol.sieves	59	56
	36b	$Cu(OTf)_2$, **3**	38	42
	36b	$Cu(OTf)_2$, **3**, 4Å mol.sieves	70	59
9 → **5**	**36a**	$Cu(OTf)_2$, **3**	35	13
	36a	$Cu(OTf)_2$, **3**, 4Å mol.sieves	63	45
	36b	$Cu(OTf)_2$, **3**	43	70
	36b	$Cu(OTf)_2$, **3**, 4Å mol.sieves	58	81
14 → **6**	**36b**	$Cu(OTf)_2$, **3**	48	52

(iii) Use of copper amino acid based complexes.

Feringa has probed similar allylic oxidations using a copper(II)-(*S*)-proline based complex (C. Zondervan and B.L. Feringa, Tetrahedron: Asymmetry, 1996, **7**, 1895-1898). Moderate enantiocontrol (up to 60% e.e.) was observed using the naturally available (*S*)-proline, **37**, as the chiral ligand. Using the more Lewis basic ligand, (*S*)-thiaproline, caused both the yield and stereoselectivity to be reduced. Surprisingly the addition of an additive, anthraquinone, resulted in an increase in the facial selectivity, but not when *t*-butyl hydroperoxide was used as the oxidant. Further investigations into the mechanism revealed that there was a non-linear effect between the enantiomeric excess of the chiral ligand used (*e.g.*, proline) and the product formed, although no definitive deductions as to the nature of the active catalyst were made.

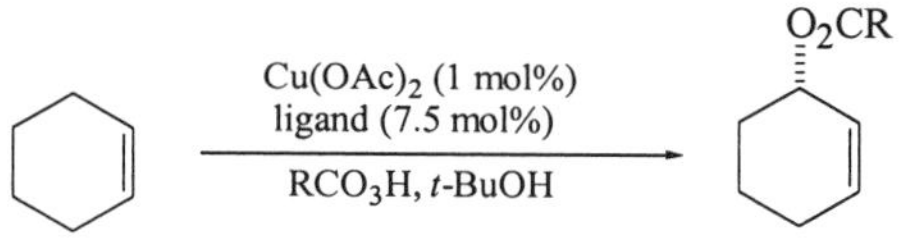

Entry	ligand	% yield	% e.e.
1	(*S*)-proline **37**	89	45
2	(*S*)-thiaproline	25	28
3	**37** and anthraquinone (4eq.)	80	60

Feringa has also investigated the use of substituted proline variants on this oxidation procedure (M.T. Rispens, C. Zondervan, B.L. Feringa, Tetrahedron: Asymmetry, 1995, **6**, 661-664). (*S*)-Proline, **37**, was found to be the most reliable ligand giving better facial control. Any attempt at protecting the nitrogen lone pair only served to lower the enantiomeric excess. Whereas, by using the strained four-membered azetidine-2-carboxylic acid, moderate levels of enantioselectivity returned.

Entry	ligand	% conversion	% e.e.
1	proline	38	51
2	α-methyl-proline	11	20
3	α-benzyl-proline	37	26
4	*N*-benzyl-proline	18	7
5	azetidine-2-carboxylic acid	37	35
6	pipecoline-2-carboxylic acid	7	8

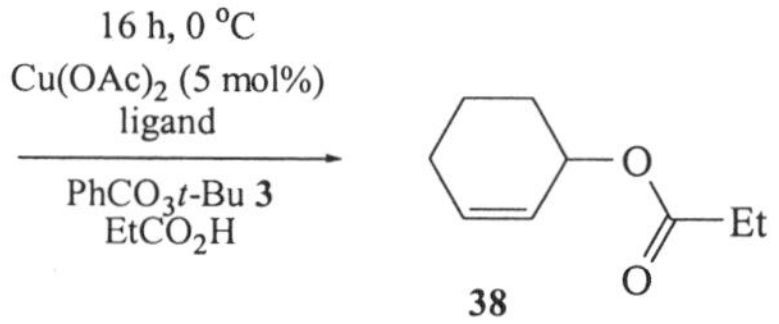

Andersson has synthesised a number of bicyclic α-amino acids based on the proline skeleton using a diastereoselective Diels-Alder approach (M.J. Sodergren and P.G. Andersson. Tetrahedron Lett., 1996, **37**, 7577-7580). Only the bicyclic [2.2.1] amino acid **39** gave better stereocontrol than the original (*S*)-proline **37**. These catalysts were very sensitive to the substitution pattern of the ligand; simple addition of an extra methylene unit in **40** had a dramatic effect on the facial control, causing it to be reduced from 60% to 11% e.e. (entry 1 versus entry 2).

39 **40**

2 (or **9**) → **4** (or **5**): $Cu(OAc)_2$ (5 mol%), $PhCO_3t$-Bu **3**, $PhCO_2H$, benzene

Entry	substrate	ligand	% yield	% e.e.
1	**2**	**39**	54	(*S*)-60
2	**2**	**40**	44	(*S*)-11
3	**9**	**39**	63	(*S*)-65
4	**9**	**40**	30	(*S*)-15

Feringa has also shown that the enantioselectivity was dependent on the structural nature of perester oxidant (M.T. Rispens, C. Zondervan, B.L. Feringa, Tetrahedron: Asymmetry, 1995, **6**, 661-664). This clearly indicates that the rate of perester decomposition has to fit smoothly and efficiently into the catalytic cycle, otherwise the stereoselectivity is compromised.

9 → **38**: 16 h, 0 °C, $Cu(OAc)_2$ (5 mol%), (*S*)-proline **37**, $R^1O_3CR^2$, $EtCO_2H$

Entry	perester	% conversion	% e.e.
1	*t*-butylperoxyacetate	86	43
2	*t*-butylperoxypropanoate	9	43
3	1,2-dimethylpropylperoxypivaloate	6	52
4	di(*t*-butylperoxy)di-cyclohexanoate	23	37
5	*t*-butylperoxybenzoate	43	45

This has led Muzart to further investigate the rate of addition of the perester to the reaction mixture (A. Levina, F. Henin and J. Muzart,

J. Organomet. Chem., 1995, **494**, 165-168). It was concluded that slow addition of the perester oxidant caused a decrease in chemical yield and evidently affects the catalytic turnover. The stereoselectivity remains constant (within experimental error) when an excess of the oxidant *t*-BuO_2COPh was used (entry 1 and 4-5). The optimum conditions were found to require 0.4 mmol of the oxidant *t*-BuO_2COPh, added in a single portion.

9 → [Cu{(*S*)-proline}$_2$], $PhCO_2H$, MeCN, reflux → **5** (O_2CPh)

Entry	$PhCO_3t$-Bu **3** (mmol)	% yield	% e.e.
1	4.0	71	38
2	0.4	76	47
3	0.4 × 2	63	43
4	0.4 × 3	48	44
5	0.4 × 4	61	40

Muzart has probed the use of a series of copper(I)-proline-like catalysts on the oxidation of cyclohexene, **9** (A. Levina, J. Muzart, Tetrahedron: Asymmetry, 1995, **6**, 147-156). They have shown that there was a direct correlation between the co-ordinating ability of the chiral ligand (to the copper atom) and the enantioselectivity observed. By lowering the co-ordination ability of the nitrogen atom through delocalisation (into either an *exo*- or *endo*-carbonyl group) in **41** and **42**, the facial selectivity was lost. Whereas, by simple methylation to give an *N*-methyl protected proline **43** some facial control returned.

9 → Cu_2O (2 mol%), **37**, **41**-**43** (10 mol%), $PhCO_3t$-Bu **3** (70% aq.) → **5** (O_2CPh)

CO_2H / N / H — **37**; 59%; 45% e.e.

CO_2H / N / Boc — **41**; 62%; 0% e.e.

O / CO_2H / N / H — **42**; 62%; 0% e.e.

CO_2H / N / Me — **43**; 27%; 15% e.e.

A series of cyclic and acyclic alkenes were then systematically screened against the optimum conditions using (*S*)-proline **37**, as the chiral ligand. They have concluded that cyclic alkenes gave better facial oxidation than those of the corresponding acyclic alkenes. Furthermore, the more conformationally rigid the alkene the better the stereocontrol as shown overleaf.

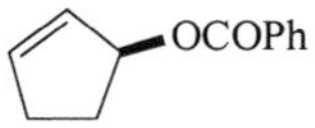

4

39%; 54% e.e.

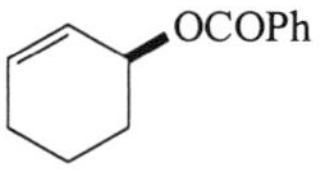

5

27%; 23% e.e.

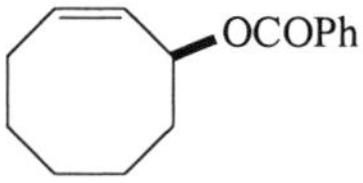

44

32%; 4% e.e.

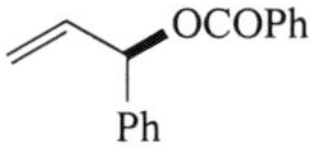

16

77%; 0% e.e.

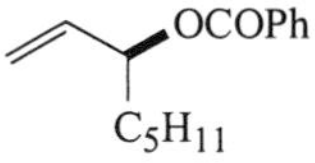

18

23%; 9% e.e.

They have further investigated the structural nature and acidity of the capturing carboxylic acid (RCO_2H). There were two distinct trends; (i) the more acidic the carboxylic acid, the worse the facial control [CH_3CO_2H (entry 3) > $ClCH_2CO_2H$ (entry 2) > CF_3CO_2H (entry 1)]; (ii) the more sterically demanding the carboxylic acid the better control [t-$BuCO_2H$ (entry 4) > *c*-$C_6H_{11}CO_2H$ (entry 5)]. Whereas, the addition of a further co-ordinating substituent in the carboxylic acid (such as a pyridinyl group (entry 7 and 8) had little effect on the facial preference.

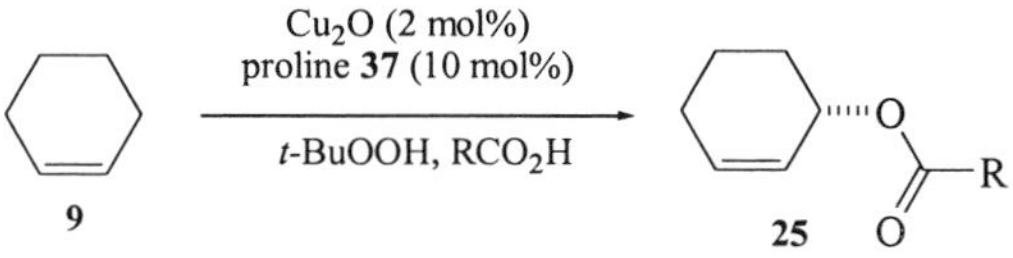

9 → 25: Cu_2O (2 mol%), proline **37** (10 mol%), *t*-BuOOH, RCO_2H

Entry	R	% yield	% e.e.
1	CF_3	0	0
2	$ClCH_2$	43	38
3	CH_3	44	41
4	*t*-Bu	7	52
5	*c*-C_6H_{11}	37	43
6	Ph	36	39
7	4-pyridyl	21	25
8	3-pyridyl	39	32

2. Enantioselective C-H oxidation using manganese(III)-salen complexes.

Katsuki has thoroughly investigated the enantioselective C-H oxidation of substituted pyrrolidines as a route to enantiomeric enriched pyrrolidinones (T. Punniyamurthy, A. Miyafuji and T. Katsuki, Tetrahedron Lett., 1998, **39**, 8259-8298).

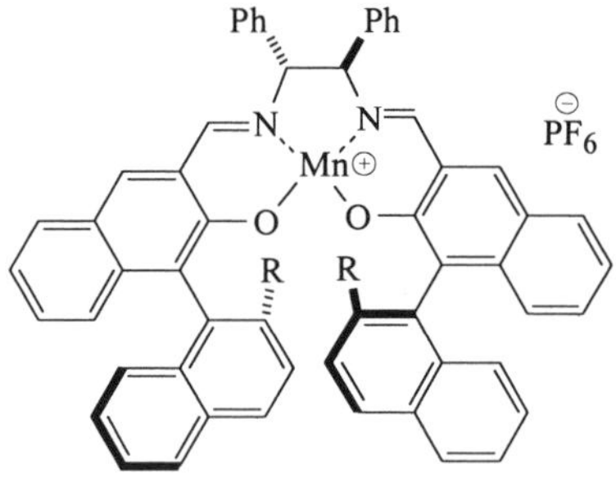

45; R=Ph

They have thoroughly screened a variety of substituted Mn(III) salen complexes and have shown that the tetra-naphthyl derivative **45**; R=Ph was the most facially selective. These hemi-aminals were oxidised *in-situ* to give directly the pyrrolidinones **48** to prevent possible epimerisation.

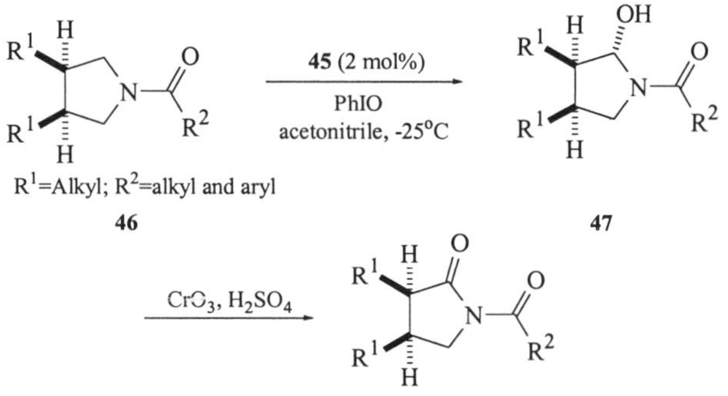

45 (2 mol%), PhIO, acetonitrile, -25°C

R^1=Alkyl; R^2=alkyl and aryl

46 **47**

CrO_3, H_2SO_4

48

This Mn(III) salen catalysis **45** was further screened against a series of *meso*-pyrrolidines and was shown to be efficient over a wide range of substrates. The required enantiomerically enriched pyrrolidinones (*e.g.*, **51**) were directly synthesised by simple deprotection of the resulting *exo*-amide in **50** through basic hydrolysis (LiOH in THF/H_2O). The yields were excellent and no loss of diastereoisomeric purity was observed.

1. **45** (2 mol%), PhIO
2. CrO_3, H_2SO_4

49 → **50**

65% yield, 76% e.e.

LiOH, THF, H_2O (1:1) → **51**

quantitative yield

Katsuki has further extended this methodology towards functionalisable bicyclic pyrrolidinones **52-55** (T. Punniyamurthy, and T. Katsuki, Tetrahedron, 1999, **55**, 9439-9454). The enantiofacial control was particularly impressive (up to 99% e.e.) when considering the presence of other co-ordinating substituents. This selectivity was shown to be sensitive to the substitution pattern in the adjacent bicyclic ring, such as the presence of a double bond in the cyclohexenyl framework (*e.g.*, **55**) which causes a substantial lowering of the stereoselectivity relative to the saturated cyclohexyl case **54**.

52: 68%, 78% e.e.

53: 35%, >99% e.e.

54: 70%, 88% e.e.

55: 29%, 64% e.e.

Katsuki has extended this protocol towards the synthesis of lactols **57-60** by the allylic oxidation of substituted tetrahydrofurans, such as **56** (A. Miyafuji and T. Katsuki, Tetrahedron, 1998, **54**, 10339-10348). The stereoselectivity was shown to be similar to that of other saturated heterocycles like pyrrolidines. However, the presence of the less nucleophilic *endo*-cyclic oxygen (within the original substituted tetrahydrofuran ring) allowed the lactols to be easily isolated - no further oxidation to the lactone was required. This enantioselectivity was high for a wide range of cyclic ethers and the better selectivity was observed with less conformationally flexible substituted tetrahydrofurans than the larger more flexible rings like a tetrahydropyran. This selectivity was surprisingly dependent on the structural nature and the polarity of the reaction solvent and chlorobenzene was found to be the best.

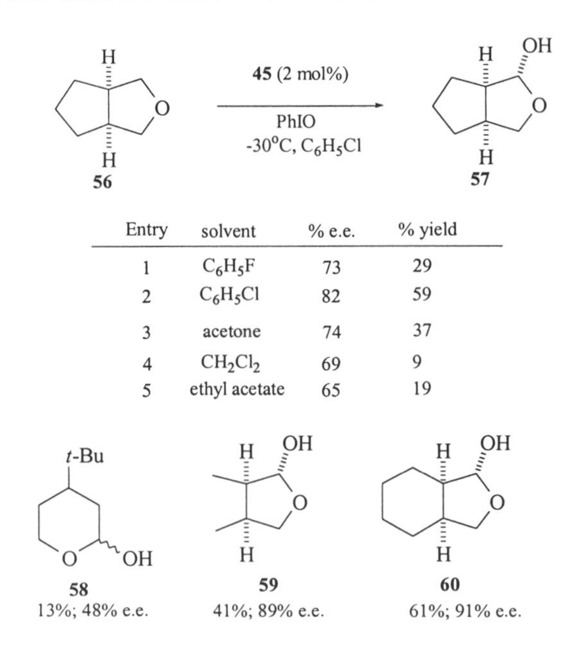

Entry	solvent	% e.e.	% yield
1	C_6H_5F	73	29
2	C_6H_5Cl	82	59
3	acetone	74	37
4	CH_2Cl_2	69	9
5	ethyl acetate	65	19

3. Use of molecular oxygen, catalysed by chiral monoaza-crown ethers.

Brussee has reported an elegant procedure for the enantioselective hydroxylation of 2-methylindanone **62** using molecular oxygen (O_2) as the oxidant (E.F.J. de Vries, L. Ploeg, M. Colao, J. Brussee and A. van der Gen, Tetrahedron: Asymmetry, 1995, **6**, 1123-1132). These reactions were performed in the presence of a catalytic quantity of chiral phase transfer reagent; monoaza crown ethers **61a** or **61b**. Hydroxylation certainly proceeds *via* direct addition of molecular oxygen to the corresponding enolate. The trimethylphosphite $P(OMe)_3$ reagent was required to reduce the intermediate peroxide to give directly the α-hydroxy ketone **63**. The enantioselectivity was generally poor, but values up to 52% have been observed using **61a**; R=H.

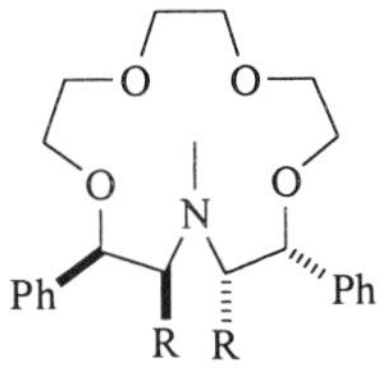
O
O
O
O
N
Ph
Ph
R R

61a

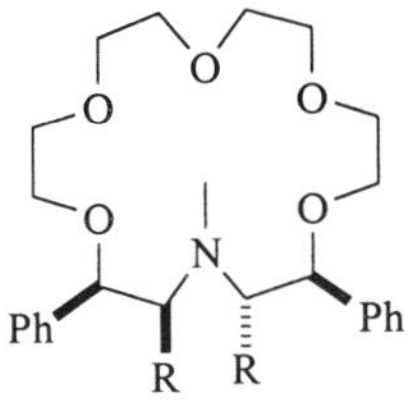
O
O
O
O
O
N
Ph
Ph
R R

61b

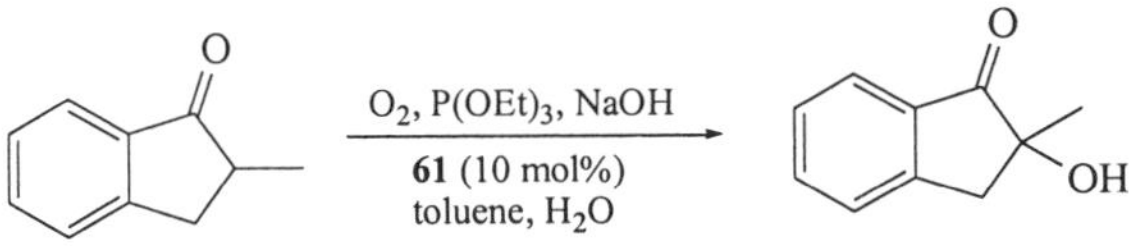
O
O_2, $P(OEt)_3$, NaOH
61 (10 mol%)
toluene, H_2O
O
OH

62 **63**

Entry	ligand	% yield	% e.e.	configuration
1	**61a**; R= H	93	52	(*R*)-
2	**61a**; R= Me	85	2	(*S*)-
3	**61a**; R= C_2H_5	85	6	(*S*)-
4	**61b**; R= H	91	6	(*R*)-
5	**61b**; R= Me	94	2	(*S*)-
6	**61b**; R= C_2H_5	75	3	(*S*)-

4. Baeyer-Villiger Oxidation

The synthesis of enantiomerically pure lactones using a Baeyer-Villiger oxidation of racemic, *meso*- and/or achiral ketones has attracted significant attention in recent years (M. Renz and B. Meunier, Eur. J. Org. Chem., 1999, 737-750 and G. Strukul, Angew. Chem., Int. Ed. Engl., 1998, **37**, 1198-1209). The majority of these reports deal with the use of biological catalysts, such as enzymes (E. Schoffers, A. Golebiowski and C.R. Johnson, Tetrahedron, 1996, **52**, 3769-3826 and S.M. Roberts, J. Chem. Soc., Perkin Trans. 1, 1998, 157-169). Some have also involved the use of substrate control (G. Fronza, C. Fuganti, G. Pedrocchi-Fantoni, V. Perozzo, S. Servi and G. Zucchi, J. Org. Chem. 1996, **61**, 9362-9367), whereas others have used auxiliary control (T. Sugimura, Y. Fujiwara and A. Tai, Tetrahedron Lett., 1997, **38**, 6019-6022).

The use of metal catalysts to promote this Baeyer-Villiger transformation is well known, but most are concerned with the synthesis of racemic lactones through the use of achiral catalysts (A.M.F. Phillips and C. Romao, Eur. J. Org. Chem., 1999, 1769-1770; M.D.T. Frisone, F. Pinna and G. Strukul, Organometallics, 1993, **12**, 148-156 and C. Bolm, G. Schlingloff and K. Weickhardt, Tetrahedron Lett., 1993, **34**, 3405-3408). Very little attention has been paid to the use of enantiomerically pure catalysts.

(i) Enantioselective oxidation involving achiral ketones.

Kanger has recently reported the use of a modified Sharpless titanium catalyst [Ti(O-*i*-Pr)$_4$, (-)-diethyl tartrate (DET), **66**, and *t*-BuOOH] in the enantioselective oxidation of 3-substituted cyclobutanones **64** (T. Kanger, K. Kriis, A. Paju, T. Pehk and M. Lopp, Tetrahedron: Asymmetry, 1998, **9**, 4475-4482). They have investigated the use of ketal-protected diethyltartrates like **67** as chiral ligands. The enantiocontrol was modest giving enantiomeric excesses up to 40% e.e. They have also noted that this simple protection caused the configuration of the lactone to be reversed, presumably due to the OH groups in the tartrate back-bone **66** directing the stereocontrol, whereas using the complementary chelating CO_2Et groups in **67** lead to the other enantiomer.

64 → **65**: Ti(O-*i*-Pr)$_4$, *t*-BuOOH, chiral ligand

Entry	ligand	config.	e.e. (%)	yield (%)
1	**66**	(*S*)-	33	23
2	**67**	(*R*)-	40	14

Bolm also chose a substituted cyclobutanone **69** to test the utility of an oxazoline-based copper(II) catalyst **68** (1 mol %) in this Baeyer-Villager oxidation using molecular oxygen (O_2) as the oxidant (C. Bolm, G. Schlingloff and F. Bienewald, J. Mol. Catal., 1997, **117**, 347). Better selectivity was observed with the *p*-nitro-substituted copper complex **68** in water-saturated benzene using pivalaldehyde as the oxygen acceptor (co-reducing agent). For example, oxidation of the achiral 4-

phenylcyclobutanone **69** under their general reaction conditions gave the (*S*)-lactone **70** (44% e.e.) in moderate yield.

68

68 (1 mol%)
O_2, *t*-BuCHO

69

70; 44% e.e.

Strukul has investigated the use of platinum(II) based catalysts involving hydrogen peroxide as the oxidant (C. Paneghetti, R. Gavagnin, F. Pinna and G. Strukul, Organometallics, 1999, **18**, 5057-5065). A variety of C_2-symmetric diphosphine-based chiral ligands were screened against a range of achiral substrates like 4-phenylcyclohexanone **72**. The highest enantioselectivity was observed with the atropisomeric BINAP catalyst **71** giving the ε-lactone (-)-**73** in 53% e.e.

71

72

H_2O_2, **71** (5 mol%)

(-)-**73**; 53% e.e.

(ii) Desymmetrisation through oxidation of *meso*-ketones.

Reports into the desymmetrisation of *meso*-ketone using a metal catalysed Baeyer-Villiger reaction are rare. Strukul has also probed the use of their platinum based BINAP catalyst, **71**, towards the desymmetrisation of *meso*-2,6-dimethyl-cyclohexanone **74** (C. Paneghetti, R. Gavagnin, F. Pinna and G. Strukul, Organometallics, 1999, **18**, 5057-5065). The selectivity was good giving the ε-lactone (-)-*syn*-**75**

71 (5 mol%)
H_2O_2

68 (1 mol%)
O_2, *t*-BuCHO

syn-**74**

(-)-*syn*-**75**; 79% e.e.

76

77; 91% e.e.

in 79% e.e.

Bolm has further extended the scope of their oxazoline-based copper(II) catalyst **68** towards the desymmetrisation of a tricyclic ketone **76** (C. Bolm, G. Schlingloff and F. Bienewald, J. Mol. Catal., 1997, **117**, 347). The stereoselectivity was excellent giving the γ-lactone **77** in 91% e.e. It is worthy of note, that the newly formed lactone **77** now contains four fully defined stereocentres.

(iii) Kinetic resolution through oxidation of racemic ketones.

The kinetic resolution of racemic cyclic ketones is by far the most popular Baeyer-Villiger metal catalysed reaction. Kanger has shown that Sharpless' titanium catalyst [Ti(O-*i*-Pr)$_4$, (-)-DET, **66**, and *t*-BuOOH] can be efficiently used to resolve racemic 3-substituted cyclobutanones such as **78** (T. Kanger, K. Kriis, A. Paju, T. Pehk and M. Lopp, Tetrahedron: Asymmetry, 1998, **9**, 4475-4482). These kinetic resolutions were particularly efficient giving γ-lactones (*e.g.*, **79**) with chemical yields up to 40% (from a theoretical maximum of 50%) and with good enantioselectivity (up to 75% e.e.).

O H H HO Br Ti(O-*i*-Pr)$_4$ *t*-BuOOH **66** O O H H HO Br

(*rac*)-**78** **79**; 40%; 75% e.e.

Strukul has also probed the resolution of a variety of cyclic ketones using their platinum BINAP catalyst **71** (A. Gusso, C. Baccin, F. Pinna and G. Strukul, Organometallics, 1994, **13**, 3442-3452). The stereoselectivity was shown to be varied and also dependent on ring size of the cyclic ketone. They found that substituted cyclohexanones (*e.g.*, **80**; 45%) gave better enantioselectivity than the corresponding cyclopentanone **84** (58%) and the larger the *C*(2)-substituent, the worse the control (*e.g.*, **85**; 12%).

H_2O_2, **71**

(*rac*)-**80** → (*S*)-**81**; 45% e.e.

(*rac*)-**82**;
R = CH_3, C_5H_{11} and *t*-Bu

83 (*S*)-16% e.e. **84** (*S*)-58% e.e. **85** 12% e.e.

68
O_2, *t*-BuCHO

(*rac*)-**86** → (*R*)-**87**; 41%; 65% e.e.

Bolm has synthesised related ε-lactones using their copper(II) oxazoline based catalyst **68** (C. Bolm, G. Schlingloff and K. Weickhardt, Angew. Chem. Int. Ed. Engl., 1994, **33**, 1848-1849). These kinetic resolutions were particularly efficient, for example 2-phenyl substituted cyclohexanone **86** gave the ε-lactone (*R*)-**87** (65% e.e.) in good yield (41%). They have further extended this methodology towards the synthesis of γ-lactones (C. Bolm and G. Schlingloff, J. Chem. Soc., Chem. Commun., 1995, 1247-1248). It was found that the positional C-O bond selectivity was rather poor and a considerable amount of the unusual less substituted lactone **90** [at position C(4)] was formed. However, more surprisingly the enantioselectivity for these unusual lactones (*e.g.*, **90**) were much higher than that of their more common constitutional counterparts, such as **89**. The overall yield was good indicating a good turnover.

68 (mol%)
O_2, *t*-BuCHO

(*rac*)-**88** → **89**; 67% e.e. + **90**; 92% e.e. (75:25)

overall yield = 61%

5. Heteroatom Oxidation

(i) Catalytic enantioselective oxidation of sulfides.

Enantiomerically pure sulfoxides have been shown to be useful synthetic intermediates (H.B. Kagan and R. Rebiere, Synlett, 1990, 643; G.H. Posnner, Acc. Chem. Res., 1987, **20**, 72 and G. Solladie, Synthesis, 1981, 185). Previous methods have relied on Andersen's sulfinate resolution procedure (K.K. Andersen, Tetrahedron Lett., 1962, **3**, 93), whereas, recently direct asymmetric oxidation of prostereogenic sulfides has been the preferred method. This procedure is rather attractive because of its simplicity, however, it is rather limited to the choice of substrates, namely the use of alkyl aryl sulfides. There are currently two procedures namely; Davis' oxidation involving chiral camphor-based oxaziridines derivatives (F.A. Davis, R.T. Reddy, W. Han and P.J. Carroll, J. Am. Chem. Soc., 1992, 114, 1428) and Kagan's modified Sharpless oxidation involving diethyl tartrate (DET) and a substituted peroxide (P. Pitchen and H.B. Kagan, Tetrahedron Lett., 1984, **25**, 1049). Both reactions are stoichiometric in the chiral component, but attempts at making Sharpless' procedure catalytic have resulted in a slower reaction with much lower enantioselectivity (K. Nakajima, C. Sasaki, M. Kojima, T. Aoyama, S. Ohba, Y. Saito and J. Fujita, Chem Lett., 1987, 2189).

Uemura has partially solved this catalytic problem using (*R*)-binaphthol **91** instead of (+)-DET, **66**, as the chiral ligand (N. Komatsu, M. Hashizume, T. Sugita and S. Uemura, J. Org. Chem., 1993, **58**, 4529-4533). Thorough investigation into the optimum reaction conditions have revealed that oxidation with 70% aqueous *t*-butyl hydroperoxide in an oxygen (O_2) containing atmosphere in the presence of Ti(O-*i*-Pr)$_4$ gave the best selectivity. The oxidation of a variety of substituted alkyl aryl sulfides **92** has revealed that this reaction appears to be particularly efficient. A low concentration of the procatalyst Ti(O-*i*-Pr)$_4$ (2.5 mol%) in the presence of the chiral ligand (*R*)-binapthol **91** (5 mol%) is required to give the corresponding sulfoxide (*R*)-**93** with superb enantioselectivity (96% e.e.). The yields were unfortunately low, due to the unwanted conversion of the sulfoxide to the sulfone **94**. However, they have elegantly shown that the origin of this near perfect selectivity was in part due to the kinetic resolution and the removal of the minor enantiomer of the sulfoxide **93** to the achiral sulfone **94**. In fact, the enantiomeric excess of the sulfoxide **93** increases from 66% to 96% e.e. under the reaction conditions.

92 → (2.5 mol% $Ti(O\text{-}i\text{-}Pr)_4$, 5 mol% **91**, H_2O, TBHP) → (*R*)-**93**

91

Entry	R	% yield	% e.e.
1	Ph; **92a**	28	96
2	*p*-Tolyl; **92b**	44	96
3	*p*-BrC_6H_4; **92c**	39	96

R= *p*-tolyl (*R*)-**93b**; 66% e.e. + R= *p*-tolyl (*S*)-**93b** → (2.5 mol% $Ti(O\text{-}i\text{-}Pr)_4$, 5 mol% **91**, H_2O, TBHP) → R= *p*-tolyl (*R*)-**93b**; 96% e.e. + R= *p*-tolyl **94**

Rosini has studied this reaction using a catalyst formed *in-situ* by reacting $Ti(O\text{-}i\text{-}Pr)_4$ with enantiomerically pure 1,2-diphenylethane diol **95** (M.I. Donnoli, S. Superchi and C. Rosini, J. Org. Chem., 1998, **63**, 9392-9395). Their initial investigation was centred on inducing enantiocontrol as a result of facial oxidation of the sulfide rather than by kinetic resolution of the resulting sulfoxide. They argued that the unwanted sulfone was formed by direct complexation of the sulfoxide to the titanium catalyst. By lowering the amount of titanium in the catalyst to 0.1-0.5 equivalents relative to the chiral diol ligand **95**, slow sulfone formation gave the required sulfoxide **95** in good yield with reasonable enantiomeric excess. The selectivity was modest with either an electron donating or a withdrawing substituent on the adjacent aryl ring (entry 2 and 3). The more sterically demanding 2-naphthyl substituent gave similar control to that of the 'bench-mark' phenyl methyl sulfide **92a**. Altering the alkyl substituent to a more sterically demanding group gave a slight increase in the facial control. A benzyl substituent (entry 8) gave the best control (e.e. >99%; yield =73%).

(*R*,*R*)-**95** (10 mol%.), Ti(O-*i*-Pr)$_4$ (5 mol%), H_2O, TBHP (200 mol%), CCl_4, 0 °C

Ar–S–R **92** → Ar–S(=O)–R **93**; (*R*,*R*)-**95**: HO, OH, Ph, Ph

Entry	Ar	R	% yield	% e.e.
1	Ph	Me; **92a**	63	80
2	*p*-MeO-C_6H_4	Me; **92d**	55	69
3	*p*-NO_2-C_6H_4	Me; **92e**	74	44
4	2-naphthyl	Me; **92f**	65	73
5	Ph	Et; **92g**	71	70
6	Ph	*n*-Bu; **92h**	69	80
7	Ph	^{i}Pr; **92i**	60	22
8	Ph	$PhCH_2$;**92j**	73	>99

Bolm has recently extended this procedure using 10 mol% of Ti(O-*i*-Pr)$_4$ with *Schering's* atropisomeric diols (*S*,*S*)- and (*R*,*S*)-**96** (C. Bolm and O.A.G. Dabard, Synlett., 1999, 360-362). In all cases studied, the diol (*S*,*S*)-**96** gave far better stereocontrol in the oxidation of phenyl methyl sulfide, **92a**, than its diastereoisomeric partner (*S*,*R*)-**96**. The yields were always high, presumably indicating the presence of water in the reaction mixture slows the rate of kinetic resolution to the unwanted sulfone.

Ph–S–Me **92a** → TBHP (200 mol%), **96**, 20 mol%, Ti(O-*i*-Pr)$_4$; 10 mol%, THF, H_2O (300 mol%) → Ph–S$^{\oplus}$(–O$^{\ominus}$)–Me

(*S*)-**93a**; 92% e.e.; 76% [using (*S*,*S*)-**96**]

(*R*)-**93a**; 15% e.e.; 36% [using (*R*,*S*)-**96**]

HO, HO (*S*,*S*)-**96** HO, HO (*R*,*S*)-**96**

Bolm has further extended this methodology using a chiral vanadium complex (C. Bolm and F. Bienewald, Synlett., 1998, 1327-1328). One of the more interesting features of this procedure is the efficiency of this vanadium catalyst; only 0.01 to 1 mol% was required. Many of these imine based chiral ligands, such as (*S*)-**98** are readily available by the addition of an amino alcohol to a substituted salicylaldehyde. The stereoselectivity was modest leading to a variety of single diastereoisomeric sulfoxides **99**. The electronic and steric nature of the aryl substituents appears to play an important role in this oxidation process. A more sterically demanding substituent such as *t*-Bu lowers the selectivity, as does the presence of an electron withdrawing and donating R^1 substituent on the aryl ring.

97 → **99**: $VO(acac)_2$(1 mol%), (*S*)-**98** (1.5 mol%); H_2O_2 (30%), DCM, RT

Entry	R^1	% yield	% e.e.
1	Ph	84	85
2	*p*-CH_3O-C_6H_4	60	57
3	*o*-NO_2-C_6H_4	75	62
4	*t*-Bu	67	46
4	$C(CH_3)_2CH_2OH$	62	47

(*S*)-**98**

Bonchio has reported the use of a chiral zirconium (IV) catalyst based on a chiral tetradenate amino alcohol **100** (M. Bonchio, G. Licini, F Di Furia, S. Mantovani, G. Modena and W.A. Nugent, J. Org. Chem., 1999, **64**, 1326-1330). The enantioselective oxidation of tolyl methyl sulfide **92b** was good, however the yield was particularly low due to the competing kinetic resolution of the required sulfoxide **93b** to give the unwanted corresponding achiral sulfone **94**.

p-Tol-S-Me **92b** → (**100** (4 mol%), $Zr(O$-*n*-$Bu)_4$ (4 mmol%), dichloroethane) → (*R*)-**93b**; 86% e.e.; 25% + **94**; 75%

(ii) Titanium mediated enantioselective oxidation of sulfides.

The majority of these reports into the enantioselective oxidation of prostereogenic sulfides utilise Sharpless' titanium-mediated oxidation process involving $Ti(O$-*i*-$Pr)_4$, $PhCMe_2CO_2H$ and (*R*,*R*)- or (*S*,*S*)-DET, **66**, which was independently developed by Kagan and Modena (K.-U.

Baldenius and H.B. Kagan, Tetrahedron: Asymmetry, 1990, **1**, 597 and P. Bendazzoli, F. Di-Furia, G. Licini and G. Modena, Tetrahedron Lett., 1993, **34**, 2975). This procedure utilises a stoichiometric quantity of Ti(O-*i*-Pr)$_4$ in the presence of two equivalents of the chiral ligand DET, **66**.

$PhCMe_2CO_3H$ (2 eq.)
Ti(O-*i*-Pr)$_4$ (1 eq.)
(+) or (-) DET **66** (2 eq.)
H_2O (1 eq.)

101

Entry	R	temp	% yield	ratio *anti-*:*syn-*	% e.e. *anti-*	% e.e. *syn-*	tartrate
1	Et; **101a**	-20°C	66	3:1	65	65	(+)
2	Bn; **101b**	-20°C	63	3:1	66	64	(+)
3	*t*-Bu; **101c**	-37°C	60	1:1	92[a]	88[a]	(-)

[a]one recrystallisation leads to optical purity

Page has used this methodology in the enantioselective oxidation of a series of acyl substituted 1,3-dithianes (P.C.B. Page, M.T. Gareh and R.A. Porter, Tetrahedron: Asymmetry, 1993, **4**, 2139-2142). The diastereoselective control within the oxidation step was rather non selective - ranging from a ratio of 1:1 to 3:1 in favour of the major *anti*-diastereoisomer **101**. It is worth noting the unusual selectivity difference between the related substituted esters **101**; R=Et, Bn and *t*-Bu. The more sterically demanding *t*-butyl ester (92 % e.e. - entry 3) gave the best selectivity, presumably due to the additional chelation to titanium in the stereochemical-determining step.

102

$PhCMe_2CO_3H$
Ti(O-*i*-Pr)$_4$, (-)-DET **66**
H_2O, CH_2Cl_2, -20 °C

103; >99% e.e.; 71%

104

61% complete
Ti(O-*i*-Pr)$_4$, (+)-DET **66**, TBHP
DCM, H_2O, -20 °C

105; 62% e.e.

This protocol has been applied to the synthesis of two biologically active molecules. Firstly, Pitchen utilised this methodology towards the synthesis of a potent biological inhibitor based on the enantioselective oxidation of a substituted imidazole **102** giving the required sulfoxide **103** in perfect stereocontrol (>99% e.e.) and in good yield (P. Pitchen, C.J. France, I.M. McFarlane, C.G. Newton and D.M. Thompson, Tetrahedron Lett., 1994, **35**, 485-488). Secondly, Panetta has reported the synthesis of sulfoxide **105**, which has been used as a potent anti-inflammatory agent for bowel disease (M.L. Phillips and J.A. Panetta, Tetrahedron: Asymmetry, 1997, **8**, 2109-2114).

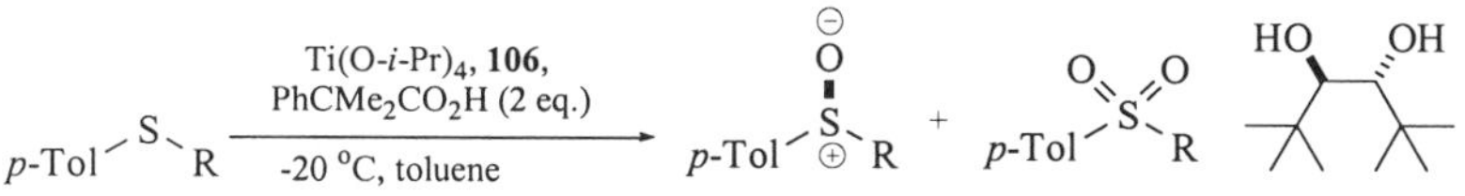

$Ti(O\text{-}i\text{-}Pr)_4$, **106**, $PhCMe_2CO_2H$ (2 eq.), -20 °C, toluene

92 → **93**; 95% e.e.; 42% + **94**; 40%; **106**

Imamoto has attempted to extend the scope of this titanium-mediated process by using a different C_2-symmetric diol **106** as the chiral ligand (Y. Yamanoi and T. Imamoto, J. Org. Chem., 1997, **62**, 8560-8564). The selectivity was modest for *p*-tolyl methyl sulfide **92a**, giving the sulfoxide **93a** with superb enantiomeric excess (up to 95% e.e). The yield was slightly compromised by further oxidation to give the sulfone **94** in 40% yield. However, this complementary reaction is necessary since the oxidation of the sulfoxide **93a** to the sulfone **94** increased the enantioselectivity through kinetic removal of the unwanted enantiomer of the sulfoxide **93a**.

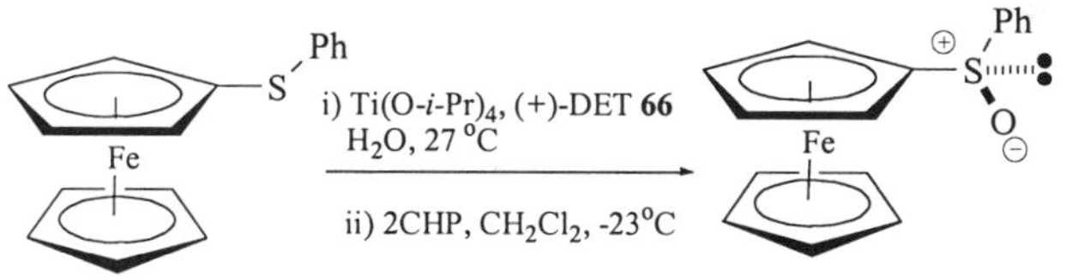

i) $Ti(O\text{-}i\text{-}Pr)_4$, (+)-DET **66**, H_2O, 27 °C; ii) 2CHP, CH_2Cl_2, -23°C

107 → **108**; >99% e.e.; 78%

Kagan has further investigated the use of a ferrocenyl phenyl sulfide **107** as a substrate for his oxidation protocol (P. Diter, O. Samuel, S. Taudien and H.B. Kagan, Tetrahedron: Asymmetry, 1994, **5**, 549-552). The enantioselectivity was excellent giving the (*R*)-sulfoxide **108** with an excellent 99% enantiomeric excess. The chemical yield was surprisingly good (78%), presumably indicating that this bulky ferrocenyl substituent aids the facial selectivity of the oxidation. In addition to this the electron withdrawing nature of the ferrocenyl group prevents further oxidation to the achiral sulfone.

92 → **93**: i) $Ti(O\text{-}i\text{-}Pr)_4$, (+)-DET **66**; ii) H_2O, 16 °C then -22 °C; iii) CHP, CH_2Cl_2, -22°C

Entry	R^1	R^2	% yield	% e.e.
1	*p*-tolyl; **92b**	Me	75	>99.5
2	*p*-MeO-C_6H_4; **92d**	Me	78	99.5
3	*p*-NO_2-C_6H_4; **92e**	Me	51	99.3
4	2-naphthyl; **92f**	Me	81	77.5
5	1-naphthyl; **92g**	Me	91	91.2
6	*p*-tolyl; **92h**	Et	82	86.6
7	*p*-tolyl; **92i**	*n*-butyl	64	38.2

Kagan has investigated a variety of C(4)-substituents in the adjacent aryl ring within a number of related sulfides **92** to probe the steric and electronic effects of this oxidation procedure (J-M. Brunel, P. Diter, M. Duetsch and H.B. Kagan, J. Org. Chem., 1995, **60**, 8086-8088). This method appears to be insensitive to the electronic nature of the 4-aryl substituent – an electron withdrawing and donating substituent behaves similarly giving perfect enantioselectivity (entry 1-3). However, a much lower yield was shown with sulfides with an electronically withdrawing group, such as *para*-nitro group (entry 3), which could be attributed to the less nucleophilic nature of the sulfide. The less sterically demanding 2-naphthyl substituent gave lower selectivity than its near neighbour 1-naphthyl (entry 4 versus entry 5). Whereas, increasing the size of the alkyl portion in **92** served only to lower both the enantioselectivity as well as the yield (entry 1 versus entry 6 and 7).

92a → Na_2MoO_4, H_2O_2, **109** → (*R*)-**93a**; 60% e.e.; 98%

109 ⇌ Na_2MoO_4, H_2O_2 → metal capped apohost

Carofiglio has elegantly shown the use of a molybdenum β-cyclodextrin-based oxidising reagent **109** for the enantioselective oxidation of phenyl methyl sulfide **92a** (M. Bonchio, T. Carofiglio, F. Di Furia and R. Fornasier, J. Org. Chem., 1995, **60**, 5986-5988). The stereoselectivity was modest (up to 60% e.e.), whereas, the yield was impressively nearly quantitative. It is worth noting that this reagent does not over oxidise the required sulfoxide **93a** to the unwanted sulfone, which was evidently problematic with Sharpless' modified procedure.

By comparison Sharpless' asymmetric dihydroxylation (AD) reaction has been used to dihydroxylate prostereogenic allylic sulfides (P.J. Walsh, P. Tong Ho, S.B. King and K.B. Sharpless, Tetrahedron Lett., 1994, **35**, 5129-5132). The presence of the additional sulfur atom in allylic sulfide did not affect the enantioselectivity and furthermore was not oxidised under these reaction conditions (see chapter 1).

p-Tol–S–Me (**92b**) → [**110** (Ph(Me)C(Me)OOH), $Ti(O\text{-}i\text{-}Pr)_4$ (5 mol%)] → *p*-Tol–S(O)–Me (**93b**; 71% e.e.) + *p*-Tol–SO_2–Me (**94**)

93b:**94** 20:80 (overall yield 88%)

An obvious area for extension is the use of enantiomerically pure peroxides such as **110** (W. Adam, M.N. Korb, K.J. Roschmann and C.R.Saha-Moller, J. Org. Chem., 1998, **63**, 3423-3428). Korb has shown the concept of this strategy by synthesising the enantiomerically enriched sulfoxide **93b**. The stereoselectivity was good (up to 71% e.e.), but sadly the yields were poor due to over oxidation to the corresponding sulfone **94**. The presence of the $Ti(O\text{-}i\text{-}Pr)_4$ catalyst was paramount, without it no stereoselectivity was observed giving exclusively the racemic sulfoxide **93b** in a low 6% yield.

(iii) Kinetic resolution of sulfoxides.

Skarzewski has investigated the enantioselective oxidation of a series of sulfides as a general method for the preparation of C_2-symmetric sulfoxides using a vanadium-based catalyst with ligand **112** (J. Skarzewski, E. Ostrycharz and R. Siedlecka, Tetrahedron: Asymmetry, 1999, **10**, 3457-3461). The enantioselectivity was excellent and evidently relied on the reagent control associated with catalyst to give the C_2-symmetric bis-sulfoxide **113** as the major diastereoisomer. This bis-sulfoxide can only be formed by direct kinetic resolution of the intermediate sulfoxide **114**. An electron donating substituent in the adjacent aryl ring (in **111/114** – entry 2 and 4 versus entry 1) appears to give better control than an electron withdrawing substituent (entry 3). This is presumably due to a more nucleophilic sulfide demanding a much tighter transition state. The overall diastereoselectivity could be further increased to 80% d.e. by ensuring the sulfide groups have a 1,5-relationship. However, the enantioselectivity diminished rapidly to 20% e.e., which may indicate some internal chelation (substrate control) from intermediate mono-sulfoxide **114** in the kinetic resolution step.

111 → **113**: $VO(acac)_2$ (1.5 mol%), **112** (1.5 mol%), H_2O_2, CH_2Cl_2, 0 °C

Entry	R	% yield	% d.e.	% e.e.
1	H	41	60	95
2	2-OMe	95	65	>95
3	2-COOMe	12	84	20
4	4-Me	56	78	92

112 (NO_2, Ph, OH, N, OH, *i*-Pr)

114

115; >80% d.e.; 25% e.e.; 92%

Aggarwal has similarly used a Sharpless based method to synthesise a related *trans*-1,3-dithiane 1,3-dioxide **101a** (V.K. Aggarwal, B.N. Esquivel-Zamora, G.R. Evans and E. Jones, J. Org. Chem., 1998, **63**, 7306-7310). The stereoselectivity was excellent under Modena's protocol, giving the bis(sulfoxide) **102a** in 97% e.e. The yield was surprisingly good. However, there was some competitive formation of the mono-sulfoxide (8%) and sulfoxide-sulfone (19%). Simple decarboxylation of this intermediate bis(sulfoxide) **102a** gave the required C_2-symmetric *trans*-1,3-dithiane 1,3-dioxide **116** in good yield.

CHP(4 eq.), (+)-DET **66** (2 eq.) Ti(O-*i*-Pr)$_4$ (0.5 eq.), -22 °C, 24 h; NaOH, H_2O

101a → **102a** 60% (>97% e.e.) → **116** 83% >99% e.e.

Aggarwal has also extended this procedure towards the synthesis of a synthetically valuable C_2-symmetric chiral ketene equivalent **119** by oxidation of the dithiolane **117** (V.K. Aggarwal, Z. Gultekin, R.S. Grainger, H. Adams and P.L. Spargo, J. Chem. Soc., Perkin Trans. 1, 1998, 2771-2781). The stereoselectivity was excellent (>98% e.e.) and this intermediate *trans*-bis(sulfoxide) **118** was efficiently converted to the required methylene-1,3-dithiolane 1,3-dioxide **119**.

CHP (4 eq.) (+)-DET **66** (2 eq.) Ti(O-*i*-Pr)$_4$ (0.5 eq.)

117 → **118** 68%, >98% e.e. → **119**

Licini has additionally investigated a similar kinetic resolution procedure towards the synthesis of a sulfoxide **121** (P. Bendazzoli, F. Di Furia, G. Licini and G. Modena, Tetrahedron Lett., 1993, **34**, 2975-2978). The stereocontrol was surprisingly good for all chiral sulfoxides **121**, **122**

TBHP, Ti(O-*i*-Pr)$_4$, (+)-DET **66**, CH_2Cl_2, 0 °C

120

Ratio **120**:Ti:(+)-DET:TBHP 8:1:4:2

121; 94% e.e.; 84% + **122**; 99% e.e.; 16%

meso-**123**; 0.7% + **124**; 88% e.e.; 2.7%

and **124** formed. This presumably indicates that the initial enantioselective oxidation of the bis(sulfide) **123** was very facially selective leading to the sulfoxide **121** in high enantiomeric excess. Furthermore the yield was excellent which indicates the subsequent chemoselective oxidation to give the bis(sulfoxide) **122** and **123**, and the sulfoxide-sulfone **124** was much slower than the original oxidation to give the sulfoxide **121**.

As an alternative, the synthesis of enantiomerically pure sulfoxides using enzymatic processes has attracted a great deal of interest in recent years. Generally, the enantiocontrol is excellent (up to >99% e.e.), but was very dependent on the substrate used. It is worth noting that further oxidation to the unwanted sulfone can be prevented using these methods (S. Colonna, N. Gaggero and M. Leone, Tetrahedron, 1991, **47**, 8385-8398; S. Colonna, N. Gaggero, A. Bertinotti, G. Carrea, P. Pasta and A. Bernardi, J. Chem. Soc., Chem. Commun., 1995, 1123-1124 and V. Alphand, N. Gaggero, S. Colonna, P. Pasta and R. Furstoss, Tetrahedron, 1997, **53**, 9695-9706).

Second Edition of Rodd's Chemistry of Carbon Compounds,
Volume V, Topical Volumes
Asymmetric Catalysis, edited by M.Sainsbury

Chapter 3

REACTIONS INVOLVING METALLOCARBENES

DAVID. M. HODGSON, PAUL A. STUPPLE and DAVID C. FORBES

The field of catalytic asymmetric reactions involving carbene-like processes is currently dominated by transformations of diazo compounds. For reviews see: M. P. Doyle, M. A. McKervey and T. Ye, "Modern Catalytic Methods for Organic Synthesis with Diazo Compounds", John Wiley & Sons, Inc., New York, 1997; M. P. Doyle in "Catalytic Asymmetric Synthesis", 2nd edn, Ojima, I., Ed., VCH Publishers, New York, 2000, Chapter 5; M. P. Doyle, "Catalytic Asymmetric Synthesis", Ojima, I., Ed., VCH Publishers, New York, 1993, Chapter 16.1 – 16.3 in "Comprehensive Asymmetric Catalysis", Vol. II, E. N. Jacobsen, A. Pfaltz, H. Yamamoto, Eds., Springer-Verlag, Berlin, 1999; M. P. Doyle, D. C. Forbes, "Chemistry for the 21st Century: Transition Metal Catalyzed Reactions", S. G. Davies, S. Murahashi, Eds., Blackwell Science, Oxford, 1999, Chapter 14; M. P. Doyle, D. C. Forbes, Chem. Rev., 1998, **98**, 911; H. M. L. Davies, Eur. J. Org. Chem., 1999, **9**, 2459; H. M. L. Davies, "Advances in Cycloaddition", M. E. Harmata, Ed., JAI Press Inc, Greenwich, CT, Vol. 5, 1999; M. A. Calter, Curr. Org. Chem., 1997, **1**, 37; M. P. Doyle, M. A. McKervey, Chem. Commun. 1997, 983; M. P. Doyle, Aldrichim. Acta 1996, **29**, 3; M. P. Doyle, Chem. Rev., 1986, **86**, 919. These reactions are generally believed to proceed *via* transient metallocarbenes arising from the reaction of the diazo (usually α–diazocarbonyl) substrate with a metal (often copper or rhodium)-ligand system. The metal species used is believed to effect formation of the metallocarbene by acting as a Lewis acid and accepting electron density from the diazo carbon at a vacant coordination site on the metal. This is then followed by back donation of electron density from the metal to the carbene carbon with concomitant loss of N_2. As the intermediate metallocarbene can be considered to be electrophilic at carbon, it can accept electron density from a suitable donor XY with eventual loss of the the metal-ligand complex which therefore functions as a catalyst.

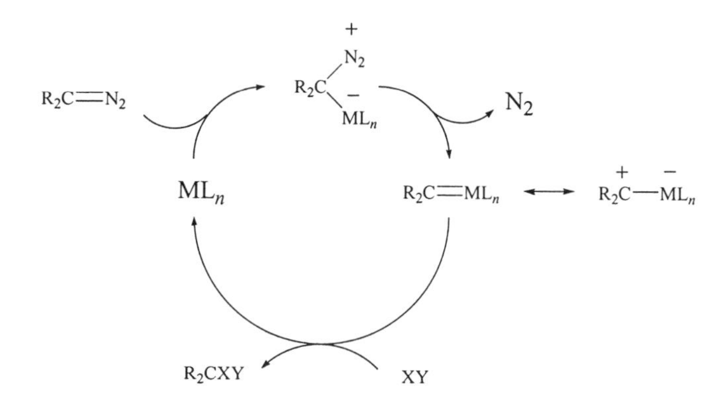

Specific metal-ligand combinations can allow the metallocarbene to discriminate between potentially competing carbene reactions, such as cyclopropanation and C-H insertion, and/or paths to different stereo–isomers by controlling the stereoselectivity of a reaction. If in such a transformation achiral starting materials are used to generate a product that contains a stereocentre, or stereocentres, then asymmetric induction is possible with a metal-ligand catalyst where the ligand is chiral (and non-racemic); this is because in the presence of such ligands the transition states leading to the enantiomeric products become diastereomeric. Significant progress has been made using chiral, non-racemic transition metal-based catalysts in transformations of diazocarbonyl compounds involving enantioselective X-H (X = C, Si) insertions (see Section 1), cyclopropanations with alkenes (see Section 2), cyclopropenations with alkynes (see Section 3), and progress is beginning to be made in enantioselective ylide transformations (see Section 4).

1. Asymmetric Insertion Reactions

Asymmetric metal carbene transformations have proven to be highly versatile when considering the insertion of a carbene moiety into carbon-hydrogen and heteroatom-hydrogen bonds. Systems of high bond polarity, including O-H, N-H and S-H, proceed as formal insertions, though they are better understood as ylide transformations. Bonds of low polarity, such as C-H and Si-H, are inserted by the metal carbene in a concerted fashion whereby the bond-breaking/bond-forming occurs in one step.

$$R_2C{=}ML_n \quad + \quad X{-}H \quad \longrightarrow \quad R_2C\genfrac{}{}{0pt}{}{X}{H} \quad + \quad ML_n$$

Recently, examples of high selectivities have been reported in both intermolecular and intramolecular insertion reactions. Both processes are highly dependent upon the electronic and steric environment of both the metal carbene and the targeted carbon-hydrogen bond. For intramolecular insertion reactions, there is a high preference for formation of five-membered rings and a dramatic increase in reactivity with either increasing alkyl substitution, tertiary C-H being highest (D. F. Taber, "Comprehensive Organic Synthesis: Selectivity, Strategy, and Efficiency in Modern Organic Chemistry", B. M. Trost, I. Fleming, Eds., Pergamon Press, New York, 1991), or with adjacent heteroatomic substitution (P. Wang, J. Adams, J. Am. Chem. Soc., 1994, **116**, 3296). Regardless of the mode of insertion, conformational influences imposed on either carbon-hydrogen bond or metal carbene can greatly influence product formation (M. P. Doyle *et al.*, J. Am. Chem. Soc., 1993, **115**, 958).

a. Enantiocontrol

i. Intramolecular: diazoesters

McKervey and co-workers were the first to report asymmetric induction in intramolecular C-H insertion reactions (M. Kennedy *et al.*, J. Chem. Soc., Chem. Commun., 1990, 361). Using an α-diazo-β-ketosulfone in the presence of a chiral dirhodium(II) prolinate catalyst, an enantiomeric excess of 12% was achieved (eq 1).

SO_2Ph; N; O–Rh; O–Rh; 4

$PhSO_2$; O; N_2; CH_3 → (CH_2Cl_2, >90%) O; SO_2Ph; CH_3; ~12% ee (1)

Although this level of stereocontrol was low, Hashimoto, Ikegami and co-workers subsequently have reported higher levels of enantiocontrol using their phthalimide-derivatized phenylalanate dirhodium(II) catalysts with a series of α-diazo-β-ketoesters (eq 2) (S. Hashimoto, N. Watanabe, S. Ikegami, Tetrahedron Lett., 1990, **31**, 5173; S. Hashimoto *et al.*, Tetrahedron Lett., 1993, **34**, 5109).

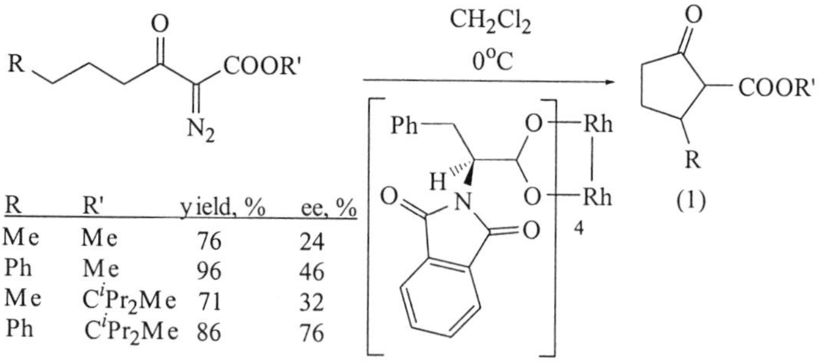

CH_2Cl_2, 0°C

COOR'

N_2

Ph, H, N, O, O, O, Rh, Rh, 4

(1) (2)

R	R'	yield, %	ee, %
Me	Me	76	24
Ph	Me	96	46
Me	C^iPr_2Me	71	32
Ph	C^iPr_2Me	86	76

The highest levels of enantioselectivity observed within this series were with phenyl substituted α-diazo-β-ketoesters (R' = CH_3, 46% ee; R' = $C^iPr_2CH_3$, 76% ee). Other catalysts have been employed with this system, such as Ito's ferrocenecarboxylate derivative of a dirhodium(II) catalyst, which resulted in lower observed enantioselectivities for (1) (M. Sawamura *et al.*, Bull. Chem. Soc. Jpn., 1993, **66**, 2725).

Exceptionally high levels of enantiocontrol have been reported in a number of intramolecular C-H insertion reactions using chiral dirhodium(II) carboxamidate catalysts. Capitalizing on Spero and Adams's work (J. Adams, D. M. Spero, Tetrahedron, 1991, **47**, 1765) which demonstrated that the oxygen heteroatom enhanced the reactivity of adjacent C-H bonds, Doyle and co-workers reported on highly enantioselective insertions to form alkoxy-substituted γ-lactones (M. P. Doyle *et al.*, J. Am. Chem. Soc., 1991, **113**, 8982). Just 0.1 mol % of catalyst was sufficient to cause complete diazo decomposition of (2) and furnish the desired lactones in good yield (eq 3).

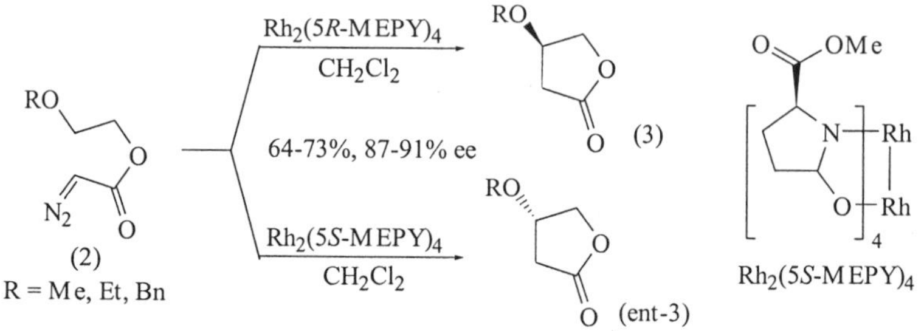

RO, N_2, O, (2), R = Me, Et, Bn

$Rh_2(5R\text{-MEPY})_4$, CH_2Cl_2 → (3)

64-73%, 87-91% ee

$Rh_2(5S\text{-MEPY})_4$, CH_2Cl_2 → (ent-3)

OMe, N, Rh, Rh, 4

$Rh_2(5S\text{-MEPY})_4$

(3)

Both (3) and (ent-3) using $Rh_2(5R\text{–MEPY})_4$ and $Rh_2(5S\text{-MEPY})_4$ were prepared using this methodology. The dirhodium(II) catalyst $Rh_2(5S\text{-MEPY})_4$ was found to effect the highest level of enantiocontrol with these systems in the initial 1991 report. Improvements in enantiocontrol have since been disclosed. For example, (ent-3) (R=Et) has been reported in 92% ee with Rh_2(4*S*-MEOX)$_4$ and 94% ee with Rh_2(4*S*-MPPIM)$_4$.

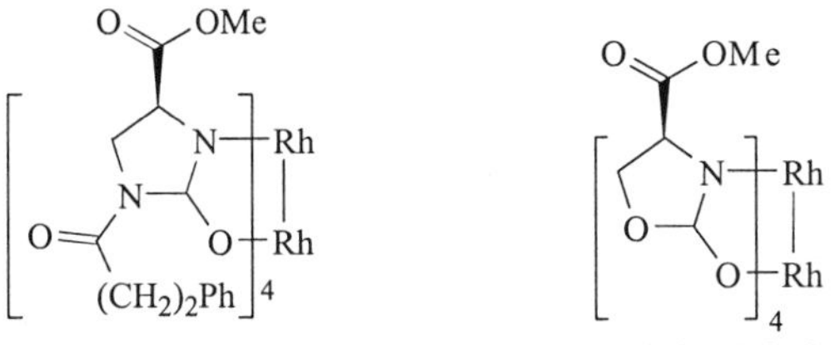

Rh_2(4*S*-MPPIM)$_4$

Rh_2(4*S*-MEOX)$_4$

Polyethylene-bound dirhodium(II) complexes have been prepared and shown to be effective reagents in both asymmetric intramolecular C-H insertion reactions and intramolecular cyclopropanation reactions (M. P. Doyle *et al.*, J. Org. Chem., 1992, **57**, 6103). Construction of the polymer-bound catalyst began with esterification of 2-pyrrolidone-5-(*S*)-carboxylic acid with a modified polyethylene oligomer. This modified oligomer was then subjected to ligand exchange using Rh_2(5*S*-MEPY)$_4$.

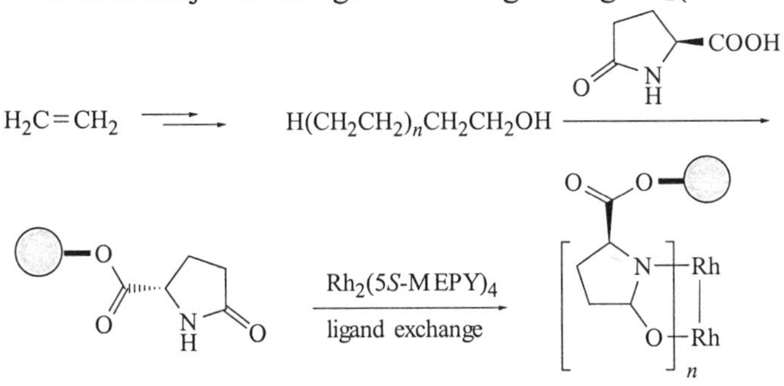

$H_2C{=}CH_2 \rightarrow H(CH_2CH_2)_nCH_2CH_2OH \rightarrow$

Rh_2(5*S*-MEPY)$_4$, ligand exchange

PE-Rh_2(5*S*-PYCA)$_4$

The levels of enantiocontrol observed with the polymer-bound catalyst were less than those reported using Rh_2(MEPY)$_4$ catalysts alone. However, the modified catalyst could be recovered and reused seven times without any appreciable loss in either conversion or stereocontrol. The rationale for loss in reactivity and stereocontrol was attributed to the displacement of chiral ligand from the oligomer-bound catalyst.

H_3CO–CH2CH2–O–C(=O)–CH=N_2 $\xrightarrow{Rh_2L_4}$ lactone (H_3CO)

Rh_2(5*S*-MEPY)$_4$ 62%, 91% ee
PE-Rh_2(5*S*-PYCA)$_4$ 47%, 69% ee

As with many metal carbene transformations, catalyst selection has an important bearing on the efficiency of the transformation (eq 4). This element of diversity was demonstrated in Sulikowski's examination of catalysts suitable for C-H insertion with (4) (H.-J. Lim, G. A. Sulikowski, J. Org. Chem., 1995, **60**, 2326; H.-J. Lim, G. A. Sulikowski, Tetrahedron Lett., 1996, **37**, 5243). Chiral dirhodium(II) carboxamidates were found to be ineffective, but the bis(oxazoline) copper(I) catalyst gave the desired product in moderate levels of enantiocontrol (eq 4). Although a detailed screening of catalysts was not reported, Burgess and Sulikowski found that both catalyst and solvent influence diastereoselectivity in product formation (K. Burgess *et al.*, Angew. Chem., Int. Ed. Engl., 1996, **35**, 220).

(4) $\xrightarrow[\text{CHCl}_3]{\text{CuOTf, bis(oxazoline) } (H_3C, CH_3; {}^iPr)}$ product (COOMe, H, N, O, CH_3, CH_3) 48% ee (4)

Davies has reported that asymmetric intramolecular C-H insertion of aryldiazoacetates is very dependent on the site of the C-H insertion. The highest enantioselectivities being obtained for insertion into methine C-H bonds (H. M. L. Davies *et al.*, Org. Lett., 2001, **3**, 1475).

ii. Intramolecular: diazoacetamides

Using acylic diazoacetamides, a modest level of enantiocontrol was found with intramolecular C-H insertion reactions (M. P. Doyle *et al.*, Tetrahedron Lett., 1992, **33**, 7819). Observed were competitive cyclizations to yield both the β-lactam and γ-lactam (eq 5). Exclusive formation of the γ-lactam was achieved only when an activating group was employed (R = OEt). The level of enantiocontrol varied with catalyst choice, but was found to be best with Rh_2(4*S*-MEOX)$_4$.

R–CH2CH2–N(^{t}Bu)–C(=O)–CH=N_2 $\xrightarrow[\text{CH}_2\text{Cl}_2]{Rh_2(4S\text{-MEOX})_4}$ γ-lactam (R, N–^{t}Bu, O) + β-lactam (R, N–^{t}Bu, O) (5)

R = Et, ^{i}Pr, OEt
82-97%, 65-80%ee

Switching to cyclic diazoacetamides, where the formation of β-lactams dominates, an improvement in enantiocontrol was observed (M. P. Doyle, A. V Kalinin, Synlett, 1995, 1075). Using $Rh_2(5S\text{-MEPY})_4$, enantioselectivities in the mid to high 90's were observed in several systems. Ring size as well as conformational issues influenced both regio- and enantioselection.

96% ee　　97% ee

iii. Intermolecular C-H Insertion Reactions

Davies and co-workers have reported on high levels of stereocontrol using methyl aryldiazoacetates and a range of both unactivated and activated alkanes in the presence of dirhodium(II) tetrakis(*S*-(*N*-dodecylbenzenesulfonyl)prolinate) ($Rh_2(S\text{-DOSP})_4$) (H. M. L. Davies, T. Hansen, M. R. Churchill, J. Am. Chem. Soc., 2000, **122**, 3063; H. M. L. Davies, E. G. Antoulinakis, J. Organomet. Chem., 2001, **617**, 47; H. M. L. Davies, P. D. Ren, J. Am. Chem. Soc., 2001, **123**, 2070). These intermolecular C-H insertion reactions utilize diazoacetates substituted with electron-donating groups. Both cyclic and acyclic systems have been employed using this methodology (eqs 6-8).

Ar, N_2, COOMe; $(CH_2)_n$; $Rh_2(S\text{-DOSP})_4$, reflux; MeOOC, Ar, $(CH_2)_n$; 55-96%, 60-89% ee (6)

Ar, N_2, COOMe; O; $Rh_2(S\text{-DOSP})_4$, reflux; MeOOC, Ar, O; 50-82% y, 52-76% ee, low de (7)

SO_2Ar, N, O, Rh, O, Rh, 4; Ar = 4-$C_{12}H_{25}C_6H_4$; $Rh_2(S\text{-DOSP})_4$

Ar, N_2, COOMe; R, OTBS; $Rh_2(R\text{-DOSP})_4$, 24 °C; MeOOC, Ar, R, OTBS; 35-72% y, 74-92% ee, 96-98% de (8)

These combinations of electron-withdrawing and electron-donating groups bound to the diazo carbon have been shown to be extremely useful from a synthetic standpoint and have been found to be exceptionally chemoselective.

A novel application of this methodology is in the preparation of *syn*-aldol products using activated C-H insertion reactions (eq 9). Diazo decomposition of methyl phenyldiazoacetate using Rh_2(*S*-DOSP)$_4$ in the presence of tetraalkoxysilanes afford β-hydroxy esters with high diastereoselectivities and enantioselectivities (H. M. L. Davies, E. G. Antoulinakis, Org. Lett., 2000, **2**, 4153).

Ph(N$_2$)C–COOMe + $(EtO)_4Si$ —[Rh_2(*R*-DOSP)$_4$, 2,2-dimethylbutane, 23 °C]→ MeOOC, Ph / H_3C, OSi(OEt)$_3$ (9)

70% y, 95% ee, >90% de

iv. Intermolecular X-H Insertion Reactions

Despite the potential of asymmetric X-H insertions to afford α-substituted carboxylic acid derivatives where X ≠ C, there have been few reports focusing on enantiocontrol in intermolecular X-H insertion reactions. The research groups of Davies and Moody have been most successful in this area and have reported on several systems where enantiocontrol has been realized (eqs 10 and 11). Using dihrodium(II) carboxylates and carboxamidates in the presence of methyl phenyl–diazoacetate, Moody, Doyle and co-workers have obtained enantio–selectivities up to 47% using dimethylphenylsilane (R. T. Buck *et al.*, Tetrahedron Lett., 1996, **37**, 7631); Hashimoto and co-workers have recently used related catalytsts incorporating *N*–phthaloyl-amino acids as chiral bridging ligands to obtain up to 74% ee (S. Hashimoto, *et al.*, Tetrahedron: Asymmetry, 2000, **11**, 3855). Davies and co-workers reported much higher enantioselectivities (92% ee) using vinyl diazoacetates with a chiral dirhodium(II) prolinate catalyst (H. M. L. Davies *et al.*, Tetrahedron Lett., 1997, **38**, 1741).

Ph(N$_2$)C–COOMe + $PhMe_2SiH$ —[Rh_2(*R*–MEPY)$_4$, CH_2Cl_2, 70%]→ Ph(SiMe$_2$Ph)CH–COOMe (10)

45-47% ee

MeOOC(N$_2$)C–cyclohexenyl + $PhMe_2SiH$ —[Rh_2(*S*-DOSP)$_4$, pentane, -78 °C, 77%]→ MeOOC(SiMe$_2$Ph)CH–cyclohexenyl (11)

92% ee

Chiral auxiliaries have also been examined in both Si-H and O-H insertion reactions with limited to moderate success (E. Aller *et al*., J. Org. Chem., 1995, **60**, 4449). Other than these studies, no significant levels of enantiocontrol have been reported.

b. Diastereocontrol

Diazocarbonyl compounds substituted with groups on the diazo carbon other than hydrogen generally undergo C-H insertion reactions with lower enantiocontrol. Not only enantiocontrol but also diastereocontrol is a factor in such cases. However, with proper choice of catalyst high levels of stereocontrol can be achieved. McKervey and co-workers have explored this area and reported on the stereoselective formation of chromanones using chiral dirhodium(II) prolinates (M. A. McKervey, T. Ye, J. Chem. Soc., Chem. Commun., 1992, 823. 105; T. Ye, C. F. Garcia, M. A. McKervey, J. Chem. Soc., Perkin Trans. 1, 1995, 1373). Enantioselectivities of up to 82% and diastereoselectivities of up to 86% were obtained (eq 12). Thus far, for reactions of disubstituted diazomethanes, carboxylate ligated dirhodium(II) complexes provide the highest levels of enantio– and diastereocontrol.

CH_2Cl_2, $[Rh_2(\text{N-}SO_2Ph\text{-prolinate})_4]$, >90% (12)

R	% ee (cis)	c:t
CH_3	82	75:25
Ph	62	89:11
vinyl	79	93:7

Asymmetric C-H insertion reactions using acyclic diazoesters provide a convenient entry into the syntheses of 2-deoxyxylolactones (M. P. Doyle, A. B. Dyatkin, J. S. Tedrow, Tetrahedron Lett., 1994, **35**, 3853). 1,3-Dichloro-2-propanol has been used as starting material in the preparation of several disubstituted γ-lactones using the $Rh_2(MEPY)_4$ catalysts (eq 13).

R = Me, Et, Bn; $Rh_2(5R\text{-MEPY})_4$, CH_2Cl_2, 65-81%, 86% de; 89-97% ee + 45-50% ee (13)

Either epimer can be prepared depending upon which configured catalyst is employed. High levels of both enantioselectivity and diastereoselectivity were observed in these examples. Use of $Rh_2(MEOX)_4$ provided a higher level of enantiocontrol (96% ee) at the expense of diastereocontrol (90:10). An alternative route has been subsequently reported which employs a cyclic prochiral acetal template (M. P. Doyle *et al.*, J. Org. Chem., 1999, **64**, 8907). A level of 94% ee in high yield as a single diastereomer was reported using $Rh_2(MEPY)_4$ catalysts. The desired 2-deoxyxylolactone was furnished upon simple hydrogenolysis of the benzyl acetal (eq 14).

$Rh_2(5S\text{-MEPY})_4$, CH_2Cl_2, 70%; hydrogenolysis, quant; 94% ee (14)

Cyclic diazoacetates afford bicyclic lactones when subjected to intramolecular C-H insertion reactions. This type of cyclization can furnish as many as four isomers when considering both enantiocontrol and diastereocontrol. An examination of catalysts and substrates in asymmetric C-H insertion reactions by Doyle, Müller and co-workers (M. P. Doyle *et al.*, J. Am. Chem. Soc., 1994, **116**, 4507; P. Müller, P. Polleux, Helv. Chim. Acta, 1994, 77, 645) revealed that high levels of stereoselectivity can be obtained using unsubstituted cycloalkyl diazoacetates and $Rh_2(MACIM)_4$ as catalyst (eq 15).

n = 1, 2, 3; $Rh_2(4S\text{-MACIM})_4$, CH_2Cl_2, 62-75%; 96-97% ee, 98% de (15)

Using substituted cyclic diazoacetates adds yet another element of stereocomplexity with these types of cyclizations. It has been found that prochiral 4-substituted cyclohexyl diazoacetates yield fused lactones in high enantiomeric excess and the fusion mode (cis or trans) is dependent upon the relative configuration of the reactant. Observed with each system was a high preference for insertion to occur into the equatorial C-

H bond using $Rh_2(MEOX)_4$. *Cis*-fused lactones resulted from cyclization onto *cis* substituted cyclohexyl diazoacetates (eq 16) whereas a *trans*-fused product was observed with the corresponding *trans* configured diazoacetate (eq 17). From a broad selection of dirhodium(II) carboxamidates, the $Rh_2(MEOX)_4$ catalysts afforded fused lactones with the highest level of diastereocontrol and enantiocontrol.

$Rh_2(4S\text{-MEOX})_4$, CH_2Cl_2 (16)

98% ee, >99:1 dr

$Rh_2(4S\text{-MEOX})_4$, CH_2Cl_2 (17)

95% ee, 90:10 dr

A preference towards formation of the cis-fused lactone (5) (63:37 dr) was observed with $Rh_2(MACIM)_4$, in high enantioselectivity.

(5)

93% ee

c. Regiocontrol

Both enantiocontrol and regiocontrol need to be considered with asymmetric C-H insertion processes that do not contain activated C-H bonds. However, it was demonstrated with dirhodium(II) carboxamidates and suitably modified diazoesters that regioselectivity can be achieved for the formation of the five-membered ring lactones in ≥95% (eq 18). Doyle and co-workers have reported exceptionally high levels of enantiocontrol in the formation of β-substituted γ-butyrolactones (M. P. Doyle *et al.*, J. Org. Chem., 1995, **60**, 6654).

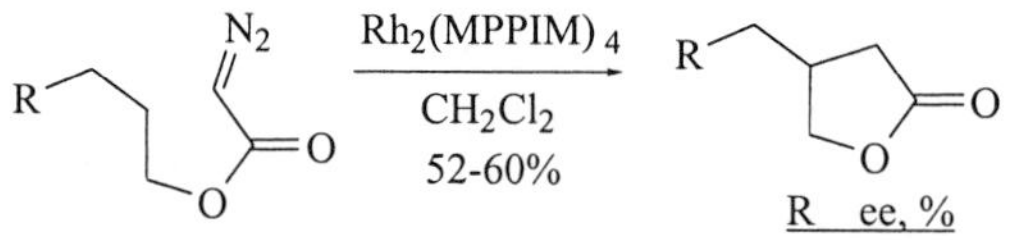

Rh$_2$(MPPIM)$_4$, CH$_2$Cl$_2$, 52-60% (18)

R	ee, %
Ph	91
Me	96
^{i}Pr	95

Disubstituted γ-lactones (6) and (7) were prepared using this methodology and found to exhibit levels of enantiocontrol in the mid-90's (M. P. Doyle *et al*., Tetrahedron Lett., 1995, **36**, 7579). High levels of both diastereocontrol and regiocontrol were also observed with these systems.

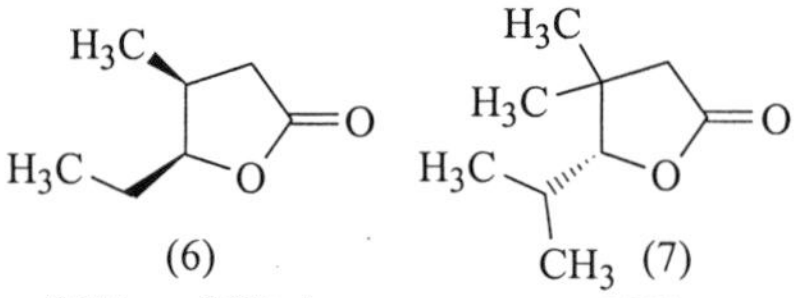

(6) 99% ee, 96% de

(7) 95% ee

Starting with cinnamic acid derivatives, a series of lignan lactones have been prepared using this asymmetric C-H insertion methodology (J. W. Bode *et al*., J. Org. Chem., 1996, **61**, 9146).

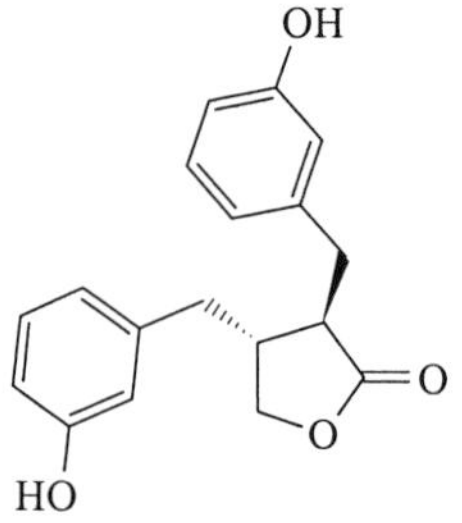

(+)- and (-)-enterolactone

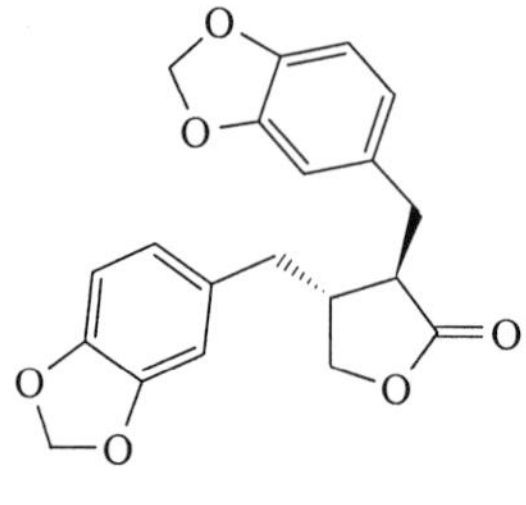

(+)- and (-)-hinokinin

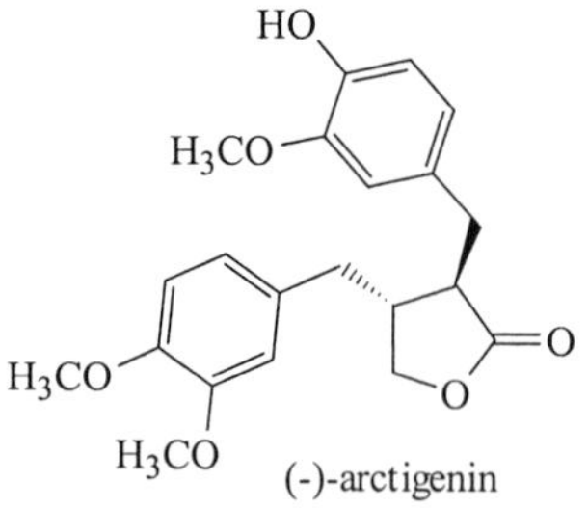

(-)-arctigenin

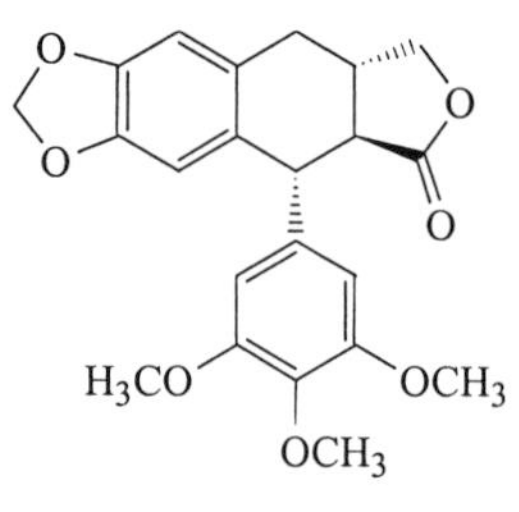

(+)-isodeoxypodophyllotoxin

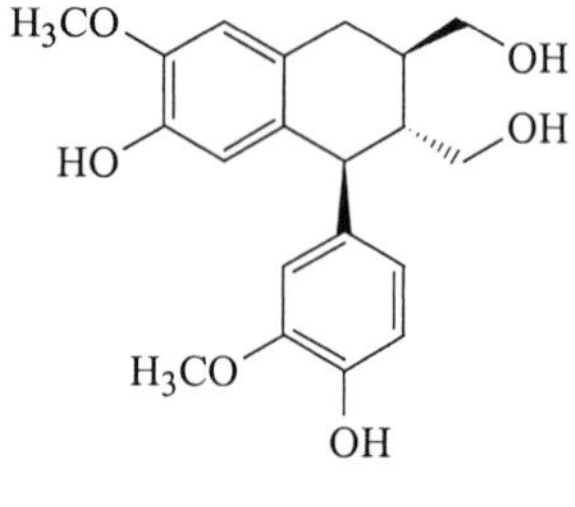

(+)-isolauricerisinol

A total of five natural products of noteworthy biological and medicinal properties were prepared using $Rh_2(4S\text{-MPPIM})_4$ as catalyst in the key cyclization step in high overall yield and stereocontrol (R. S. Ward, Chem. Soc. Rev., 1982, **11**, 75; R. S. Ward, Tetrahedron, 1990, **46**, 5029; S. S. C. Koch, A. R. Chamberlin, Studies Nat. Prod. Chem., 1995, **16**, 687).

d. Enantiomer Differentiation

Processes involving enantiomer differentiation reveal the remarkable effectiveness of metal carbene technology (M. P. Doyle, A. V. Kalinin, D. G. Ene, J. Am. Chem. Soc., 1996, **118**, 8837). That is, an *S*-configured chiral catalyst yielding a specific product will form the diastereomer upon use of the corresponding *R*-configured catalyst. The same holds true when employing either the *R*- or *S*-configured substrate in the presence of a chiral non-racemic catalyst (M. P. Doyle, A. V. Kalinin, Russ. Chem. Bull., 1995, **44**, 1729). Ideally, high levels of enantiomeric excess of either diastereomer can be obtained solely based upon selection of epimeric reagent. Several examples of this concept have been reported and are illustrated in equations 19-22. Two examples of enantiomer differentiation by the chiral catalyst are shown in equations 19/20 and 21/22 where diazo decomposition of diazoester affords γ-lactones under dirhodium(II) carboxamidate catalysis.

CH_3 O N_2 O

$Rh_2(4R\text{-MPPIM})_4$, CH_2Cl_2, 88%, 98% product selectivity — (19)

$Rh_2(5S\text{-MEPY})_4$, CH_2Cl_2, 95%, 94% product selectivity — (20)

$Rh_2(4R\text{-}MEOX)_4$, CH_2Cl_2, 86%, 87% product selectivity (21)

$Rh_2(4S\text{-}MEOX)_4$, CH_2Cl_2, 77%, 87% product selectivity (22)

With (*S*)-2-octyl diazoacetate, either the β-lactone (8) or γ-lactone (9) is obtained selectively as the major product depending upon catalyst choice.

$Rh_2(4R\text{-}MEOX)_4$, CH_2Cl_2, 53%, 85% product selectivity (8)

$Rh_2(4S\text{-}MPPIM)_4$, CH_2Cl_2, 72%, 84% product selectivity (9)

Low yields and selectivities are commonly attributed to a mismatch in catalyst and diazocarbonyl compound. However, as demonstrated by the examples reported thus far, the appropriate match in catalyst and substrate configuration ensures production of single diastereomers/regioisomers of high enantioselectivity in C-H insertion reactions.

2. Asymmetric Cyclopropanation Reactions

Metal carbenes are perhaps best known for cyclopropanation transformations, where a formal addition occurs across a carbon-carbon double bond. For reviews see those cited at the beginning of this chapter and A. Pfaltz, Acc. Chem. Res., 1993, **26**, 339; V. K. Singh, A. DattaGupta, G. Sekar, Synlett, 1997, 568; D. J. Agar, I. Prakash, D. R.

Schaad, Chem. Rev., 1996, **96**, 835; R. Schumacher, F. Dammast, H.-U. Reißeg, Chem. Eur. J., 1997, **3**, 614; V. K. Singh, A. DattaGupta, G. Sekar, Synthesis, 1997, 137. From a stereochemical perspective, up to three contiguous stereocenters can be formed in one step and thus issues of diastereocontrol and enantiocontrol need to be considered. As with asymmetric C-H insertions, the source of chirality originates from a catalyst which is bound to the transferring carbene moiety. During the past decade, significant advances have been made on catalytic asymmetric cyclopropanation reactions using a host of metal centers. Given the volume of information currently available on the development and utilization of cyclopropanation processes which provide detailed accounts in differences of reactivity and rationale of catalyst selection, the following review will focus on only those transformations utilizing chiral non-racemic metal catalysts.

$$R_2C{=}ML_n \quad + \quad = \quad \longrightarrow \quad \triangleright CR_2 \quad + \quad ML_n$$

A vast array of chiral non-racemic cyclopropane compounds of high medicinal and biological importance have been prepared using this approach which utilizes diazocarbonyl compound, alkene and suitably modified chiral catalyst. A tremendous amount of research has focused on the development of chiral, non-racemic catalysts to be used in asymmetric cyclopropanation processes. These methods have successfully been applied toward a number of intermolecular and intramolecular processes across a broad spectrum of olefinic substrates. Conformational issues are also present in cyclopropanation reactions. There is a strong preference for cyclopropanation even if the possibility of a C-H insertion or ylide-type transformation is present. Because of the relative inactivity of the C-H bonds when compared to electron rich alkenes and the reversibility associated with the initial intermediates found with ylide-type transformations, cyclopropanation processes with proper selection of catalyst are highly chemoselective (J. L. Maxwell *et al.*, Organometallics, 1992, **11**, 645; G. Maas *et al.*, Tetrahedron, 1993, **49**, 881).

a. Intermolecular

The first reported example of an enantioselective cyclo–propanation using a homogeneous chiral catalyst was in 1966 by Nozaki and co-workers (H. Nozaki *et al.*, Tetrahedron Lett., 1966, 5239).

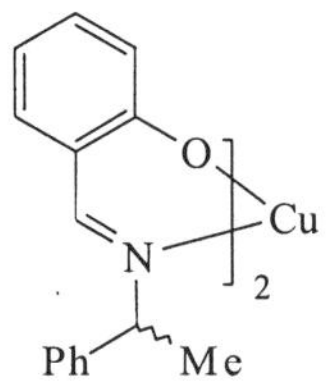

Although the initial search for asymmetric induction resulted in less than 10% enantiomeric excess, this single report set in motion extensive research efforts not only in the area of metal carbene transformations but those in the field of asymmetric catalysis. Described was reaction of ethyl diazoacetate with styrene in the presence of a copper(II) complex consisting of a salicylaldimine template constructed from salicylaldehyde and enantiopure 1-amino-1-phenylethane.

Aratani subsequently reported on a modified salicylaldimine design with sterically larger substituents which resulted in a substantial improvement in enantiocontrol (T. Aratani, Y. Yoneyoshi, T. Nagase, Tetrahedron Lett., 1975, 1707). Most notable is the enantioselectivity observed in the reaction of ethyl diazoacetate with either 2,5-dimethyl-2,4-hexadiene (to provide chrysanthemic acid), or isobutylene (eq 23) (T. Aratani, Y. Yoneyoshi, T. Nagase, Tetrahedron Lett., 1977, 2599).

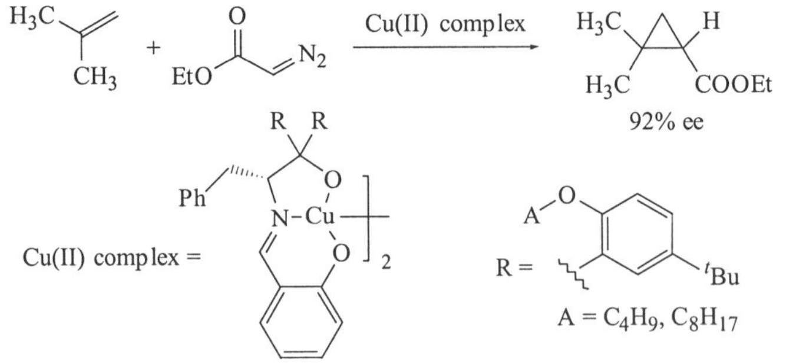

Cu(II) complex → 92% ee (23)

Cu(II) complex = … R = … $A = C_4H_9, C_8H_{17}$

From a historical standpoint, the development of other chiral catalysts to influence enantiocontrol in catalytic cyclopropanation reactions from the late 1960's through the mid-1980's was evident; however, major advances beyond those initially reported with the Aratani catalyst have occurred only during the past decade and a half.

i. Diazoacetates

In 1986, Pfaltz and co-workers provided the next major advancement in chiral catalyst design since the development of the Aratani catalyst. They described for the first time the construction and implementation of a chiral semicorrin ligand suitable for coordination with copper(II) (H. Fritschi, U. Leutenegger, A. Pfaltz, Angew. Chem., Int. Ed. Engl., 1986, **25**, 1005). The research groups of Masamune, Evans and Pfaltz then constructed a series of structurally similar chelating ligands from available amino alcohols which all have been successfully utilized in cyclopropanation reactions (R. E. Lowenthal, S. Masamune, Tetrahedron Lett., 1991, **32**, 7373; D. A. Evans *et al.*, J. Am. Chem. Soc., 1991, **113**, 726; A. Pfaltz, Acc. Chem. Res., 1993, **26**, 339).

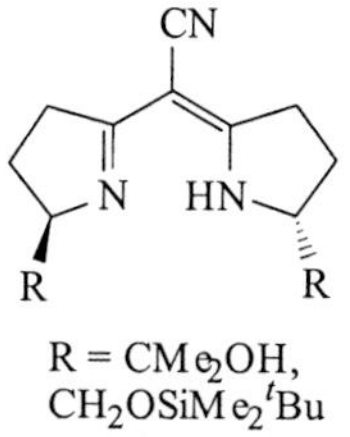

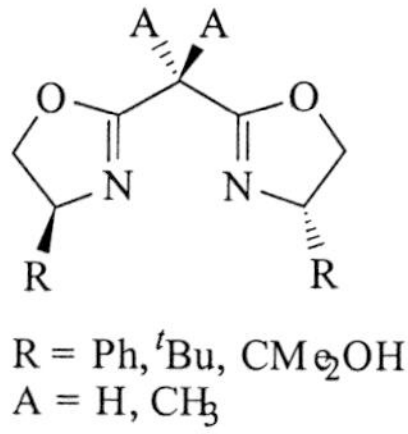

R = CMe_2OH, $CH_2OSiMe_2{}^tBu$

R = Ph, ^tBu, CMe_2OH
A = H, CH_3

The C_2-symmetric bis(oxazoline) ligands have been applied in a number of transformations. However, those focusing on intermolecular cyclopropanation reactions with diazoacetates have been met with limited success. When considering optimal catalyst/ligand combinations for all alkenes or diazo compounds, C_2-symmetric bis(oxazoline) ligands in the presence of a copper center are the most effective protocols to date. Observed are exceptionally high levels of enantiocontrol (mid to high 90's) and moderate levels of diastereocontrol (27:73 dr (c:t)). In order to achieve high levels of both enantiocontrol and diastereocontrol at the same time, it is necessary to employ bulky diazoesters such as 2,6-di-*tert*-butyl-4-methylphenyl diazoacetate.

Catalyst/ligand combinations which effect high levels of both enantiocontrol and diastereocontrol using a broad spectrum of diazocarbonyl compounds and alkenes without recourse to substrate modifications are few. Focusing on only monosubstitued alkenes, two methods have been reported. Buono's chiral copper(I) iminodiazaphospholidine complex and Nishiyama's chiral ruthenium(II) 2,2-bis(2-oxazolin-2-yl) or pybox catalyst are exceptionally good examples in stereocontrol for intermolecular cyclopropanation reactions of monosubtituted alkenes and diazoacetates (J. M. Brunel, *et al.*, J. Am.

Chem. Soc., 1999, **121**, 5807; S.-B. Park, *et al.*, Tetrahedron: Asymmetry, 1995, **6**, 2487). Diastereoselectivities of up to 96% (2:98 dr (c:t)) in 94% ee were observed upon reaction of ethyl diazoacetate with styrene in the presence of CuOTf and Buono's ligand (10).

Cu(I) or Ru(II) complexes

70-90%

R= alkyl, aryl

(10)

(pybox)$RuCl_2$(ethene)

ii. Aryl and Vinyldiazoesters

Chiral dirhodium(II) catalysts have been successfully utilized in intermolecular cyclopropanation reactions with diazocarbonyl compounds possessing groups attached to the diazo carbon other than hydrogen. Davies and co-workers have tested their vinyl and aryldiazo compounds for enantiocontrol in several cyclopropanation reactions (H. M. L. Davies *et al.*, J. Am. Chem. Soc., 1996, **118**, 6897). These studies revealed not only a temperature dependency, but a solvent effect when considering enantiocontrol using chiral dirhodium(II) prolinate catalysts and monosubstituted alkenes (eq 24). Optimal reaction conditions were observed when diazo decomposition was performed in pentane at -78°C.

pentane (24)

Ar **a** = 4-$^{t}BuC_6H_4$

b = 4- $C_{12}H_{25}C_6H_4$

Ar	R	A	temp, °C	ee, %
a	PhCH=CH	Ph	25	90
b	PhCH=CH	Ph	–78	98
b	PhCH=CH	OAc	–78	95
a	Ph	Ph	25	85

Davies has successfully applied catalytic asymmetric cyclopropanation to the solid-phase, where resin-bound alkenes were used as the limiting agents in reactions with aryldiazoacetates in excess and carbene side products remained in the liquid phase (H. M. L. Davies, T. Nagashima, J. Am. Chem. Soc., 2001, **123**, 2695) .

Doyle and McKervey have also examined enantiocontrol for cyclopropanation reactions with aryl diazoacetates (M. P. Doyle, *et al.*, Tetrahedron Lett., 1996, **37**, 4129), and found similar selectivity enhancements when using nonpolar solvents (eq 25). Solvent dependency has been attributed to the alignment of prolinate ligands associated with the dirhodium(II) complex. This conclusion stems from the observation that this solvent effect is not evident with the rigid dirhodium(II) carboxamidates and is opposite with the Ikegami and Hashimoto catalyst.

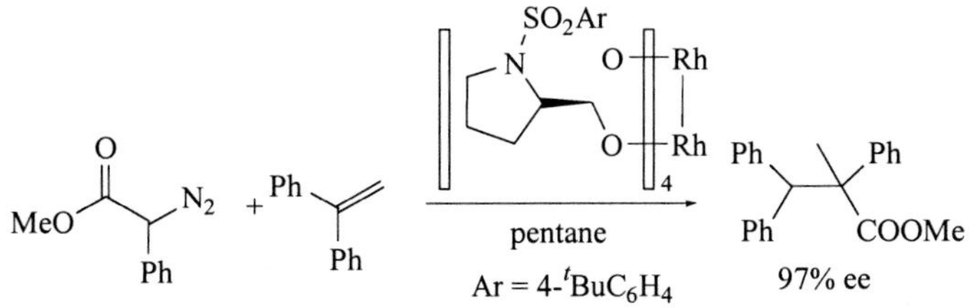

pentane

Ar = 4-^{t}BuC_6H_4

97% ee

(25)

Davies and co-workers have employed this methodology using their vinyldiazoacetates and dienes to prepare cycloheptadienes (H. M. L. Davies, Z.-Q. Peng, J. H. Houser, Tetrahedron Lett., 1994, **35**, 8939) and bicyclic dienes (H. M. L. Davies, H. D. Smith, O. Korkor, Tetrahedron Lett., 1987, **28**, 1853) enantioselectively via a formal [3+4]-cycloaddition (H. M. L. Davies, T. J. Clark, H. D. Smith, J. Org. Chem., 1991, **56**, 3817). Formation of *cis*-1,2-divinyl cyclopropanes makes possible facile [3,3]sigmatropic rearrangement to afford the annulated products stereoselectively. Factors that influence enantiocontrol as well as alternative chiral dirhodium(II) catalysts for diazo decomposition in the presence of other substrates have been reported. (H. M. L. Davies, Tetrahedron, 1993, **49**, 5203).

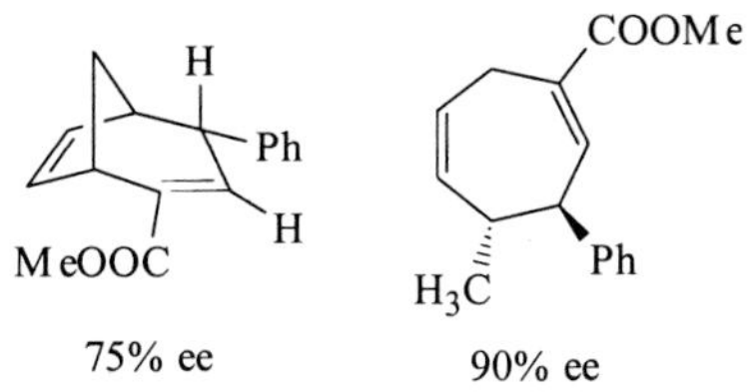

75% ee

90% ee

b. Intramolecular

In contrast to the diversity associated with intermolecular cyclopropanation reactions, there does exist an optimal ligand/catalyst combination to effect high levels of stereocontrol in the corresponding intramolecular transformation. Few methodologies have been so effectively optimized than those involving carbon-carbon double bonds tethered to diazocarbonyl compounds using chiral metal complexes. Indeed, metal carbene technology has afforded bicyclic compounds in high yield and enantiocontrol for virtually all allylic systems examined (M. P. Doyle *et al.*, J. Am. Chem. Soc., 1995, **117**, 5763). Enantioselectivities using their homoallylic counterparts are only moderately reduced. Extension of this methodology to remote carbon-carbon double bonds has been recently shown to occur with high enantiocontrol (M. P. Doyle *et al.*, J. Am. Chem. Soc., 2000, **122**, 5718). Dirhodium(II) carboxamidates are in general the catalysts of choice. However, recent reports reveal the compatibility of other catalyst systems with this mode of cyclization to effect exceptionally high levels of enantiocontrol.

i. Diazoacetates: allylic and homoallylic

Allylic diazoacetates, when subjected to diazo decomposition in the presence of chiral dirhodium(II) complexes, produce bicyclic lactones in high yield. Levels of enantiocontrol generally exceed 90% with the $Rh_2(MEPY)_4$ catalysts.

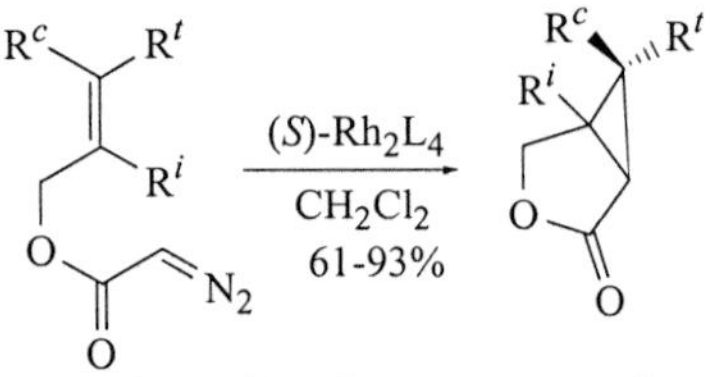

A broad selection of allylic diazoacetates have been converted to their corresponding bicyclic lactones using this protocol. If enantio–selectivities drop while using this catalyst series, such as in the case of trans-substituted allyl diazoacetates, a change in ligand design generally improves enantiocontrol (M. P. Doyle *et al.*, J. Am. Chem. Soc., 1991, **113**, 1423). A catalyst loading as low as 0.1 mol % is effective for complete diazo decomposition and scale-up offers no obvious difficulties. Other catalyst systems are relatively ineffective with allylic diazoacetates, such as, the copper(I) bis(oxazoline) and ruthenium(II) pybox catalysts (M. P. Doyle *et al.*, J. Chem. Soc., Chem. Commun., 1997, 211).

On switching to the one carbon homologue - homoallylic diazoacetates - then δ-lactones can be prepared stereoselectively. Intramolecular cyclopropanation using these diazocarbonyl compounds afford the desired cyclopropanes in high yield and, using $Rh_2(MEPY)_4$ catalysts, with up to 90% enantioselectivity (S. F. Martin, C. J. Oalmann, S. Liras, Tetrahedron Lett., 1992, **33**, 6727). Even though substrate generality afforded cyclopropane compounds at a reduced level of enantiocontrol (10-15% lower ee) when compared to their allylic counterparts, substitution at R^i and R^t did not have as profound an influence on enantiocontrol using $Rh_2(MEPY)_4$ catalysts.

$Rh_2(5S\text{-}MEPY)_4$, CH_2Cl_2, 61-93%

ii. Diazoacetates: macrocyclization

High levels of enantiocontrol can be achieved in intramolecular cyclopropanation reactions of remote carbon-carbon double bonds (M. P. Doyle, C. S. Peterson, D. L. Parker Jr., Angew. Chem., Int. Ed. Engl. 1996, **35**, 1334). Doyle and co-workers have reported enantioselectivities of up to 90% with several diazoacetate systems (eq 26).

$Cu[CH_3CN]_4PF_6$, CH_2Cl_2, 82%; 90% ee (26)

This unexpectedly high preference for macrocycle formation utilizes chiral bis(oxazoline) ligated copper(I) complexes. The metal carbenes generated are more reactive and suitable for macrocyclization (M. P. Doyle *et al.*, J. Am. Chem. Soc., 1995, **117**, 7281; M. P. Doyle *et al.*, J. Am. Chem. Soc. 1997, **119**, 8826). This preference in macrocycle formation is opposite to that observed with $Rh_2(MEPY)_4$ catalysts. Equations 27 and 28 illustrate the diversity in product formation as a function of catalyst choice in systems with both remote and allylic carbon-carbon double bonds.

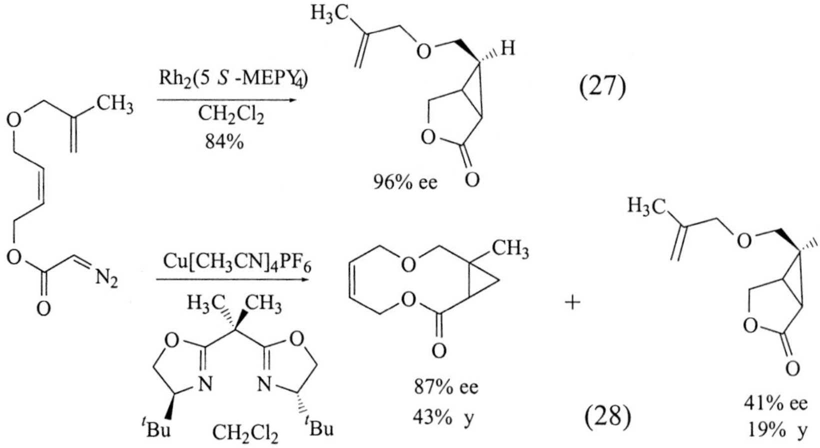

(27)

(28)

Enantioselective cyclopropanation reactions involving ring sizes of up to 20 have been reported without any changes in enantiocontrol as long as chiral bis(oxazoline) ligated copper(I) complexes are employed.

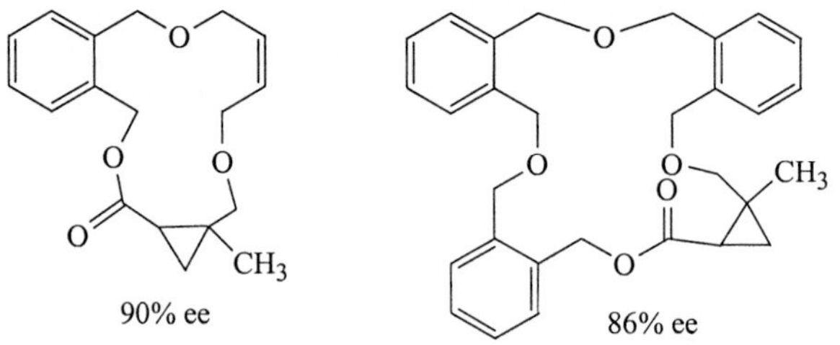

iii. Diazoacetamides

Conformational influences play an important role in the effectiveness of intramolecular cyclopropanation reactions of diazoacetamides. In addition, an uncatalysed intramolecular dipolar cycloaddition occurs in competition with catalytic decompostion (M. P. Doyle *et al.*, Tetrahedron, 1994, **50**, 4519). Cyclopropanation is favored if either substitution on nitrogen is small (R = H, Me) or when $n \geq 2$ (eq 29) (M. P. Doyle, A. V. Kalinin, J. Org. Chem., 1996, **61**, 2179). Use of $Rh_2(MEPY)_4$ catalysts resulted in the highest enantiocontrol (93% ee) and isolated yield (62%). Enantioselectivities were found to parallel those with diazoacetates.

(29)

iv. Diazoketones

With catalytically active metal complexes, diazoketones are more reactive toward diazo decomposition relative to their diazoester and diazoamide counterparts and hence, from a relative reactivity standpoint, intramolecular cyclopropanation reactions utilizing diazoketones afford cyclopropane compounds with low enantiocontrol. This has been reported by Doyle and co-workers who have observed uniformly low enantioselectivities in reactions catalyzed by chiral dirhodium(II) complexes (M. P. Doyle, M. Y. Eismont, Q.-L. Zhou, Russ. Chem. Bull., 1997, **46**, 955). The rationale for lower levels of enantiocontrol has been attributed to both the relative rates of cyclization and the conformational alignment of the metal carbene which is opposite to that found in processes involving diazoesters and diazoamides. Switching to copper catalysis has provided some advances in enantiocontrol in intramolecular cyclopropanation reactions involving diazoketones. Pfaltz and co-workers have reported on the enantioselective intramolecular cyclopropanation of (11) using a copper(II) semicorrin complex (eq 30) (R. Tokunoh, B. Fähndrich, A. Pfaltz, Synlett, 1995, 491). Extension of this methodology to other systems has been met with limited success. Variable levels of enantiocontrol were observed when other diazoketones were tested.

(11) Cu(II)* ClCH$_2$CH$_2$Cl 57% 95% ee (30)

* activated by reduction with PhNHNH$_2$

v. *Enantiomer Differentiation*

The stereospecificities associated with product formation based upon proper match of chiral catalyst and substrate in intramolecular cyclopropanation reactions parallel those found with asymmetric C-H insertion reactions. Treatment of racemic 2-cyclohexen-1-yl diazoacetate with $Rh_2(cap)_4$ afforded racemic tricyclic cyclopropane (rac-12) in 95% yield (eq 31) (M. P. Doyle *et al.*, J. Am. Chem. Soc., 1995, **117**, 11021). However, the same reaction using $Rh_2(4S\text{-MEOX})_4$ afforded tricyclic product ((1*S*, 2*R*, 6*S*)-12) in 40% yield and 94% ee (eq 32).

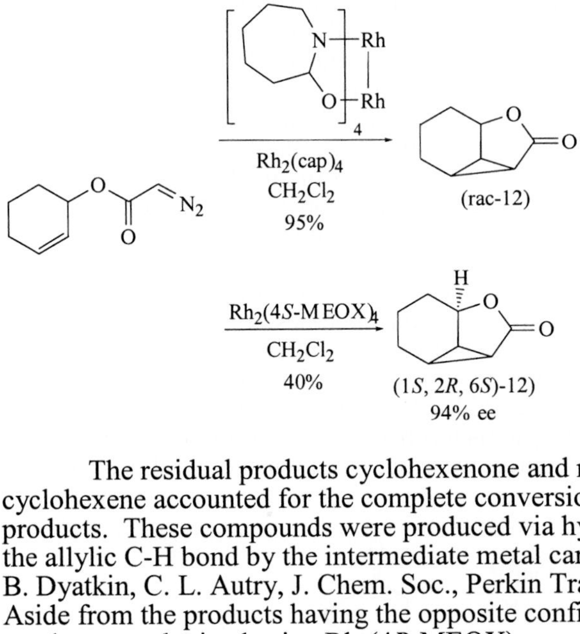

N–Rh, O–Rh (bracketed)$_4$; $Rh_2(cap)_4$, CH_2Cl_2, 95% → (rac-12) (31)

$Rh_2(4S\text{-MEOX})_4$, CH_2Cl_2, 40% → (1*S*, 2*R*, 6*S*)-12), 94% ee (32)

The residual products cyclohexenone and methylene-2-cyclohexene accounted for the complete conversion of diazoacetate to products. These compounds were produced via hydride abstraction from the allylic C-H bond by the intermediate metal carbene (M. P. Doyle, A. B. Dyatkin, C. L. Autry, J. Chem. Soc., Perkin Trans. 1, 1995, 619). Aside from the products having the opposite configuration, the same results were obtained using $Rh_2(4R\text{-MEOX})_4$.

Enantiomer differentiation with 95% ee in tricyclic product formation has been observed with methyl-substituted 2-cyclohexen-1-yl diazoacetates (13) and (14) and with 2-cyclopenten-1-yl diazoacetates (15).

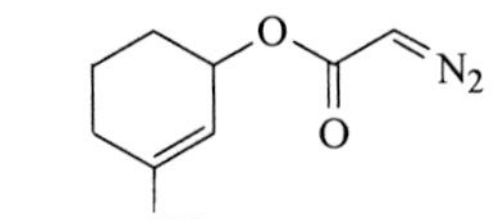

(13) 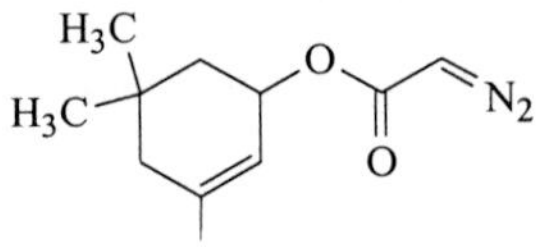 (14) 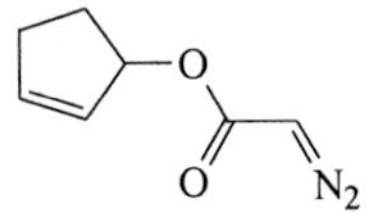 (15)

With acyclic diazoacetates, Martin and co-workers have reported a diastereoselective match/mismatch between chiral *sec* allylic diazoacetates and chiral $Rh_2(MEPY)_4$ catalysts (S. F. Martin *et al.*, J. Am. Chem. Soc., 1994, **116**, 4493). Intramolecular cyclopropanation for the formation of *endo*-(17) of greater than 20:1 (80% yield) was observed using diazoacetate (16) and $Rh_2(5S\text{-MEPY})_4$ (eq 33). Treatment of diazoacetate (16) with $Rh_2(5R\text{-MEPY})_4$ resulted in a mismatch, that is, cyclopropane (17) was obtained in only a 39% yield with a *endo/exo* ratio of only 1.0:1.5.

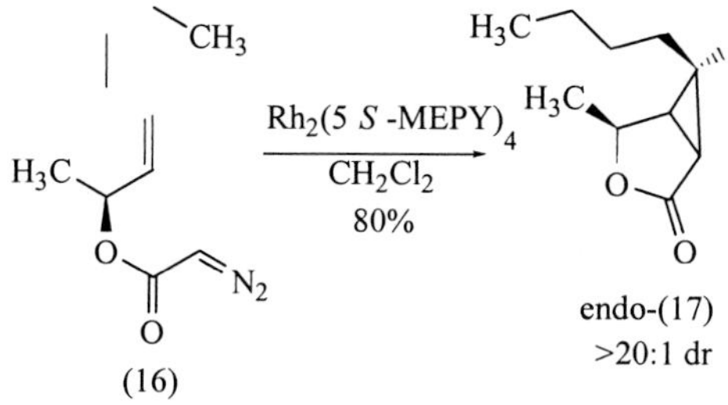

(16) $\xrightarrow[\text{CH}_2\text{Cl}_2,\ 80\%]{Rh_2(5S\text{-MEPY})_4}$ endo-(17), >20:1 dr (33)

3. Asymmetric Cyclopropenation Reactions

Excellent high levels of enantiocontrol have been achieved in both intramolecular and intermolecular cyclopropenation reactions. The synthesis of cyclopropenes occurs by metal carbene addition across a carbon-carbon triple bond under mild reaction conditions. Both copper(I) and dirhodium(II) catalysts have been shown to be effective for enantioselection in these processes. Cyclopropenes are generally stable to self-decomposition (M. S. Baird, Top. Curr. Chem., 1988, **144**, 137) and have well-defined biological effects (J. Salaun, M. S. Baird, Fr. Curr. Med. Chem., 1995, **2**, 511).

$$R_2C{=}ML_n \quad + \quad \equiv \quad \longrightarrow \quad \text{(cyclopropene)}CR_2 \quad + \quad ML_n$$

a. Intermolecular

Use of $Rh_2(MEPY)_4$ catalysts have proven to be exceptional in effecting enantiocontrol in the construction of cyclopropenes from diazoacetates (M. P. Doyle *et al.*, J. Am. Chem. Soc., 1994, **116**, 8492). Levels up to 98% ee have been reported upon reaction of several substituted alkynes and diazoesters in the presence of $Rh_2(5S\text{-MEPY})_4$ (eq 34).

$$\equiv\!\!-R \quad + \quad N_2CHCOOR' \xrightarrow[CH_2Cl_2]{Rh_2(5S\text{-MEPY})_4} \text{cyclopropene (R, COOR')} \qquad (34)$$

R	R'	yield, %	ee, %
$CH(OEt)_2$	Me	42	>98
CH_2OMe	Et	73	69
Bu^t	Et	85	57

An increase in enantiocontrol is obtained with *N,N*-dimethyldiazoacetamide (18) (eq 35). Diazoamides have been shown to be less reactive and thus more selective in metal carbene reactions. Although the isolated yields are moderate at best, the levels of enantiocontrol are exceptional.

$$\equiv\!\!-R \quad + \quad \underset{(18)}{N_2CHCONMe_2} \xrightarrow[CH_2Cl_2]{Rh_2(5S\text{-MEPY})_4} \text{cyclopropene (R, CONMe}_2) \qquad (35)$$

R	yield, %	ee, %
CH_2OMe	22	>94
Bu^t	47	89

b. Intramolecular

High levels of enantiocontrol can be achieved in intramolecular cyclopropenation reactions of remote carbon-carbon triple bonds (M. P. Doyle *et al.*, Angew. Chem., Int. Ed. Engl. 1999, **38**, 700). As shown in macrocyclic cyclopropanation reactions, selectivities are highly dependent upon choice of catalyst (eq 36).

In those cases where ring strain is further increased, the cyclopropene products have been shown to be unstable and rapidly decompose to vinyl carbenes in the presence of transition metal catalysts. Accordingly, 10- to 15-membered rings produced by cyclopropenation are stable and can be prepared in high yield. Most notable would be the construction of 10-membered ring cyclopropene (21) where enantioselectivities in excess of 99% were reported using $Rh_2(4S\text{-IBAZ})_4$ (eq 37).

ML_n / CH_2Cl_2 → (19) + (20)   (36)

ML_n	yield, %	(19):(20)	ee (19), %	ee (20), %
$Rh_2(5S\text{-MEPY})_4$	76	4:96	-	96
H_3C, CH_3 bis(oxazoline) (^{t}Bu) / $Cu[CH_3CN]_4PF_6$	54	31:69	75	46
$Rh_2(4S\text{-IBAZ})_4$	80	84:16	97	88

$Rh_2(4\,S\text{-IBAZ})_4$ / CH_2Cl_2 → (21) >99% ee   (37)

3. Rearrangements and Cycloadditions Involving Ylides

Ylides are normally reactive intermediates which are known to undergo a number of synthetically useful transformations (rearrangements and dipolar cycloadditions are considered in this Chapter), to form stable products. Ylides can be viewed as species in which a positively charged heteroatom X (such as O, I, S, Se, N) is connected to an atom (carbon is considered here) possessing an unshared pair of electrons. An attractive method of ylide formation involves the reaction of heteroatom-substituted organic compounds *via* a lone pair on the heteroatom with carbenes. Such intermediates are usually generated by transition metal catalysed decomposition of diazo functionality.

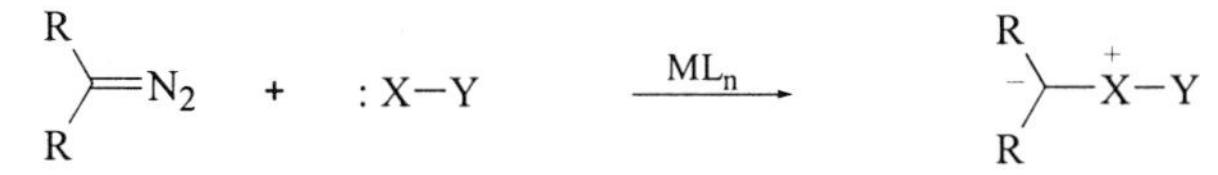

Until a few years ago it was not considered likely that a catalyst, having decomposed a diazo compound to form an ylide, was still able to exert an influence on subsequent transformations of the ylide. Recent developments have shown this not to be the case and a new era of synthetically important ylide transformations in organic chemistry has begun where appropriate metal - chiral (non racemic) ligand combinations are starting to be developed to render these transformations enantioselective. This area has recently been reviewed (D. M. Hodgson, F. Y. T. M. Pierard, P. A. Stupple, Chem. Soc. Rev., 2001, **30**, 50).

a. Oxonium Ylides

In 1966 Nozaki *et al.* reported the reaction of racemic 2-phenyl oxetane and methyl diazoacetate with a chiral (non racemic) ligand-containing copper chelate which furnished a mixture (75%) of cis and trans tetrahydrofurans (H. Nozaki *et al.*,Tetrahedron, 1968, **24**, 3655, and references cited therein). Although no enantiomeric excesses were determined, the optically active tetrahydrofurans indicated the first examples of a catalytic asymmetric ylide transformation.

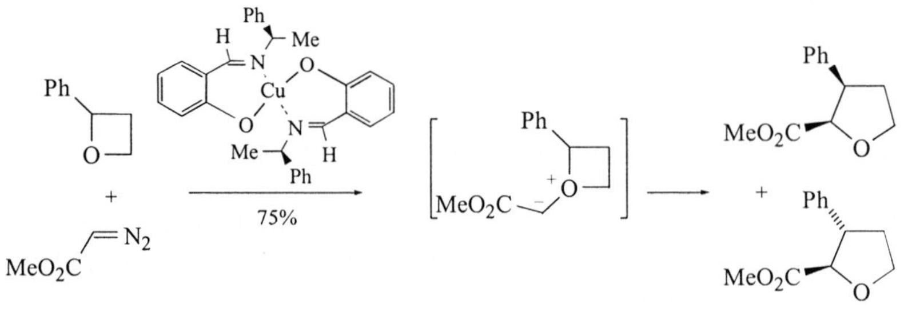

Not until the early 1990s were further asymmetric oxonium ylide transformations reported, when Katsuki *et al.* described a more detailed investigation of the oxetane ring-expansion process (K. Ito, M. Yoshitake, T. Katsuki, Heterocycles, 1996, **42**, 305). These workers first studied the reaction of excess racemic 2-phenyloxetane with *tert*-butyl diazoacetate using a copper catalyst with a C_2–symmetric reduced bipyrindine ligand. The cis- and trans- ring-expanded tetrahydrofurans were each formed in good ee (~80%) and the predominant absolute configuration at the ester-substituted stereocentre (C-2) was the same in both tetrahydrofurans.

Bu^tO_2C, N_2, Ph, O, CuOTf-L, Bu^tO_2C, Ph, O, Bu^tO_2C, Ph, O, L =, N, N, TBSO, OTBS

Doyle and co-workers have recently devised an alternative strategy to chiral oxonium ylides based on desymmetrisation of 1,3-dioxanes (M. P. Doyle *et al.*, Tetrahedron Lett., 1997, **38**, 4365). Presumably the intermediate carbenoid selectively interacts with one of the diastereotopic ethers, followed by [1,2] rearrangement to furnish the bicyclic lactone virtually exclusively, in good ee.

N_2, O, O, O, O, CO_2Me, Ph, N, N, Rh, O, Rh, O, 4, $Rh_2(4S\text{-MPPIM})_4$, L_nRh, O, O, O, O, O, O, O, O

86%, 81% ee

The most frequently observed asymmetric transformation of an oxonium ylide is [3,2] sigmatropic rearrangement, which was first reported by McKervey and co-workers in 1992 (N. McCarthy *et al.*, Tetrahedron Lett., 1992, **33**, 5983). A novel chiral Rh(II) catalyst was employed [$Rh_2(S\text{-binaphtholphosphate})_2(O_2COH)_2$], which resulted in enantioselectivities of up to 30% ee after [3,2] sigmatropic rearrangement of the intermediate oxonium ylide. Further improvement of enantiocontrol (up to 60% ee) was achieved when a range of chiral carboxylate Rh(II) catalysts were screened (N. Pierson, C. Fernandez-Garcia, M. A. McKervey, Tetrahedron Lett., 1997, **38**, 4705).

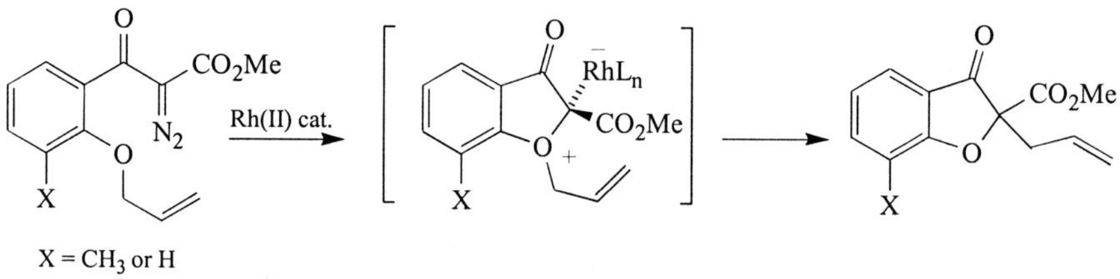

O
CO_2Me
N_2
Rh(II) cat.
O
X
$X = CH_3$ or H
RhL_n
CO_2Me
O
CO_2Me

More recently, Clark and co-workers have reported analogous enantioselective rearrangements (up to 57% ee) using a copper catalyst in combination with a C_2–symmetric diimine (J. S. Clark *et al.*, Tetrahedron Lett., 1998, **39**, 97). The presence of a phenyl ring in the tether gave slightly lower ees.

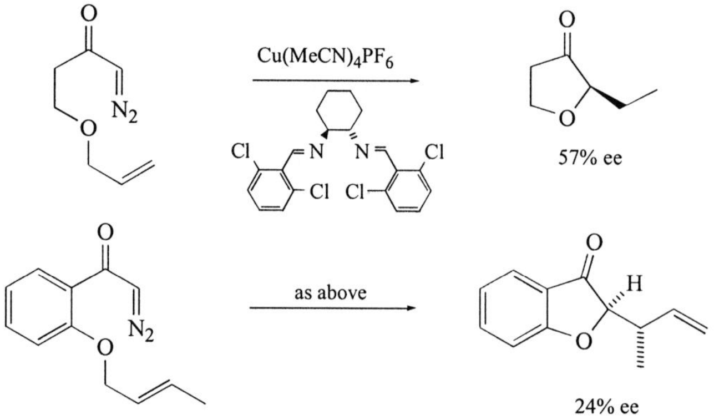

$Cu(MeCN)_4PF_6$
N_2
Cl
N
Cl
57% ee
as above
H
24% ee

Doyle *et al.* have extended this methodology to enantioselective macrocyclisation by intramolecular oxonium ylide formation - [3,2] sigmatropic rearrangement to form a ten-membered ring in 65% ee (M. P. Doyle *et al.*, J. Am. Chem. Soc., 1998, **120**, 7653).

OMe O O N$_2$ O

Cu(MeCN)$_4$PF$_6$

Me Me O O N N Bu^t Bu^t

O O OMe O

35%, 65% ee

Doyle has also reported a study of an enantioselective *inter*molecular oxonium ylide formation - [3,2] sigmatropic rearrangement - remarkably in up to 98% ee, although yields were low (M. P. Doyle *et al.*, J. Am. Chem. Soc., 1998, **120**, 7653).

EtO$_2$C N$_2$

+ Me :O Ph

CO$_2$Me O N Rh O Rh 4

Rh$_2$(*S*-MEOX)$_4$

32%

Me + H O L$_n$Rh Ph EtO$_2$C H

Me H H +O L$_n$Rh EtO$_2$C Ph

OMe CO$_2$Et Ph

98% ee

+

OMe CO$_2$Et Ph

94% ee

ratio:

85

15

Calter and Sugathapala have reported an asymmetric approach to the core of the zaragozic acids based on desymmetrisation of cyclic unsaturated acetals (M. A. Calter, P. M. Sugathapala, Tetrahedron Lett., 1998, **39**, 8813).

MeO$_2$C O N$_2$ O O

H O Rh N O Rh SO$_2$Ar 4

Rh$_2$(*S*-TBSP)$_4$

(Ar = 4-Bu^{t}C$_6$H$_5$)

O MeO$_2$C + O O

O MeO$_2$C O O

47%, 34% ee

b. Iodonium, Sulfonium and Selenonium Ylides

Doyle and co-workers have extended the utility of the [3,2] rearrangement methodology to include iodonium ylides (M. P. Doyle *et al.*, J. Am. Chem. Soc., 1998, **120**, 7653). These studies were undertaken to prove the involvement of the chiral catalyst in the product-forming intermediate. Unlike the other ylides described so far, catalyst free iodonium ylides are achiral at the onium centre, therefore enantio–selectivity cannot arise from a catalyst free, but 'configurationally constrained' ylide. In this case, Cu(I) catalysis was found to give both superior yields and asymmetric induction to Rh(II).

EtO_2C $=N_2$ + I → [EtO_2C L_nM I⁺] → EtO_2C I

CO_2Me $Rh_2(S\text{-MEOX})_4$ 12-15 % y, 39% ee

Me Me O O N N Bu^t Bu^t $Cu(MeCN)_4PF_6$ 62% y, 69% ee

In 1995 Uemura *et al.* reported the first examples of catalytic asymmetric [3,2] sigmatropic rearrangements of allylic sulfur and selenonium ylides (Y. Nishibayashi, K. Ohe, S. Uemura, J. Chem. Soc., Chem. Commun., 1995, 1245, and references cited therein). Trans-cinnamyl phenyl sulfide (and related selenides) and ethyl diazoacetate were treated with a copper(I) - bisoxazoline catalyst or $Rh_2(S\text{-MEPY})_4$ resulting in up to 20% ee for the sulfide and up to 41% ee with the corresponding selenide.

EtO_2C $=N_2$ + : Se(Ph) Ph

CO_2Me $Rh_2(S\text{-MEPY})_4$ 65%

SePh CO_2Et Ph 25% ee + SePh CO_2Et Ph 41% ee

ratio: 58 42

Hashimoto and co-workers have reported ees up to 58% using trans-cinnamyl phenyl sulfide and a sterically hindered diazoacetate (S. Hashimoto, *et al.*, Heterocycles, 2001, **54**, 623). Higher enantioselectivities (up to 64%) have been achieved by Katsuki and co-workers with the same sulfide using chiral cobalt(III) - salen catalysts and tert-butyl diazoacetate (T. Fukuda, R. Irie, T. Katsuki, Tetrahedron, 1999, **55**, 649).

Bu^tO_2C $=N_2$ + :S(Ph) (cinnamyl) → [Co(III)-salen catalyst (Ph, Ph; MeO, OMe; Co–Br)], 81% → SPh / CO_2Bu^t / Ph, 64% ee (ratio: 85) + SPh / CO_2Bu^t / Ph, 64% ee (ratio: 15)

In 1999 Aggarwal and co-workers reported low ees from rearrangement of allylic sulfur ylides derived from trimethylsilyldiazomethane (V. K. Aggarwal *et al.*, Tetrahedron Lett., 1999, **40**, 8923).

Me_3Si $=N_2$ + :S(Me) (cinnamyl) → $Rh_2(S\text{-MEPY})_4$, 94% → [transition states: Me, H, S, L_nRh, Me_3Si, Ph, H] → SMe / $SiMe_3$ / Ph, 13% ee (ratio: 49) + SMe / $SiMe_3$ / Ph, 15% ee (ratio: 51)

More recently McMillen has found that increasing the steric demands of an allyl sulfide (from allyl phenyl sulfide to allyl 2,6-dimethylphenyl sulfide) increased ee (from 14% to 52%) in the resulting homoallylic sulfide following treatment with ethyl diazoacetate and a copper(I) - bisoxazoline catalyst (D. W. McMillen *et al.*, J. Org. Chem., 2000, **65**, 2532).

c. Carbonyl Ylides

Hodgson and co-workers reported the first examples of enantioselective carbonyl ylide cycloaddition (up to 53% ee) using unsaturated α-diazo-β-keto esters in 1997 (D. M. Hodgson, P. A. Stupple, C. Johnstone, Tetrahedron Lett., 1997, **38**, 6471). In further studies it has been shown that rhodium phosphate catalysts can be superior to the more commonly utilised carboxylate and carboxamidate analogues in asymmetric transformations of diazocarbonyl compounds. Enantioselectivities of up to 90% were observed using a hydrocarbon soluble variant of Rh_2(binaphtholphosphate)$_4$. (D. M. Hodgson, P. A. Stupple, C. Johnstone, Chem. Commun., 1999, 2185).

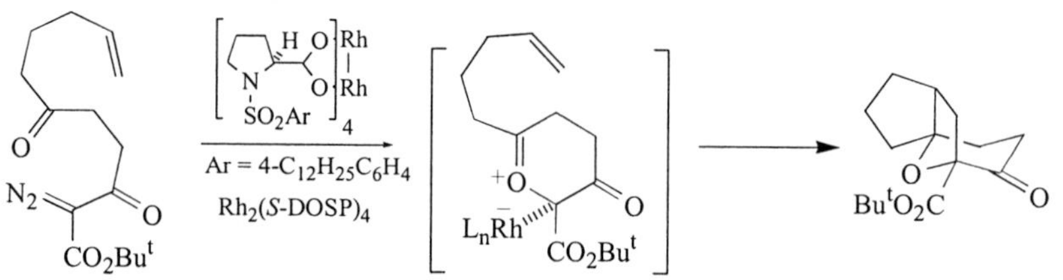

Ar = 4-$C_{12}H_{25}C_6H_4$
Rh_2(*S*-DOSP)$_4$

In 1998 a study of catalyst stereocontrol in intermolecular cycloaddition of an 'aromatic' carbonyl ylide, specifically an oxidopyrylium, with *N*-substituted maleimides was reported (H. Suga, H. Ishida, T. Ibata, Tetrahedron Lett., 1998, **39**, 3165). The principal aim of the study was to investigate the factors controlling the endo/exo selectivity in the cycloaddition step and in doing so the first examples of asymmetric induction in an oxidopyrylium cycloaddition were observed, albeit at a low level (up to 20% ee).

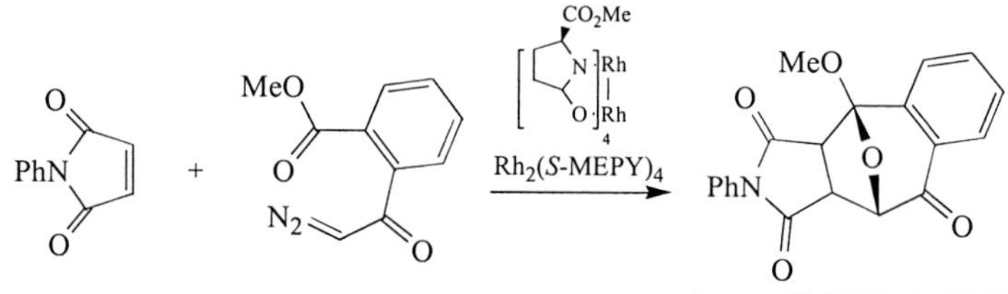

Rh_2(*S*-MEPY)$_4$

endo : exo, 11 (20% ee) : 89 (5% ee)

Hashimoto and co-workers have observed significant asymmetric induction using α-diazoketones with DMAD as the dipolarophile, where enantioselectivities of up to 92% were reported (S. Kitagaki *et al.*, J. Am. Chem. Soc., 1999, **121**, 1417).

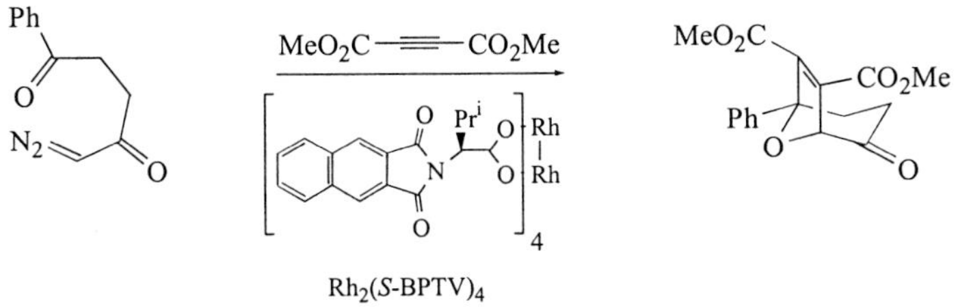

$Rh_2(S\text{-}BPTV)_4$

Further structural changes to both the carbonyl ylide and dipolarophile have since been described (S. Kitagaki *et al.*, Tetrahedron Lett., 2000, **41**, 5931). In systems analogous to those of Ikegami, cycloaddition to oxidopyryliums, was reported in up to 93% ee.

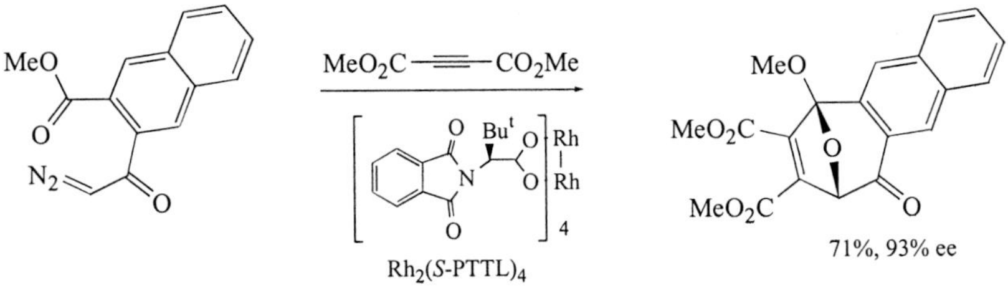

$Rh_2(S\text{-}PTTL)_4$

71%, 93% ee

Recently, Ikegami *et al.* have described a new approach to enantiocontrol in oxidopyrylium cycloadditions which involves the use of chiral (non-racemic) Lewis acids (H. Suga *et al.*, Org. Lett., 2000, **2**, 3145).

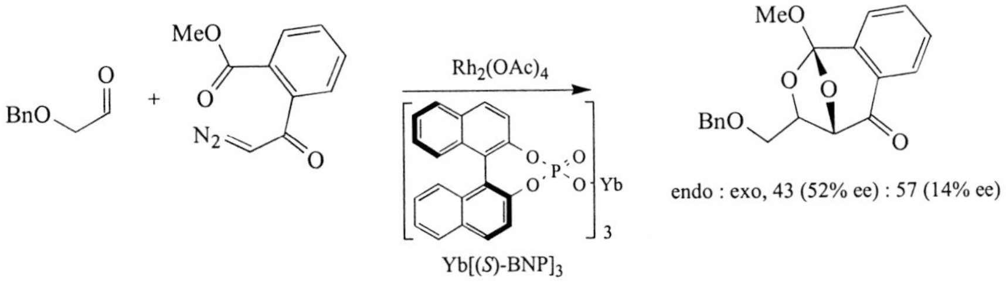

$Yb[(S)\text{-}BNP]_3$

endo : exo, 43 (52% ee) : 57 (14% ee)

Acknowledgments: The authors would like to thank the many colleagues who have contributed to the development of the chemistry described and who are cited in the references.

Second Edition of Rodd's Chemistry of Carbon Compounds,
Volume V, Topical Volumes
Asymmetric Catalysis, edited by M.Sainsbury

Chapter 4

PHASE-TRANSFER REACTIONS

B. LYGO

1. Introduction

This chapter will cover recent developments in the use of phase-transfer reactions. For the purposes of this review we will define a 'phase-transfer reaction' as a heterogeneous reaction which requires transfer of a reagent or substrate ion from one phase to another, in order for the reaction to occur rapidly. To keep the size of the chapter down the review will be restricted to selected examples of liquid-liquid (specifically aqueous-organic) and solid-liquid systems that involve the transfer of anionic species into the organic phase and have high synthetic utility. Since detailed discussions of the mechanistic issues relating to phase-transfer reactions have already been published (M.E. Halpern, "Phase Transfer Catalysis, Mechanisms and Synthesis", ACS, Washington D.C, 1997; "Handbook of Phase Transfer Catalysis", Ed.'s Y. Sasson, R. Neumann, Blackie, London, 1997; C.M. Starks, C.L. Liotta, M.C. Halpern, "Phase-Transfer Catalysis - Fundamentals, Applications, and Industrial Perspectives", Chapman & Hall, New York, 1994; "Phase Transfer Catalysis" Ed. E.V. Dehmlow, S.S. Dehmlow, VCH, Weinheim, 1993), only a brief overview of this aspect will be presented here.

Given the vast amount of published literature concerned with phase-transfer processes only selected citations have been quoted in this review, further references can be found in the above texts and in the compendium of phase-transfer reactions (W.E. Keller, "Phase-Transfer Reactions: Fluka Compendium Vol.'s 1-3", Georg Thieme Verlag, Stuttgardt, 1986, 1987, 1990). Two extensive reviews on asymmetric phase-transfer catalysis have also been published recently (M.J. O'Donnell "Asymmetric Phase-Transfer Reactions" in "Catalytic Asymmetric Synthesis, 2nd Ed.", Ed. I.

Ojima, Verlag Chemie, New York, 2000; T. Shioiri, "Chiral Phase-Transfer Catalysts" in "Handbook of Phase-Transfer Catalysis", Ed. Y. Sasson, R. Neumann, Blackie, 1997) and are highly recommended to anyone wishing to find out more about this aspect of the subject.

Readers interested in gas-solid phase-transfer processes should consult recent reviews (S.D. Naik, L.K. Doraiswamy, *Aiche J* 1998, **44**, 612; P. Tundo, G. Moraglio, F. Trotta, *Ind. Eng. Chem. Res.* 1989, **28**, 881) that cover this topic.

2. General Background

(a) Phase-transfer Reactions

In general, ion transfer between two different reaction phases is greatly accelerated by the addition of a phase-transfer catalyst. Under such conditions it has been proposed that a number of subtly distinct mechanistic schemes may be operative (see for example: S.S. Yufit, S.S. Zinovyev, *Tetrahedron* 1999, **55**, 6319; D. Landini, A. Maia, F. Montanari, *J. Chem. Soc. Chem. Commun.* 1977, 112; M. Makosza, E. Bialecka, *Tetrahedron Lett.* 1977, **2**, 1983; C.M. Starks, R.M. Owens, *J. Am. Chem. Soc.* 1973, **95**, 3613).

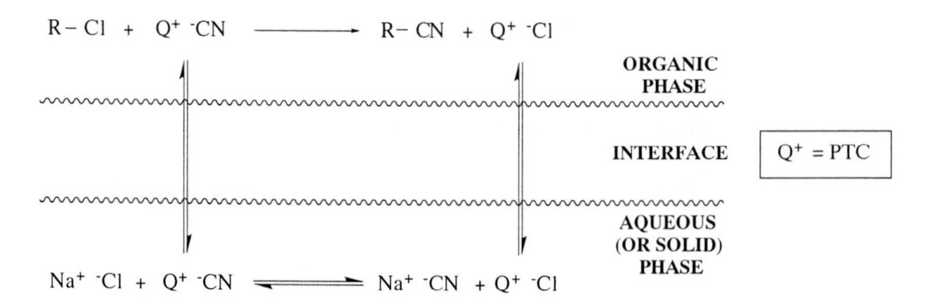

The vast majority of these processes are 'normal' phase-transfer reactions, in which the phase-transfer catalyst (PTC) facilitates reaction by solublising a reagent or substrate ion in the organic phase. This type of process is illustrated in the scheme shown above. The reactivity of the anion ($^{-}$CN) in the organic phase is usually enhanced since the catalyst-anion ion pair tends to have greater charge separation and reduced hydration compared with the precursor salt (NaCN). Consequently intrinsic reaction rates tend to be significantly higher than those obtained in homogeneous media.

Variations in this mechanistic scheme relate to whether the ion exchange process takes place mainly in the aqueous phase (as implied above), the interfacial region, or the organic phase. This can be strongly influenced by the reaction conditions, so when optimising reactions of this type a range of variables need to be considered (table 1).

Table 1. Main reaction variables in normal phase-transfer reactions

Variable	Effect on ion transfer	Effect on intrinsic reaction
agitation	√√√	
concentration of aqueous	√√	√
concentration of organic	√√	√√√
nature of organic phase	√	√√
nature of anion	√√√	√√√
nature of catalyst	√√√	√√√
nature of organic reactant		√√√
temperature	√	√√
use of co-catalyst	√√	√√

There are many approaches that can be taken towards optimising a given phase-transfer reaction. For example, solid-liquid phase-transfer reactions are generally more reactive than their liquid-liquid counterparts (F. Sirovski *et al.*, *Tetrahedron* 1999, **55**, 6363), and in many cases it is possible to use the organic substrate as one of the reaction phases, avoiding the need for an organic solvent (A. Loupy, *Top. Curr. Chem.* 1999, **206**, 153). It has also been reported that the rate of reaction in many phase-transfer catalysed reactions can be increased by the use of microwave irradiation (S. Deshayes *et al.*, *Tetrahedron* 1999, **55**, 737). However, when first developing a new phase-transfer reaction, choice of appropriate catalyst and ensuring effective mixing of the phases are probably the most important parameters to consider and these are discussed in more detail below.

Inverse phase-transfer reactions (L.J. Mathias, R.A. Vaidya, *J. Am. Chem. Soc.* 1986, **108**, 1093), in which the organic substrate is solublised in the aqueous phase by the phase-transfer catalyst have also been reported, this topic has been reviewed recently (E. Monflier, in "Recent Research Developments in Organic Chemistry Vol. 2", Ed. S. G. Pandalai, Transworld Research Network, India, 1998, p623).

(b) Agitation

The heterogeneous nature of phase-transfer reactions makes agitation an important reaction parameter (C.M. Starks, *Tetrahedron* 1999, **55**, 6261). With low levels of agitation, the interfacial area is relatively small, and so for many reaction processes, mass transfer becomes rate limiting. At higher levels of agitation, the rate of mass transfer can approach or surpass the intrinsic reaction rate. Indeed for most phase-transfer reactions there is a point at which the kinetics become independent of the degree of agitation. For mechanical stirring (half-moon Teflon blades) in conventional round-bottomed flasks, this situation typically occurs at stirring rates of 80-1200rpm (see for example: M. Makosza, *Russ. Chem. Rev.* 1977, **46**, 1151; H.H. Freedman, R.A. Dubois, *Tetrahedron Lett.* 1975, 3251), however a number of reactions, particularly those involving hydroxide ion, have been reported to benefit from substantially higher stirring rates (see for example: R. Solaro, S.D'Antone, E. Chiellini, *J. Org. Chem.* 1980, **45**, 4179). It has also been reported that ultrasonic mixing can lead to even greater increases in the rate of both solid-liquid (R. Guilet *et al.*, *C.R. Acad. Sci., Ser. IIc: Chim.* 1998, **1**, 651; W. Oppolzer, R. Moretti, C. Zhou, *Helv. Chim. Acta* 1994, **77**, 2363; A.R. Katritzky *et al.*, *Can J. Chem.* 1994, **72**, 1849; E. Diez-Barra *et al.*, *Synlett* 1992, 893; V. Ragaini *et al.*, *Ind. Eng. Chem. Res.* 1988, **27**, 1382; R.S. Davidson, A.M. Patel, A. Safdar, *Tetrahedron Lett.* 1983, **24**, 5907; R. Neumann, Y. Sasson, *J. Chem. Soc., Chem. Commun.* 1983, 616) and liquid-liquid (W.L. Xu, Y.F. Chen, F.L. Wu, *Chin. Chem. Lett.* 1999, **10**, 813; A.G. Kukovinets *et al.*, *Russ. J. Org. Chem.* 1998, **34**, 1331; J. Ezquerra, J. Alvarez-Bullia, *J. Het. Chem.* 1988, **25**, 917) phase-transfer reactions. Presumably in part due to the formation of smaller particle/droplets and the consequent increase in interfacial area.

(c) Phase-transfer Catalysts

The primary role of the phase-transfer catalyst is to rapidly transfer a reactive reagent or substrate anion into the phase containing the other reaction partner. Although a variety of compounds are capable of fulfilling this role, the most commonly employed catalysts are either quaternary ammonium salts, phosphonium salts, or crown ethers.

(i) Onium Ions

The most commonly employed onium ion phase-transfer catalysts are quaternary ammonium ions (J. Dockx, *Synthesis* 1973, 441), although phosphonium ions have also been shown to be useful in a variety of reaction processes (M.O. Wolf, K.M. Alexander, G. Belder, *Chim. Oggi* 2000, **18**, 29). The former tend to be preferred as they are generally less expensive to produce and are more stable towards hydroxide, but phosphonium salts do have the advantage of greater thermal stability. Onium ions are effective catalysts because they form relatively lipophilic ion-pairs with a wide range of counter-ions, and are typically used at 0.5-10mol% levels.

In normal liquid-liquid phase-transfer processes where the anion is transferred into an organic phase, the choice of onium ion catalyst depends upon the type of reaction taking place. Nucleophilic displacements usually require a highly activated anion, and so catalysts with high lipophilicity and poor charge accessibility (e.g. R_4NX where R=butyl or larger) are often preferred, whereas reactions involving extraction of basic anions (e.g. HO^-) seem to work best with more accessible lipophilic onium ion catalysts (e.g. $BnNEt_3X$ or $(C_8H_{17})_3NMeX$).

Charge accessibility in onium ion catalysts can influence both the rate and the regioselectivity of reaction (E.V. Dehmlow, R. Richter, *Chem. Ber.* 1993, **126**, 2765; M. Halpern, Y. Sasson, M. Rabinovitz, *Tetrahedron* 1982, **38**, 3183), for example it has been shown that the regiochemistry of alkylation of phenylacetophenone (1) is dependant on the nature of the catalyst used (table 2).

Table 2.

$$PhC(=O)CH_2Ph \xrightarrow[\text{aq. NaOH, } C_6H_6 \text{, } 34°C]{Me_2SO_4} PhC(OMe)=CHPh + PhC(=O)CH(CH_3)Ph$$

(1) → (2) + (3)

Onium ion	Ratio (2)/(3)	Conversion after 30min (%)
none	1.20	<5
Me_4N^+	0.82	21
Et_4N^+	1.68	85
$(n\text{-}Pr)_4N^+$	1.95	77
$(n\text{-}Bu)_4N^+$	1.88	54
$(n\text{-}C_5H_{11})_4N^+$	1.70	45
$(n\text{-}C_8H_{17})_4N^+$	1.53	29
$(n\text{-}Bu)_3(Me)N^+$	1.35	95
$(n\text{-}Bu)(Et)_3N^+$	1.79	100
$(n\text{-}C_8H_{17})(Et)_3N^+$	1.72	98

Similarly the extent of reaction can sometimes be controlled by the appropriate choice of catalyst, as illustrated by the alkylation of 2,5-dimethylphenol (4) with 1-bromo-3-chloropropane (E. Reinholtz *et al.*, *Synthesis* 1990, 1069). In this case, less accessible quaternary ammonium ions (e.g. $n\text{-}Bu_4N^+$ and $(n\text{-}C_8H_{17})_4N^+$) not only catalyse the alkylation reaction, but also activate the resulting bromide ion sufficiently that halogen exchange also occurs leading to formation of bromide (6) (table 3).

(ii) Crown Ethers

Crown ethers and cryptands are highly effective at solublising both organic and inorganic alkali metal salts in organic solvents. Consequently they are also effective as phase-transfer catalysts, especially for processes that involve solid-liquid phase-transfer (see for example: A. Gobbi *et al.*, *J. Org. Chem.* 1998, **63**, 5356; M. Makosza, M. Ludwikow, *Angew. Chem. Int. Ed., Engl.* 1974, **13**, 665). The most commonly employed catalysts are 18-Crown-6 (7) and dibenzo-18-crown-6 (8), the latter often being preferred due to its greater lipophilicity.

Table 3.

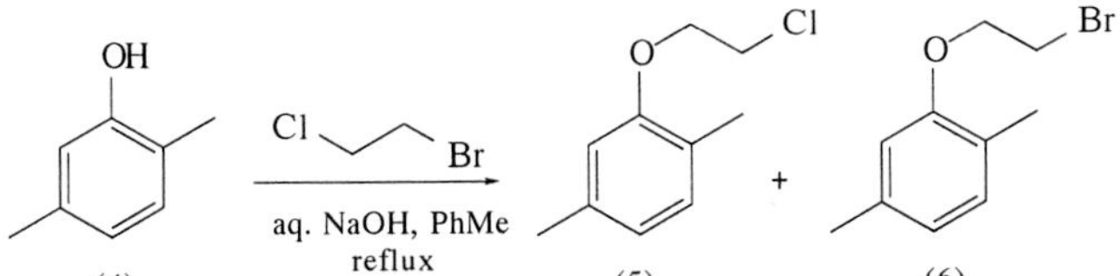

(4) $\xrightarrow[\text{aq. NaOH, PhMe, reflux}]{\text{Cl}\sim\sim\text{Br}}$ (5) + (6)

Catalyst	Ratio (5)/(6)	Combined Yield (%)
Et_4NCl	26.2	60
$BnNMe_3Br$	25.6	58
$BnNEt_3Br$	24.6	71
$BnNBu_3Cl$	2.5	76
18-crown-6	33.7	64
n-Bu_4NBr	1.9	75
n-Bu_4NHSO_4	2.1	72
$(n\text{-}C_8H_{17})_4NBr$	2.1	63

These catalysts are very effective at transferring both potassium and sodium salts into organic solvents and have been widely used in solid-liquid reactions involving alkali metal hydroxides. However they are significantly more expensive to produce that simple onium ion catalysts and so their use tends to be limited to small scale applications.

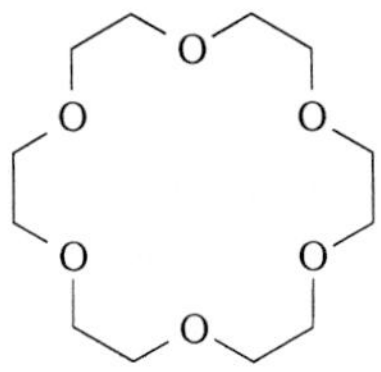

(7) 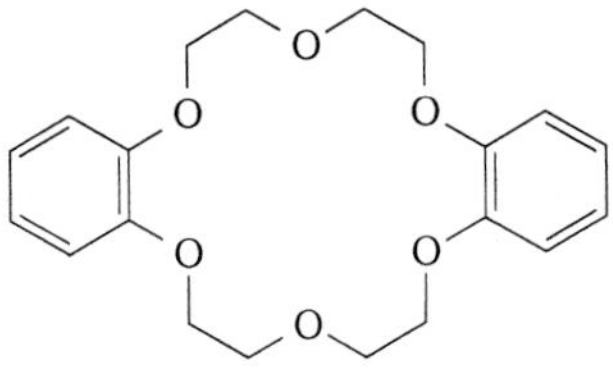 (8)

(iii) Polymer-Bound Catalysts

In principle, the attachment of conventional phase-transfer catalysts to a solid support allows straightforward retrieval and recycling of the catalyst. However, for most phase-transfer processes this necessitates the introduction of a third phase, which creates a number of additional problems. Many reactions, particularly those with a slow intrinsic reaction rate, are much slower when insoluble supported catalysts are used. In addition many polymer supports tend to break down under the high levels of agitation that are often required. Despite these problems a range of highly effective onium ion and crown ether supported catalysts have been developed and successfully applied to a range of reaction processes (see for example: R. Annunziata *et al.*, *Org. Lett.* 2000, **2**, 1737; O. Arrad, Y. Sasson, *J. Org. Chem.* 1989, **54**, 4993; M. Tomoi, W.T. Ford, "Polymeric Phase-Transfer Catalysts" in "Synthesis and Separations Using Functional Polymers", Ed. D.C Sherrington, P. Hodge, Wiley & Sons, 1988, p183). Polyethylene glycols and their derivatives have also been shown to be effective phase-transfer catalysts in their own right (see for example: G.E. Totten, N.A. Clinton, P.L. Matlock, *J. Macromol. Sci. (Part A), Chem.* 1998, **C38,** 77; B. Sauvagnat *et al.*, *Tetrahedron Lett.*, 1998, **39**, 821; A. Gobbi *et al.*, *J. Org. Chem.* 1994, **59**, 5059; J.M. Harris et al., *J. Org. Chem.* 1982, **47**, 4789).

(iv) Chiral Catalysts

In the last 20 years, a vast array of chiral phase-transfer catalysts have been reported. The majority of these are either quaternary ammonium salts derived from naturally occurring alkaloids, or chiral crown ethers derived from chiral alcohols. Since this topic has been extensively reviewed in recent years (M.J. O'Donnell "Asymmetric Phase-Transfer Reactions" in "Catalytic Asymmetric Synthesis, 2nd Ed.", Ed. I. Ojima, Verlag Chemie, New York, 2000; T. Shioiri, "Chiral Phase-Transfer Catalysts" in "Handbook of Phase-Transfer Catalysis", Ed. Y. Sasson, R. Neumann, Blackie, 1997), discussion in this review will be limited to examples of highly enantioselective processes.

3. C-C Bond Forming Reactions

(a) Nucleophilic Addition of Cyanide Ion

The reaction between alkali metal cyanides and a variety of organic electrophiles has been reported to be efficiently promoted by phase-transfer catalysis.

Primary alkyl halides have been shown to react under both liquid-liquid (C.M. Starks, R.M. Owens, *J. Am. Chem. Soc.* 1973, **95**, 3613) and solid-liquid (G. Bram, A. Loupy, M. Pedoussaut, *Bull. Chem. Soc. Fr.* 1986, 124) phase-transfer conditions to give the corresponding nitriles in good yield, however secondary and tertiary alkyl halides suffer from competing elimination reactions (C.M. Starks, C.L. Liotta, "Phase-Transfer Catalysis, Principles and Techniques", Academic Press, New York, 1978, p96). A wide variety of catalysts have been reported to effect this reaction including the commonly employed n-Bu_4NBr (M. Cariou, *Bull. Soc. Chim. Fr.* 1978, 271), Aliquat 336 (G. Bram, A. Loupy, M. Pedoussaut, *Bull. Soc. Chim. Fr.*, 1986, 124) and 18-crown-6 (F.L. Cook, C.W. Bowers, C.L. Liotta, *J. Org. Chem.* 1974, **39**, 3416). It has also been demonstrated that active quaternary ammonium salt catalysts can be generated *in situ* during reactions of this type (W.P. Reeves, M.R. White, *Synth. Commun.* 1976, **6**, 193):

Br (9) —[33% aq. NaCN; n-Bu_3N (3% w/w), 103°C, 3h]→ CN (10) 100%

Problems can arise if the resulting nitrile is sufficiently acidic so as to undergo deprotonation under the reaction conditions, as exemplified by the reaction between benzyl chloride (11) and aqueous sodium cyanide:

Cl (11) —[aq. NaCN; R_4N^+]→ CN (12) —[$PhCH_2Cl$; pH > 9.5]→

CN $\xrightarrow[\text{pH > 9.5}]{\text{PhCH}_2\text{Cl}}$ CN

(13) (14)

In such situations, slow addition of aqueous sodium cyanide has been recommended as an effective means of maintaining the pH below 9.5 and thus avoiding secondary alkylation (H. Coates, R.L. Barker, R. Guest, British Patent 1978, 1,336,883).

Acid chlorides (e.g. 15) have also be reported to undergo reaction with cyanide ion leading to formation of acylcyanides (K.E. Koenig, W.P. Weber, *Tetrahedron Lett.* 1974, 2275).

O Cl $\xrightarrow[\text{n-Bu}_4\text{NBr (10mol\%), 0°C, 1h}]{\text{aq. NaCN}}$ O CN 60%

(15) (16)

This chemistry has been extended to the preparation of acylated cyanohydrins by addition of an aldehyde (e.g. 17) or ketone to the reaction system. Quaternary ammonium salts (A.T. Au, *Synth. Commun.* 1984, **14**, 743; J.M.J. Frechet, M.D de Smet, M.J.J. Farrall, *J. Org. Chem.* 1979, **44**, 1774; J.M. Mcintosh, *Can. J. Chem.* 1977, **55**, 4200) and crown ethers (K. Okazaki, K. Nomura, E. Yoshii, *Synth. Commun.* 1987, **17**, 1021; R. Chenevert, R. Plante, N. Voyer, *Synth. Commun.* 1983, **13**, 403) appear to be equally effective in this process.

EtO(C=O)Cl, aq. NaCN, n-Bu_4NBr (cat.), CH_2Cl_2 → (18) 92%

(17)

PhCOCl, KCN, 18-crown-6 (10mol%), CH_2Cl_2 → (19) 94%

The diastereoselective addition of cyanide ion to α-amino acid derived aldehydes (e.g. 20), has also been reported (A. Faessler, G. Bold, H. Steiner, *Tetrahedron Lett.* 1998, **39**, 4925; F. Matsuda *et al.*, *Chem. Lett.* 1990 723):

BocHN, CHO (20) + (21) Cl; $BnNEt_3Cl$ (4mol%), aq. KCN, CH_2Cl_2, 0°C → BocHN, OBz, CN (22) 70% (60%d.e)

O-Silyl cyanohydrins (24) have been generated under phase-transfer conditions. In this case *via* cyanide ion addition to acyl silanes (e.g. 23) followed by Brook rearrangement (K. Takeda, Y. Ohnishi, *Tetrahedron Lett.* 2000, **41**, 4169):

$SiMe_2$t-Bu (23); aq.KCN, n-Bu_4PBr (20mol%), CH_2Cl_2 → $OSiMe_2$t-Bu, CN (24) 95%

With an appropriate substrate (e.g. 25) and by switching to solid-liquid conditions, alkylation of the resulting cyanohydrin is also possible:

O
Br SiMe$_2$t-Bu (25) — solid KCN / 18-crown-6 / CH_2Cl_2 → NC OSiMe$_2$t-Bu (26) 95%

Cyanide ring-opening of epoxides (e.g. 27) under phase-transfer conditions has also been described, and is regioselective even with substrates such as (27) (Eli Lilly Co., Patent, 1992, 5136079 (US-5136079), 92-283996.):

(27) — aq. KCN/KOH / n-Bu_4NBr (cat.), $ClCH_2CH_2Cl$ → OH CN (28)

(b) Alkylation of C-H Acids

The deprotonation and subsequent alkylation of acidic or moderately acidic C-H compounds (pKa ≤23) constitutes the largest group of phase-transfer reactions studied. These processes most commonly involve the use of simple hydroxide bases (NaOH, KOH) in conjunction with either quaternary ammonium halide or crown ether catalysts, but solid potassium carbonate is sometimes preferred for substrates containing hydrolysis-sensitive functionality.

A number of studies have suggested that the hydroxide ion may not be efficiently extracted into the organic phase by quaternary ammonium halide catalysts (see for example: M. Makosza, I. Krylowa, *Tetrahedron*, 1999, **55**, 6395; H-S. Wu *et al.*, *J. Mol. Cat. A: Chem.* 1998, **136**, 135; M. Rabinovitz, Y. Cohen, M. Halpern, *Angew. Chem. Int. Ed., Engl.* 1986, **25**, 960; A.W. Herriott, D. Picker, *J. Am. Chem. Soc.* 1975, **97**, 2345). This has led to the suggestion that many C-H acids may undergo deprotonation at the interface, and subsequently be extracted into the bulk organic phase by the phase-transfer catalyst (M. Makosza, *Pure Appl. Chem.*, 1975, **43**, 439). Other studies have found that significant extraction of hydroxide ion into the organic phase is possible (S. Cohen, A. Zoran, Y. Sasson, *Tetrahedron Lett.* 1998, **39**, 9815) and indicated that this can be strongly influenced by the nature of the quaternary ammonium counter-ion (J. de la Zerda, Y. Sasson, *J. Chem. Soc., Perkin II* 1987, 1147). Consequently it appears that the mechanistic detail of these reactions is not straightforward, and is probably strongly influenced by the nature of the substrate and catalyst involved.

In reactions where extraction of basic ions into the organic phase is necessary, the use of alcohol co-catalysts has sometimes been found to be advantageous (G. Rothenberg *et al.*, *Chem. Commun.* 2000, 1293; J. de la Zerda, Y. Sasson, *J. Chem. Soc., Perkin II* 1987, 1147; E.V. Dehmlow *et al.*, *Tetrahedron* 1985, **41**, 2927).

(i) Enolate Alkylation

Although not widely exploited, the generation and alkylation of simple ketones and aldehyde enolates under phase-transfer conditions is possible (A. Diaz-Ortiz *et al.*, *Synth Commun.* 1993, **23**, 875; Y. Goldberg, E. Abele, *Synth. Commun.* 1991, **21**, 557). For example, the alkylation of 2-methylpropanal (29) has been reported to proceed in high yield under liquid-liquid phase-transfer conditions (H.K. Dietl, K.C. Brannock, *Tetrahedron Lett.* 1973, 1273; T. Huang, S. Lin, *J. Chin. I. Ch. E* 1988, **19**, 193).

CHO
(29)
Cl
n-Bu_4NI (14mol%)
30% aq. NaOH,
PhMe, 70°C
CHO
(30)
96%

The reaction of simple ketone substrates under solid-liquid conditions has also been reported, as illustrated by the alkylation of cycloundecanone (31) (L.I. Zakharkin, G.N. Antonova, L.S. Podvisotskaya, *Izv. Akad. Nauk SSSR, Ser Khim.* 1995, 2521).

(31) $\xrightarrow[\text{dibenzo-18-crown-6, KOH, PhMe}]{\text{I(CH}_2)_4\text{CO}_2\text{t-Bu}}$ (32) 65%

More commonly, carbonyl compounds (33) incorporating additional enolate stabilisation are employed:

(33) $\xrightarrow[\text{phase-transfer catalyst, base}]{\text{R'-X}}$ (34)

Y = Ar, CO_2R, COR, CN, NC, $R_2C{=}N$, NO_2, RSO_2, RSO, RS, $(RO)_2PO$

Examples of this include α-aryl ketones (A. Jonczyk, B. Serafin, M. Makosza, *Tetrahedron Lett.* 1971, 1351), malonates (Y. Wang *et al.*, *Synth. Commun.* 1995, **25**, 1761; A.R. Kore, R.B. Mane, M.M. Salunkhe, *Bull. Soc. Chem. Belg.*, 1995, **104**, 643; S.V. Patel *et al.*, *Chem. Ind.* 1991, 24; M. Fedorynski *et al.*, *J. Org. Chem.* 1978, **43**, 4682), β-diketones (A. Choudhary, A.L. Baumstark *Synthesis* 1989, 688; G. Gelbard, S. Colonna, *Synthesis* 1977, 113), acetamidomalonates (J. Zhu *et al.*, *Synth. Commun.* 1995, **25**, 215), β-ketoesters (R.-T. Li, Q. Zhu, M.-S. Cai, *Synth. Commun.* 1997, **27**, 1351; D. Barbry, C. Faven, A. Ajana, *Org. Prep. Proced. Int.* 1994, **26**, 469; R. Deng, Y. Wang, Y. Jiang, *Synth. Commun.* 1994, **24**, 111; S. Mataka *et al.*, *J. Org. Chem.* 1986, **51**, 4618; H.D. Durst, L. Liebeskind, *J. Org. Chem.* 1974, **39**, 3271; B. Samuelsson, B. Lamm, *Acta Chem. Scand.* 1971, **25**, 1555; A. Brandstrom, U. Junggren, *Acta Chem. Scand.* 1969, **23**, 2204), cyanoesters (R.K. Singh, S. Danishefsky, *J. Org. Chem.*, 1975, **40**, 2969), isocyanoesters (S. Kotha, E. Brahmachary, *J. Org. Chem.* 2000, **65**, 1359; U. Schöllkopf, D. Hoppe, R. Jentsch, *Chem. Ber.*, 1975, **108**, 1580), iminoesters (A. Lopez *et al. Tetrahedron* 1996, **52**, 8365; M.J. O'Donnell *et al.*, *Pol. J. Chem.* 1994, **68**, 2477; J. Ezquerra *et al.*, *Tetrahedron Lett.* 1993, **34**, 8535; B. Kaptein *et al.*, *Tetrahedron Lett.* 1992, **33**, 6007; Y. Jiang *et al.*, *Tetrahedron* 1988, **44**, 5343; M.J. O'Donnell *et al.*, *Synthesis*, 1984, 313; M.J. O'Donnell, M. Eckrich, *Tetrahedron Lett.* 1978, 4625; M.J. O'Donnell, J.M. Boniece, S.E. Earp,

Tetrahedron Lett. 1978, 2641), nitroesters (E. Diez-Barra, A. de la Hoz, A. Moreno, *Synth. Commun.* 1994, **24**, 1817; M.E. Niyazymbetov, D.H. Evans, *J. Org. Chem.* 1993, **58**, 779; A. Thomas, S.G. Manjunatha, S. Rajappa, *Helv. Chim. Acta* 1992, **75**, 715; V.N. Gogte, A.A. Natu, V.S. Pore, *Synth. Commun.* 1987, **17**, 1421), nitroketones (M. Sawamura *et al.*, *J. Org. Chem.* 1996, **61**, 9090), β-ketosulfones (Z. Zhang *et al.*, *Synth. Commun.* 1989, **19**, 1167), β-ketosulfoxides (C. Xu, G. Liu, Z. Zhang, *Synth. Commun.* 1987, **17**, 1839), β-ketosulfides (R. Deng, Y. Wang, Y. Jiang, *Synth. Commun.* 1994, **24**, 1917), and β-ketophosphonates (S.M. Ruder, V.R. Kulkarni, *Synthesis* 1993, 945).

Table 4.

(35) → R-X, n-Bu_4NBr (10mol%), solid KOH, RT, 1-24h → (36) + (37)

R-X	Stochiometry (ketone:R-X:base)	Yield (%) (36) : (37)
CH_3-I	1 : 1 : 2 1 : 3 : 4	95 : 0 0 : 82
n-C_4H_9-Br	1 : 1 : 2 1 : 2 : 3	92 : 0 35 : 60
PhCH_2-Br	1 : 1 : 1 1 : 2 : 3	96 : 0 0 : 80

A wide variety of different phase-transfer conditions are effective for these substrates, the most commonly employed being either aqueous or solid hydroxide in conjunction with an organic solvent (typically dichloromethane or toluene), or solid potassium carbonate in hot acetonitrile. Under these conditions *C*-alkylation is usually favoured and it is possible to obtain good selectivity for mono-alkylation if stochiometric amounts of alkylating agent are used (for an exception to this see, E. Diez-Barra *et al.*, *Tetrahedron* 1997, **53** 3659). In most instances di-alkylation is also possible if excess alkyl halide is employed (e.g. Table 4) (A.

Aranda *et al., J. Chem. Soc., Perkin Trans. I* 1992, 3451; A. Aranda *et al., J. Chem. Soc., Perkin Trans. I* 1992, 2427)

It has also been shown that alkylation of arylacetic acids (e.g. 38) is possible if excess alkylating agent is used (C. Balo *et al., Arch. Pharm.* 1991, **324**, 967; J.A. Canicio, A. Ginebreda, R. Canela, *An. Quim., Ser. C* 1985, **81**, 181). This transformation appears to proceed *via O*-alkylation, to give an ester, followed by *C*-alkylation, and then ester hydrolysis.

MeO (38) CO_2H → MeO (39) CO_2H

Me_2SO_4 (3.2eq.), $BnNEt_3Cl$ (31mol%), solid KOH, CH_2Cl_2, reflux, 5h

In many instances the alkylation of an achiral enolate substrate leads to the generation of an asymmetric centre, and this offers the possibility of using asymmetric phase-transfer catalysis for the preparation of optically-enriched materials. Early investigations into this possibility centred around the use of quaternary ammonium catalysts derived from the cinchona and ephedra alkaloids (M.J. O'Donnell "Asymmetric Phase-Transfer Reactions" in "Catalytic Asymmetric Synthesis, 2nd Ed.", Ed. I. Ojima, Verlag Chemie, New York, 2000; T. Shioiri, "Chiral Phase-Transfer Catalysts" in "Handbook of Phase-Transfer Catalysis", Ed. Y. Sasson, R. Neumann, Blackie, 1997), and although the catalysts were effective in promoting alkylation, enantiomeric excesses were generally moderate.

(40) → (42)

(41) (10mol%), 50% aq. NaOH, PhMe, RT, 1,3-dichloro-2-butene

(42) 99% (92% e.e.)

This situation changed with the development of a highly enantioselective alkylation of indanones (e.g. 40) (U.-H. Dolling *et al.*, *J. Org. Chem.* 1987, **52**, 4745; A. Bhattacharya *et al.*, *Angew. Chem. Int. Ed., Engl.* 1986, **25**, 476; R.S.E. Conn *et al.*, *J. Org. Chem.* 1986, **51**, 4710). This process employed a cinchonine derived quaternary ammonium salt (41) and subsequently it has been reported that catalysts of this type are effective in promoting a range of related enantioselective alkylation reactions (S. Arai *et al.*, *Tetrahedron Lett.* 1999, **40**, 6785; T.B.K. Lee, G.S.K. Wong, *J. Org. Chem.* 1991, **56**, 872; W. Nerinckx, M. Vandewalle, *Tetrahedron Asymm.* 1990, **1**, 265).

Cinchona alkaloid derived quaternary ammonium salts have also been utilised in the alkylation of glycine imine ester (43). This process has attracted considerable interest since a wide variety of alkylating agents can be used and both amino acid esters (45) and free amino acids (46) can be generated from the resulting imines by simple acid hydrolysis.

$Ph_2C{=}N$–CH_2–CO_2t-Bu (43)
R-X; phase-transfer catalyst, base, solvent
$Ph_2C{=}N$–CH(R)–CO_2t-Bu (44)
15% aq. citric acid
6N HCl
H_2N–CH(R)–CO_2t-Bu (45)
H_2N–CH(R)–CO_2H (46)

Pioneering studies in this area (see for example: I.A. Esikova, T.S. Nahreini, M.J. O'Donnell, in "ACS Symposium Series 659: Phase-Transfer Catalysis - Mechanisms and Syntheses", Ed. M.E. Halpern, ACS, Washington D.C, 1997, p89; M.J. O'Donnell, S. Wu, J.C. Huffman, *Tetrahedron* 1994, **50**, 4507; M.J. O'Donnell, W.D. Bennett, S. Wu, *J. Am. Chem. Soc.* 1989, **111**, 2353) employed the *N*-benzyl derivative (47) of cinchonidine, and obtained enantioselectivities in the range 42-81% e.e. During the course of these studies, it was also demonstrated that highly enantiomerically enriched amino acids could be obtained *via* crystallisation of the intermediate imines (44).

More recently it has been reported that the corresponding *N*-anthracenylmethyl cinchona alkaloid derivatives (e.g. 48, 49, and 50) give excellent enantioselectivities with a range of different alkylating agents. As would be expected based on earlier work, it was shown that catalysts derived from cinchonidine (e.g. 48 and 50) are enantiocomplimentary to those derived from cinchonine (e.g. 49), in that they are selective for opposite enantiomers of the alkylation products (44).

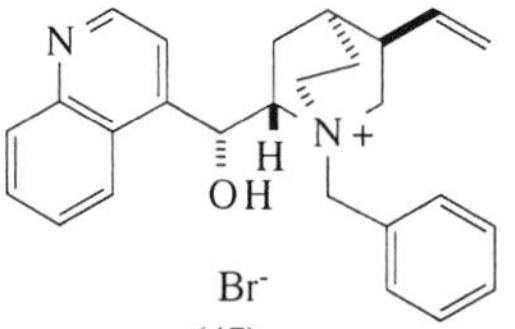

Br^- (47)

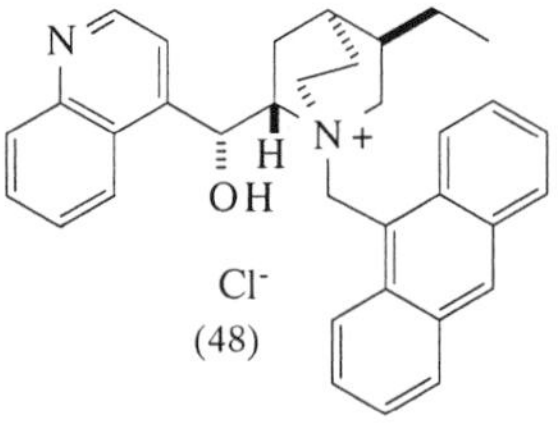

Cl^- (48)

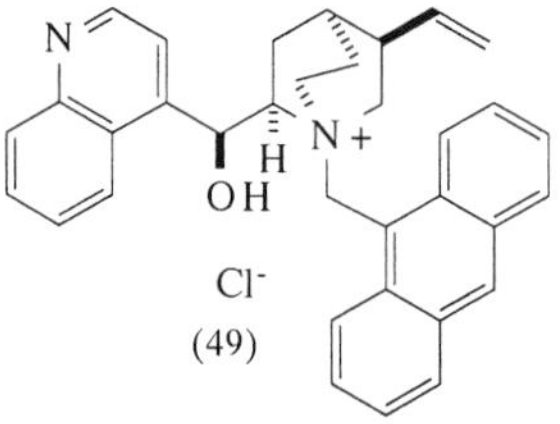

Cl^- (49)

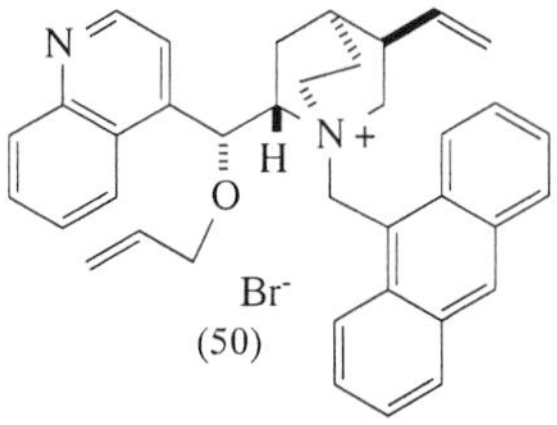

Br^- (50)

Catalysts such as (48) and (49) have been utilised under conventional liquid-liquid phase-transfer conditions using 50% aqueous potassium hydroxide and toluene at room temperature (B. Lygo, J. Crosby J.A. Peterson, *Tetrahedron Lett.* 1999, **40**, 1385; E.V. Dehmlow, S. Wagner, A. Müller, *Tetrahedron* 1999, **55**, 6335; B. Lygo, P.G. Wainwright, *Tetrahedron Lett.* 1997, **38**, 8595), typically giving imines (44) in 84-98%e.e. In these reactions the initial catalyst structures are usually modified *via* rapid *O*-alkylation (M.J. O'Donnell, S. Wu, J.C. Huffman, *Tetrahedron* 1994, **50**, 4507) during the alkylation reaction. Pre-alkylated catalysts such as (50) have been employed in solid-liquid ($CsOH.H_2O$, CH_2Cl_2) phase-transfer alkylations at low temperature (E.J. Corey, M.C. Noe, F. Xu, F. *Tetrahedron Lett.* 1998, **39**, 5347; E.J. Corey, F. Xu, F; M.C.Noe, *J. Am. Chem. Soc.* 1997, **119**, 12414), typically giving imines (44) with enantioselectivities in the range 92-99%e.e. Catalysts such as (50) have also been used in conjunction with a soluble phosphazene base in dichloromethane at low temperature (M.J. O'Donnell *et al.*, *Tetrahedron Lett.* 1998, **39,** 8775), again leading to high levels of enantioselectivity, and these latter reaction conditions have also been applied to the asymmetric alkylation of Wang resin-bound glycine imine (51) (M.J. O'Donnell, F.Delgado, R.S. Pottorf, *Tetrahedron* 1999, **55**, 6347):

$Ph_2C{=}N$ CO_2- (51) → Catalyst (69) (100mol%), CH_2Cl_2, -78°C, $PhCH_2Br$ → $Ph_2C{=}N$ CO_2- Ph, 76% e.e. (*S*)-(52)

Me, N, P, Nt-Bu, NEt_2, N, Me

Recently the binaphthyl-derived quaternary ammonium salt (53) has also been reported to be a highly effective catalysts for the alkylation of imine (43) (T. Ooi, M. Kameda, K. Maruoka, *J. Am. Chem. Soc.*, 1999, **121**, 6519), giving enantioselectivities in the range 90-96% e.e. at 0°C under liquid-liquid phase-transfer conditions (50% aq. KOH, toluene).

N^+ Br^- (53)

The asymmetric alkylation of imines derived from other amino acids (54) has also been reported. In this case it appears necessary to employ aldehyde-derived imines and to use solid-liquid phase-transfer conditions in order to obtain efficient alkylation (M.J. O'Donnell, S. Wu, *Tetrahedron Asymm.* 1992, **3**, 591).

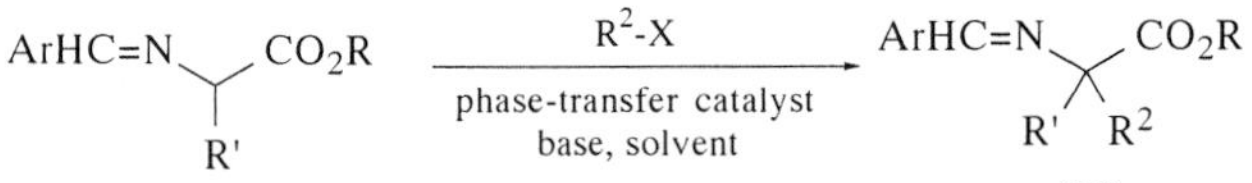

ArHC=N CO$_2$R (R') (54) —[R^2-X; phase-transfer catalyst, base, solvent]→ ArHC=N CO$_2$R (R', R^2) (55)

Catalysts of type (48), (49), (B. Lygo, J. Crosby, J.A. Peterson, *Tetrahedron Lett.* 1999, **40**, 8671) have been shown to give good enatioselectivities (up to 87% e.e.) in the alkylation of imine (54, R = *t*-Bu, R' = Me), and catalysts of type (53) reported (T. Ooi *et al.*, *J. Am. Chem. Soc.*, 2000, **122**, 5228) to give excellent selectivities (91-99%e.e.) for a range of substrates (54, R = *t*-Bu, R' = various).

Highly enatioselective (up to 92% e.e.) alkylation of imines (54) has also been achieved using solid sodium hydroxide in conjunction with both taddol (56) (Y.N. Belokon *et al.*, *Tetrahedron Asymm.*, 1998, **9**, 851) and chiral salen complexes (e.g. 57) (Y.N. Belokon, R.G. Davies, M. North, *Tetrahedron Lett.*, 2000, **41**, 7245; Y.N. Belokon *et al.*, *Tetrahedron Lett.* 1999, **40**, 6105).

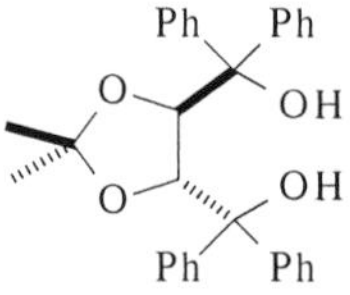

Ph Ph OH OH Ph Ph O O

(56)

N N Cu O O

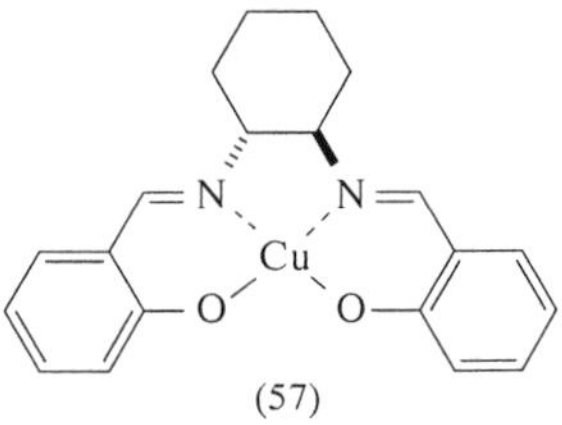

(57)

The use of amino acid imines incorporating chiral auxiliary groups has also been investigated. For example, the mono-alkylation of imines (58) (A. Lopez, R. Pleixats, *Tetrahedron Asymm.* 1998, **9**, 1967; W. Oppolzer, R. Moretti, C. Zhou, *Helv. Chim. Acta* 1994, **77**, 2363) and (59) (G. Guillena, C. Najera, *Tetrahedron Asymm.* 1998, **9**, 3935) under phase-transfer conditions, have both been reported to proceed with high levels of diastereoselectivity.

$N=CR_2$ O_2 (58)

MeN N $N=CR_2$ Me Ph (59)

Two other highly diastereoselective phase-transfer alkylation approaches to α-amino acids have been reported recently. Phenylglycine derivative (61) was obtained in high enantiomeric excess using solid-liquid phase-transfer conditions (M.J. O'Donnell *et al.*, *Heterocycles* 1997, 46, 617);

(60) → (61): EtI; n-Bu_4NI, solid KOH/K_2CO_3, CH_2Cl_2, RT, 12h — 75% (92% e.e.)

and the alanine derivative (62) was reacted with a range of alkylating agents, again under solid-liquid phase-transfer conditions (R. Chinchilla, N. Galindo, C. Najera, *Synthesis* 1999, 704):

(62) → (63): $PhCH_2Br$; n-Bu_4NBr, solid K_2CO_3, CH_3CN, RT — 75% (96% d.e.)

(ii) Alkylation of other carbanions

(64) → (65): R-X; phase-transfer catalyst, base — Y, Z = various electon stabilising groups

In addition to carbonyl compounds, a wide range of other C-H acidic compounds (64) have been alkylated under phase-transfer conditions. As might be expected, similar reaction conditions and catalysts to those described above are used for substrates of this type. Of particular note in this context are phenylacetonitriles, since these were the substrates studied in what is generally considered to be the genesis of phase-transfer *C*-alkylation chemistry (M. Makosza, B. Serafinowa, *Rocz. Chem.* 1965, **39**, 1223; M. Makosza, B. Serafinowa, *Rocz. Chem.* 1965, **39**, 1401). Since these seminal papers a large number of publications have been concerned with investigating alternative phase-transfer conditions for the alkylation of this type of substrate (see for example: T.W. Bentley *et al.*, *J. Chem. Soc., Perkin Trans. 2* 1998, 89; S. Cohen, G. Rothenberg, Y. Sasson, *Tetrahedron Lett.* 1998, **39**, 3093; R. Markovic, G. Dimitrijevic, V. Aleksic, *J. Chem. Res. (S)* 1997, 66; D. Barbry, C. Pasquier, C. Faven, *Synth. Commun.*, 1995, **25**, 3007; E. Diaz-Barra *et al.*, *Synthesis* 1989, 391; T. Balakrishnan, W. Ford, *J. Org. Chem.* 1983, **48**, 1029; R. Solaro, *J. Org. Chem.* 1980, **45**, 4179; M. Makosza, A. Jonczyk, *Org. Synth.*, 1976, **55**, 91).

Other C-H acidic substrates that have been alkylated under phase-transfer catalysed conditions include: malononitriles (E. Diez-Barra *et al.*, *Synthesis* 1989, 391; E. Diez-Barra *et al.*, *J. Chem. Soc., Perkin Trans. I* 1991, 2593; E. Diez-Barra *et al.*, *J. Chem. Soc., Perkin Trans. I* 1991, 2589), α-iminophosphonates (J.P. Genet *et al.*, *Tetrahedron Lett.* 1992, **33**, 77) α-phosphonosulfones (D.Y. Kim, K.H. Suh, *Synth. Commun.* 1998, **28**, 83), α-halosulfones (A. Jonczyk, I. Pytlewski, *Synthesis* 1978, 883; A. Jonczyk, K. Banko, M. Makosza, *J. Org. Chem.* 1975, **40**, 266), benzylsulfones (J. Golinski, A. Jonczyk, M. Makosza, *Synthesis* 1979, 461), α-sulfonylisonitriles (K. Kurosawa *et al.*, *Tetrahedron Lett.* 1982, 5335), cyclopentadiene (E. Dehmlow, C. Bollmann, *Tetrahedron Lett.* 1991, 5773; O. Nefedov, *et al.*, *Mendeleev Chem. J. (Engl. Trans)* 1986, 31, 102), nitroalkanes (P. Vanelle *et al.*, *Heterocycles* 1998, **48**, 181), indene (M. Halpern *et al.*, *New J. Chem.* 1984, **8**, 443; M. Makosza, *Tetrahedron Lett.* 1966, 4621) and phenylacetylene (E.V. Dehmlow, U. Fastabend, *Gazz. Chim. Ital.* 1996, **126**, 53).

Fischer carbene complexes (e.g. 66) have also been reported to readily undergo dialkylation under conventional liquid-liquid phase-transfer conditions (S.R. Amin, A. Sarkar, *Organomet.* 1995, **14**, 547):

OBn (OC)$_5$Cr (66) → Br (allyl bromide), n-Bu$_4$NBr (10mol%), 50% aq. NaOH, CH_2Cl_2, RT, 3h → OBn (OC)$_5$Cr (67) 80%

It is interesting to note that the starting material (66) for this reaction can also be generated under phase-transfer conditions (see under C-O bond formation)

(c) Aldol and Related Processes

It is well-known that, aldol reactions involving certain substrate combinations can be performed in a two-phase system without the need for a phase-transfer catalyst (see for example, M. Stiles, *J. Am. Chem. Soc.*, 1964, **86**, 3337):

O (68) → O_2N–C$_6$H$_4$–CHO, 0.2N NaOH, H_2O, 0°C, 2h → O OH NO_2 (69) 90%

Attempts to extend the scope of this process by employing catalysts have met with varying degrees of success. Simple aldehyde and ketone substrates (e.g. 70) have been reported to give mixtures of aldol and aldol condensation products under normal phase-transfer conditions (J. Nowicki, J. Gora, *Pol. J. Chem.* 1991, **65**, 2267; X.-P. Gu, I. Ikeda, M. Okahara, *Bull. Chem. Soc. Jpn.*, 1988, **61**, 2256; I. Aljancic-Solaja, M. Rey, A.S. Dreiding, *Helv. Chim. Acta* 1987, **70**, 1302). For substrates of this type, surfactant catalysts (e.g. n-$C_{16}H_{33}NMe_3Cl$) seem to perform better, giving high selectivity for the condensation product (72) (F. Fringuelli *et al.*, *Tetrahedron* 1994, **50**, 11499):

CHO

O OH O

Ph (70) → 0.25M aq. NaOH, RT, 2h → Ph (71) Ph + Ph (72) Ph

Catalyst	(71)	(72)
n-Bu_4NCl (10mol%)	26%	43%
n-$C_{16}H_{33}NMe_3Cl$ (10mol%)	0%	92%

In contrast, phenylacetonitrile (73) (F. Fringuelli *et al.*, *Tetrahedron* 1994, **50**, 11499) and phenylmethylsulfone (75) (C. Cardillo, D. Savoia, A. Umani-Ronchi, *Synthesis* 1975, 453) have been reported to give only condensation products when reacted with aromatic aldehydes irrespective of the catalyst employed:

CHO

CN

Ph CN (73) → 0.25M aq. NaOH, RT → Ph (74) Ph

Catalyst	(74)
n-Bu_4NCl (10mol%)	80%
n-$C_{16}H_{33}NMe_3Cl$ (10mol%)	86%

CHO

PhO_2S-CH_3 (75) → $BnNEt_3Cl$ (20mol%), 50% aq. NaOH, RT, 2-6h → PhO_2S Ph (76) 86%

The nitroaldol (Henry) reaction has also been effected under a range of phase-transfer conditions (see for example: D. Lazzari *et al.*, *Org. Prep. Proc. Int. Ed.* 1999, **31**, 543; R. Ballini, G. Bosica, *J. Org. Chem.* 1997, **62**, 425; D. Simoni, *et al.*, *Tetrahedron Lett.* 1997, **38**, 2749; W. Kobertz, C.R. Bertozzi, M.D. Bednarski, *J. Org. Chem.* 1996, **61**, 1894). Since this process often only requires catalytic amounts of base, many alternative processes that simply employ catalytic amounts of a quaternary ammonium salt (e.g. R_4NOH and R_4NF) have been developed. Polymer-supported quaternary ammonium salts appear particularly useful in this respect since they can readily be removed from the reaction mixture (see for example: C. Jousse, D. Desmaele, *Eur. J. Org. Chem.* 1999, 909; M. Caldarelli, J. Habermann, S.V. Ley, *J. Chem. Soc., Perkin. Trans. 1* 1999, 107; R. Ballini, P. Astolfi, *Liebigs Ann. Chem.* 1996, 1879; R. Ballini, G. Bosica, *J. Org. Chem.* 1994, **59**, 5466).

Recently the use of chiral quaternary ammonium salt (50) to influence the diastereoselectivity of addition of nitromethane to amino acid derived aldehydes (e.g. 77) was reported (E.J. Corey, F.Y. Zhang, *Angew. Chem. Int. Ed. Engl.* 1999, **38**, 1931).

BocHN CHO Ph (77) — CH_3NO_2, catalyst (49) (10mol%), KF, CH_2Cl_2, -10°C, 12h → BocHN, OH, NO_2, Ph (78) 88% (80% d.e.) (0% d.e. obtained with n-Bu_4NF)

Aldol-like reactions involving more complex substrates have also been investigated. For example, azidoester (79) has been reported to react with arylaldehydes (Y. Murakami *et al., Chem. Pharm. Bull.* 1997, **45**, 1739);

N_3 CO_2Et (79) — OMe, CHO; dicyclohexyl-18-crown-6 (10mol%), KF, CH_3CN, RT, 2d → N_3, CO_2Et, OMe OH (80) 39%

and glycine imine esters (e.g. 81) have also been employed in aldol-like reactions (W. Shengade, Z. Changyou, J. Yaozhong, *Synth. Commun.* 1986, **16**, 1479):

Cl, N, CO_2Et (81) — Cl, CHO; $BnNEt_3Cl$ (10mol%), solid K_2CO_3, CH_3CN → H N, CO_2Et, O, Cl, Cl 95% (82)

Attempts to influence the stereoselectivity of this latter process using chiral quaternary ammonium salts has met with limited success (C.M. Gasparski, M.J. Miller, *Tetrahedron* 1991, **47**, 5367), but a promising homogeneous asymmetric variant of this reaction has recently been reported (M. Horikawa, J. Busch-Petersen, E.J. Corey, *Tetrahedron Lett.* 1999, **40**, 3843).

Imines (e.g. 83) have also been employed as the electrophilic component in related processes (V. Dryanska, D. Tasheva, *Synth. Commun.* 1992, **22**, 63):

$Ph_2C{=}N\frown CN$ (83) —[PhCH=NPh; $BnNEt_3Cl$ (5mol%), 33% aq.NaOH, PhH, RT]→ (84) 62%

Ph₂C=N, CN, Ph, NHPh

In situ imine formation has been investigated, and this has been exploited in a remarkable multi-step process which involves subsequent addition of trichloromethide followed by hydrolysis to give aryl amino acids (e.g. 86) (D. Landini, F. Montanari, F. Rolla, *Synthesis*, 1979, 26).

$CHCl_3$ (85) —[PhCHO, NH_3; $BnNEt_3Cl$ (10mol%), KOH, LiCl, CH_2Cl_2, H_2O]→ (86) NH_2, CO_2H 66%

(d) Darzens Condensation

(87) Cl, O, R —[R'-CHO; phase-transfer catalyst, base]→ (88) R', O, O, R

Darzens condensations involving both α-chloroester and α-chloroketone substrates have been investigated using a range of phase-transfer conditions (see for example: M.-H. Munos *et al.*, *Chem. Pharm. Bull.* 1994, **42**, 1914; S. Gladaili, F. Soccolini, *Synth. Commun.* 1982, **12**, 355; M. Fedorynski *et al.*, *J. Org. Chem.* 1978, **43**, 4682).

Attempts to utilise chiral crown ether (see for example, P. Bako, *et al.*, *Tetrahedron Asymm.* 1999, **10**, 4539) and chiral quaternary ammonium salts (e.g. 41) (see for example: S. Arai *et al.*, *Tetrahedron* 1999, **55**, 6375; S. Arai *et al.*, *Chem. Commun.* 1999, 49; S. Arai, T. Shioiri, *Tetrahedron Lett.* 1998, **39**, 2145) in this process have met with limited success, however it has been shown that high levels of stereoselectivity can be obtained with certain substrate combinations (e.g. 89 and 90):

(89) + (90) CHO → (91)

catalyst (41) (10mol%)
solid $LiOH.H_2O$
$n\text{-}Bu_2O$, RT, 84h

86% (86%e.e. 85%d.e.)

Darzens condensations involving α-haloacetonitriles (see for example: J.P. Jayachandran, T. Balakrishnan, M.L. Wang, *J. Mol. Cat. A* 2000, **152**, 91; A. Jonczyk, A. Kwast, M. Makosza, *J. Chem. Soc., Chem. Commun.* 1977, **902**; A. Jonczyk, M. Fedorynski, M. Makosza, *Tetrahedron Lett.* 1972, 2395) and α-halosulfones (A. Jonczyk, K. Banko, M. Makosza, *J. Org. Chem.* 1975, **40**, 266) have also been described. Application of chiral phase-transfer catalysts (41) in the latter process has been shown to generate epoxysulfones (e.g. 93) in up to 81%e.e. (S. Arai, T. Ishida, T. Shioiri, *Tetrahedron Lett.* 1998, **39**, 8299):

PhO_2S Cl (92) → (93) SO_2Ph

CHO
cat.alyst (41) (10mol%)
KOH, PhMe, RT, 2h

70% (81%e.e.)

(e) Michael Additions

X Y (94) → (95)

Z
phase-transfer catalyst
base

X, Y, Z = various electon stabilising groups

Most of the carbanion species discussed above have also been reacted with a wide range of Michael acceptors, using similar reaction conditions to those employed for alkylation reactions (see for example: D. Tzalis, P. Knochel, *Tetrahedron Lett.* 1999, **40**, 3685; E. Diez-Barra *et al.*, *Tetrahedron*, 1998, **54**, 1835; C.D. Mudaliar, K.R. Nivalkar, S.H. Mashraqui, *Org. Prep. Proc. Int.* 1997, **29**, 584; A. Lopez *et al.* *Tetrahedron* 1996, **52**, 8365; A. Loupy, A. Zaparucha, *Tetrahedron Lett.* 1993, **34**, 473; R.M. Kellogg *et al.*, *Tetrahedron Lett.* 1991, **32**, 3727; E.V. Dehmlow, E. Kunesch, *Liebigs Ann. Chem.* 1985, 1904; G. Bram *et al.*, *Tetrahedron Lett.* 1985, 26, 4601; D. Loganathan, T. Varghese, G.K. Trivedi, *Org. Prep. Proc. Int.* 1984, **16**, 115). Amongst this body of work, a large number of studies on the application of chiral phase-transfer catalysts have been reported, and a number of highly stereoselective processes developed.

The first example of a highly enantioselective Michael addition involving phase-transfer conditions utilised binaphthyl-derived crown ethers (e.g. 96) (D.J. Cram, G.Y.D. Sogah, *J. Chem. Soc., Chem. Commun.* 1981, 625):

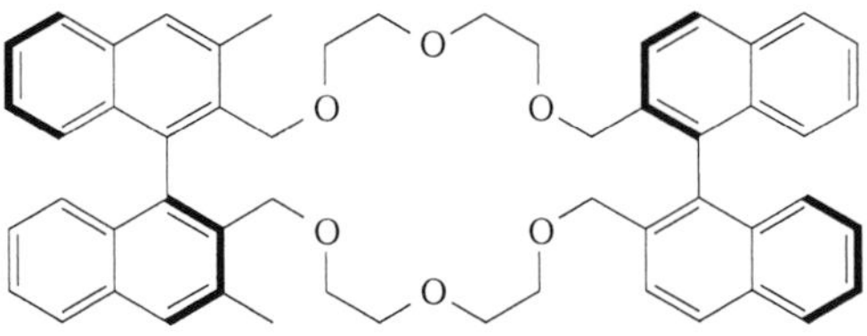

(96)

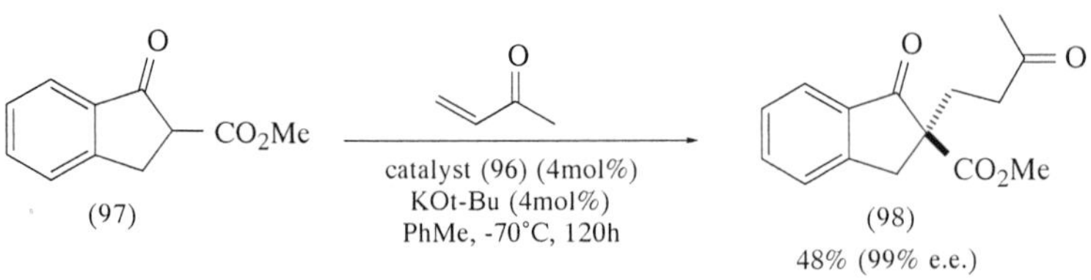

(97) CO_2Me → catalyst (96) (4mol%), KOt-Bu (4mol%), PhMe, -70°C, 120h → (98) CO_2Me, 48% (99% e.e.)

Another early example of a highly enantioselective Michael addition involved the use of crown ether (100) in the addition of methyl phenylacetate (99) to methyl acrylate (S. Aoki, S. Sasaki, K. Koga, *Tetrahedron Lett.* 1989, **30**, 7229)

(100)

Ph CO_2Me (99) → methyl acrylate, KOt-Bu (cat.), PhMe, -78°C, 3h → Ph, CO_2Me, CO_2Me (101) 95% (79% e.e.)

More recently crown ethers derived from both terpenes (E. Brunet *et al.*, *Tetrahedron Asymm.* 1994, **5**, 935) and carbohydrates (L. Toke *et al.*, *Tetrahedron* 1998, **54**, 213) have been reported to be similarly effective in this latter transformation.

As in the case of asymmetric alkylation processes, the earliest example of the use of a quaternary ammonium catalyst in a highly enantioselective Michael addition involved the use of indanone substrates (e.g. 40) in conjunction with the cinchona alkaloid-derived salt (41) (A. Bhattacharya *et al., Angew. Chem. Int. Ed. Engl.* 1986, **25**, 476; R.S.E. Conn *et al.*, *J. Org. Chem.* 1986, **51**, 4710).

Cl, Cl, O, MeO (40) → catalyst (41) (5.6mol%), 50% aq. NaOH, PhMe, RT, 30min → Cl, Cl, O, O, MeO (102) 95% (80% e.e.)

This chemistry has also been extended to other types of substrate (e.g. 103 and 106) (W. Nerinckx, M. Vandewalle, *Tetrahedron Asymm.* 1990, **1**, 265). Here it was reported that the dihydro catalyst (104, R = Et), generally gave slightly higher levels of enantiomeric excess than the corresponding unsaturated system (104, R = $CH{=}CH_2$):

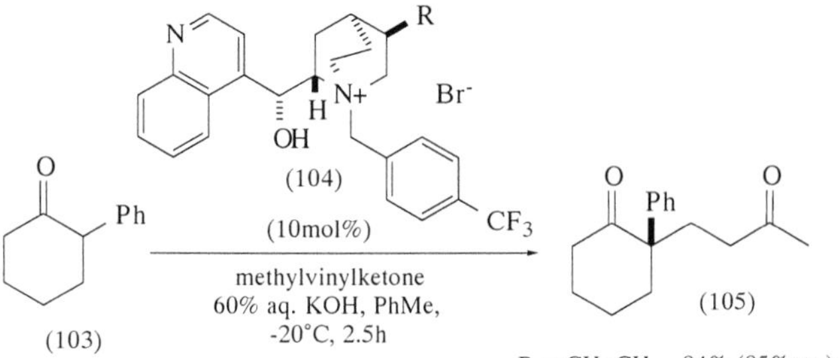

R = $CH=CH_2$, 84% (85%e.e.)
R = CH_2CH_3, 62% (87%e.e.)

With substrates such as (106), addition of 18-crown-6 to the reaction mixture after the initial Michael addition was complete, was found to give the Robinson annulation product (107) in high enantiomeric excess, and this process has subsequently been applied to the synthesis of (+)-triptoquinone (K. Shishido *et al.*, *J. Org. Chem.* 1994, 59, 406).

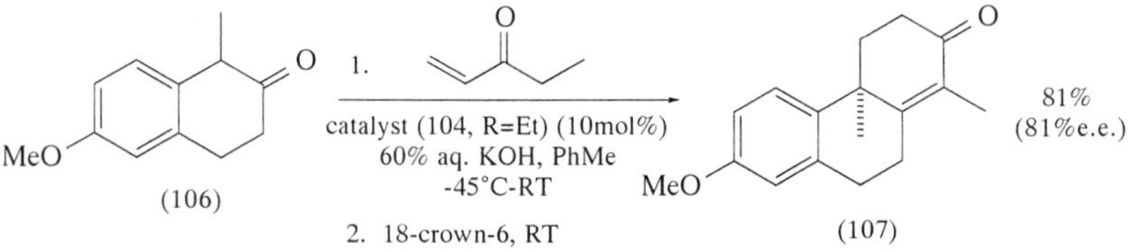

N-Anthracenylmethyl substituted cinchona alkaloid catalysts (e.g. 48) have been reported to be effective in promoting a number of enantioselective Michael additions (F.-Y. Zhang, E. J. Corey, *Org. Lett.* 2000, **2**, 1097; E.J. Corey, M.C. Noe, F. Xu, *Tetrahedron Lett.* 1998, **39**, 5347), including the addition of acetophenone (70) to enone (109):

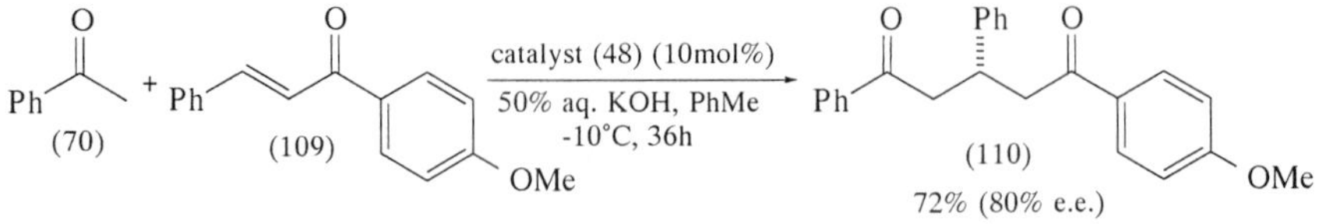

Another recent example of a highly enantioselective Michael addition involves the addition of 2-nitropropane (111) to chalcone (72) (P. Bako, Z.

Bajor, L. Toke, *J. Chem. Soc., Perkin Trans 1* 1999, 3651; P. Bako, T. Kiss, L. Toke, *Tetrahedron Lett.* 1997, **38**, 7259):

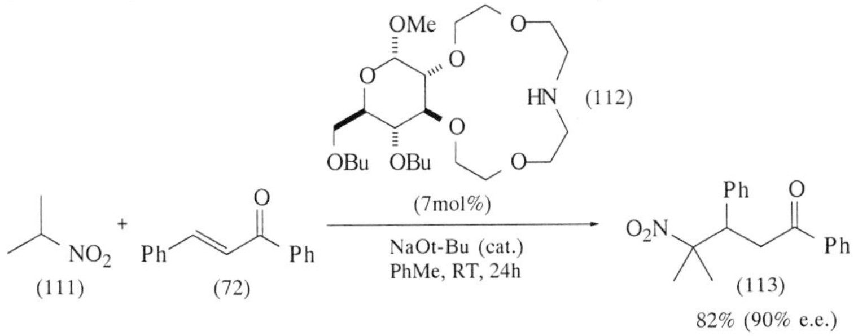

(111) + (72) → [(112) (7mol%), NaOt-Bu (cat.), PhMe, RT, 24h] → (113) 82% (90% e.e.)

The phase-transfer catalysed conjugate addition of stabilised anions to α-haloenones has also been reported. In this study it was found that either cyclic enol ethers (e.g. 115) (S. Arai, *et al.*, *Tetrahedron Lett.* 1998, **39**, 9739);

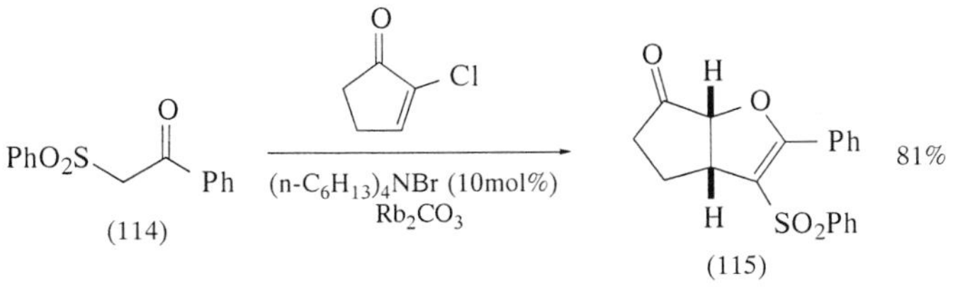

(114) → [$(n\text{-}C_6H_{13})_4NBr$ (10mol%), Rb_2CO_3] → (115) 81%

or cyclopropanes (e.g. 117) (S. Arai, *et al.*, *J. Org. Chem.* 1998, **63**, 9572) can be formed depending upon the nature of the substrates involved and the conditions employed.

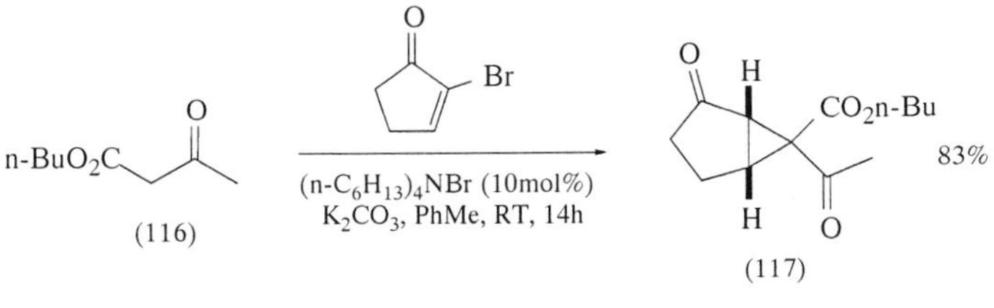

(116) → [$(n\text{-}C_6H_{13})_4NBr$ (10mol%), K_2CO_3, PhMe, RT, 14h] → (117) 83%

Use of chiral quaternary ammonium phase-transfer catalysts reactions of this type has also been investigated (S. Arai *et al.*, *Tetrahedron Lett.*

1999, **40**, 4215) but the enantioselectivities obtained were generally moderate.

(f) Wittig and Wadsworth-Emmons reactions

A number of Wittig reactions utilising phase-transfer conditions have been described. In most cases the ylide is generated in the reaction mixture from the corresponding phosphonium salt (e.g. 118), so there is no requirement for an additional phase-transfer catalyst. Reaction of the phosphonium salt with hydroxide ion to give phosphine oxides usually competes, and this has been shown to result in lower yields of alkene when less reactive carbonyl compounds are employed (Table 5) (W. Tagaki *et al.*, *Tetrahedron Lett.* 1974, 2587).

Table 5

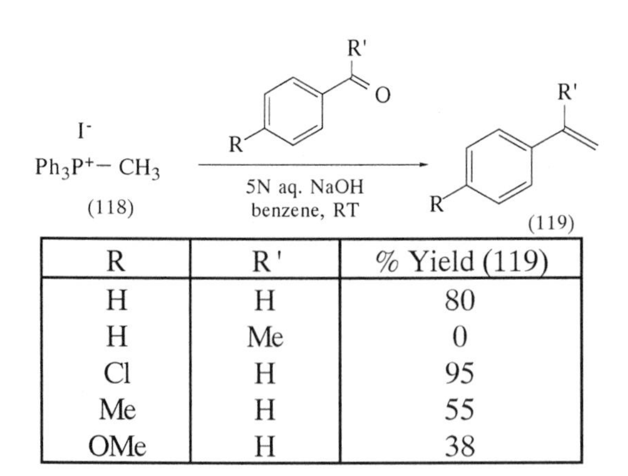

R	R'	% Yield (119)
H	H	80
H	Me	0
Cl	H	95
Me	H	55
OMe	H	38

Despite this potential shortcoming, many examples of sucessful phase-transfer Wittig reactions have been reported (see for example: J.J. Hwang, *J. Mol. Cat. A* 1999, **142**, 125; N. Daubresse, C. Francesch, C. Rolando, *Tetrahedron* 1998, **54**, 10761; E. Dehmlow, S. Barahona-Naranjo, *J. Chem. Res. (S)* 1981, 142; S. Yanagida, K. Takahashi, M. Okahara, *J. Org. Chem.* 1979, **44**, 1099; G. Markl, A. Merz, *Synthesis* 1973, 295).

It has also been demonstrated that, in favourable cases, phosphonium salts can be generated *in situ* and this strategy has been applied in the synthesis of pyrethroid intermediates (e.g. 121) (R. Galli *et al.*, *Tetrahedron* 1984, **40**, 1523):

OHC CO_2Et (120) $\xrightarrow[\text{BnNEt}_3\text{Cl (2mol\%), 40\% aq. NaOH, benzene, 30°C}]{\text{Ph}_3\text{P, CH}_3\text{SCHCl}_2}$ Cl MeS CO_2Et (121) 68%

Phase-transfer olefination reactions involving phosphonates (e.g. 122) have also been reported. Examples of this type of reaction proceeding in the absence of an additional catalyst are also known (M. Graff *et al.*, *Tetrahedron Lett.* 1986, **16**, 149), and an interesting application of this chemistry involves the double addition of phosphonate (122) to glutaraldehyde hydrate (J. Villieras, M. Rambaud, *Synthesis* 1983, 300):

$(EtO)_2P(O)CH_2CO_2Et$ (122) $\xrightarrow[\text{H}_2\text{O, K}_2\text{CO}_3]{\text{HO, O, OH}}$ HO, CO_2Et (123) 62%

However for most two-phase olefination reactions involving phosphonates, the addition of a quaternary ammonium ion or crown ether catalyst is deemed advantageous (see for example: S. Ohba, T. Kosaka, T. Wakabayashi, *Synth. Commun.* 1995, **25**, 3421; S.K. Davidsen, C.H. Heathcock, *Synthesis* 1986, 842; P.A. Aristoff, *Synth. Commun.* 1983, **13**, 145; M. Mikolajczky *et al.*, *Synthesis* 1975, 278; C. Piechucki *Synthesis* 1974, 869), as illustrated by the preparation of alkene (125) (E.N. Durantini, *Synth. Commun.* 1999, **29**, 4201):

$(EtO)_2P(O)CH_2CN$ (124) $\xrightarrow[\text{n-Bu}_4\text{NCl (2mol\%), 50\% aq.NaOH, CH}_2\text{Cl}_2\text{, RT, 20min}]{\text{O}_2\text{N-C}_6\text{H}_4\text{-CHO}}$ O_2N, CN (125) 91%

(g) Elimination Reactions

Eliminations are an integral part of condensation reactions and several examples of this type of process have been discussed above. In addition, a variety of dehydrohalogenation reactions have been achieved under phase-transfer conditions (see for example: M. Makosza, A. Chesnokov, *Tetrahedron* 2000, **56**, 3553; F.S. Sirovski, *Org. Process. Res. Dev.* 1999, **3**, 437; D. Mason, S. Magdasi, Y. Sasson, *J. Org. Chem.* 1991, **56**, 7229; M. Halpern *et al.*, *J. Org. Chem.* 1985, **50**, 5088).

A particularly useful example of this latter process involves the double dehydrohalogenation of 1,2-dibromides (e.g. 126) to generate alkynes (see for example: T. Kulinski, A. Jonczyk, *Synthesis* 1992, 757; E. Dehmlow, M. Lissel, *Tetrahedron* 1981, **37**, 1653):

Br

EtO Br OEt (126) —— $(n\text{-}C_8H_{17})_4NBr$ (1mol%), solid KOH, petroleum ether, 80°C, 6h ——> EtO OEt (127) 79%

(h) Carbene Additions

X_3CH (128) —— phase-transfer catalyst, base ——> X_2C: (129)

Dihalocarbenes (129) appear to be readily generated from the appropriate haloform (128) under basic phase-transfer conditions. In most cases the resulting carbenes are reacted with alkenes to give cyclopropanes (see for example: I. Crossland, *Org. Synth. Col. Vol. VII* 1990, 12; M. Fedorynski *et al.*, *Tetrahedron* 1997, **53**, 1053; E. Dehmlow, H. Raths, J. Soufi, *J. Chem. Res.* (S) 1988, 334; R. Le Goaller *et al.*, *Synth. Commun.* 1982, **12**, 1163; E. Dehmlow, M. Lissel, *Tetrahedron Lett.* 1976, 1783; T. Hiyama *et al.*, *Tetrahedron Lett.* 1975, 3013; G. Joshi, N. Singh, L. Pande, *Tetrahedron Lett.* 1972, 1461; C. Starks, *J. Am. Chem. Soc.* 1971, **93**, 195; M. Makosza, W. Wawrzyniewicz, *Tetrahedron Lett.* 1969, 4659) and this constitutes a highly effective method for the synthesis of cyclopropanes. Reactions involving trihalomethides are also known and can sometimes compete with cyclopropanation. This was highlighted in a recent study (M. Fedorynski, *Tetrahedron* 1999, **55**, 6329) which demonstrated that with olefins (130) and (133), the course of reaction is strongly influenced by the nature of the catalyst employed:

(130) $\xrightarrow[\text{50\% aq.NaOH}]{CHCl_3}$ (131) + (132)

Catalyst	(131)	(132)
$BnNEt_3Cl$ (cat.)	0%	65%
Me_4NHSO_4 (cat.)	68%	5%

(133) $\xrightarrow[\text{50\% aq.NaOH}]{CHBr_3}$ (134) + (135) + (136)

Catalyst	(134)	(135)	(136)
benzo-15-crown-6 (cat.)	56%	5%	9%
Ph_4AsCl (cat.)	7%	82%	9%

4. C-X Bond Forming Reactions

Phase-transfer alkylation of heteroatom nucleophiles has also been widely investigated using a range of reaction conditions. This section outlines some of the more common reaction processes of this type.

(a) C-O

$$\underset{(137)}{R-OH} \xrightarrow[\text{phase-transfer catalyst, base}]{R'\text{-}X} \underset{(138)}{R-O-R'}$$

Phase-transfer reactions have been employed in the alkylation of a variety of common O-H functions (137) including alcohols (see for example: S. Chatti, M. Bortolussi, A. Loupy, *Tetrahedron Lett.* 2000, **41**, 3367; D. Seeback *et al.*, *Org. Synth. Col. Vol. VII* 1990, 41; R.M. Nouguier, M. Mchich, *Tetrahedron* 1988, *44*, 2477; R.M. Nouguier, M. Mchich, *J. Org. Chem.* 1987, **52**, 2995; G. Mouzin *et al.*, *Synthesis* 1983, 117; H.H Freedman, R.A. Dubois, *Tetrahedron Lett.* 1975, 3251; A. Merz, *Angew. Chem. Int. Ed. Engl.* 1973, **12**, 846), phenols (see for example:

M.-L. Wang, C.-Y. Yang, *Tetrahedron* 1999, **55**, 6275; J.R. Carrillo, E. Diez-Barra, *Synth. Commun.* 1994, **24**, 945; E. Dahan, S. Biali, *J. Org. Chem.* 1991, **56**, 7269; M. Lissel, S. Schmidt, B. Neumann, *Synthesis* 1986, 382; A. McKillop, J. Fiaud, R. Hug, *Tetrahedron* 1974, **30**, 1379), carboxylic acids (see for example: H.S. Wu, J.F. Tang, *J. Mol. Cat. A* 1999, **145**, 95; M.J. O'Donnell *et al.*, *Synthesis* 1991, 989; J. Barry *et al.*, *Synthesis* 1985, 40; C.L. Liotta *et al.*, *Tetrahedron Lett.* 1974, 2417), hydroperoxides (J. Moulines *et al.*, *Synth. Commun.* 1990, **20**, 349), and hemiketals (S.N. Suryawanshi, A. Mukhopadhyay, A.S. Bhakuni, *Synth. Commun.* 1990, **20**, 687). Some examples of this chemistry are outlined below.

O-Alkylation with glycosyl halides has been used in the synthesis of a variety of glycosides (see for example: D. Carriere *et al.*, *J. Mol. Cat. A* 2000, **154**, 9; P. Lewis, S. Kaltia, K.J. Wahala, *J. Chem. Soc., Perkin Trans. 1* 1998, 2481; J.M. Kim, R.J. Roy, *J. Carbohydr. Chem.* 1997, **16**, 1281; R.R. Schmidt, *Angew. Chem. Int. Ed. Engl.* 1986, 212), and this methodology has also been combined with *in situ* Wittig reaction (N. Daubresse *et al.*, *Synthesis* 1998, 157):

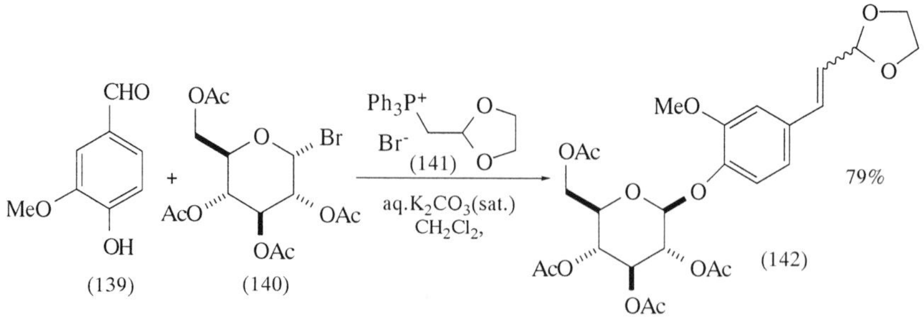

Macrocyclisation of insoluble ω-halo carboxylates under solid-liquid phase-transfer conditions has been shown to be a highly effective means of preparing macrolides since it avoids the need for high dilution conditions (Y. Kimura, S.L. Regen, *J. Org. Chem.* 1983, **48**, 1533). This strategy was recently utilised in the synthesis of Yuzu lactone (144) (L. Rodefeld, W. Tochtermann, *Tetrahedron* 1998, **54**, 5893).

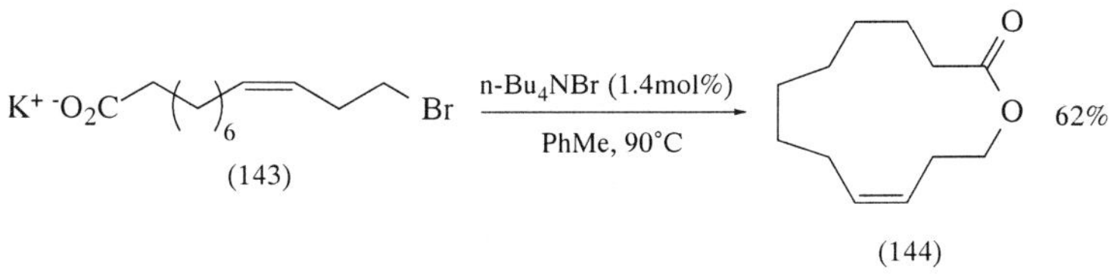

O-Alkylation of Fischer carbene complexes (e.g. 145) has also been reported (Q.H. Zheng, J. Su, *Synth. Commun.* 2000, **30**, 177; T.R. Hoye, K. Chen, J.R. Vyvyan, *Organomet.* 1993, **12**, 2806):

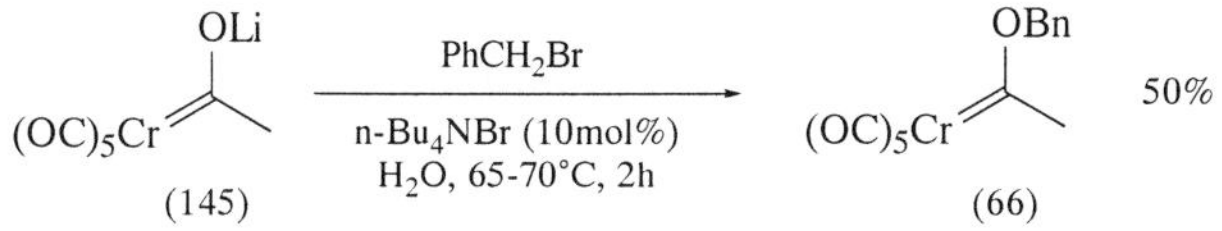

The reaction of acid chlorides with aqueous base under phase-transfer conditions has been reported to be an effective method of preparing a wide variety of symmetrical anhydrides (e.g. 147) (D. Plusquellec *et al.*, *Tetrahedron* 1988, **44**, 2471).

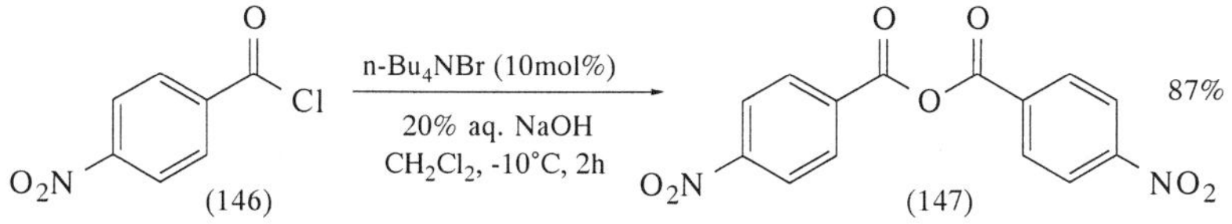

Hydrolysis of N-trifluoroacetates (e.g. 148) has also been achieved under similar reaction conditions (D. Albanese *et al.*, *J. Chem. Soc., Perkin Trans. 1* 1997, 247).

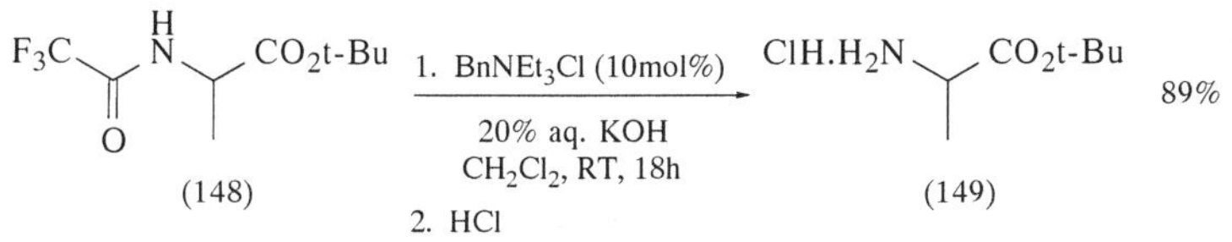

(b) C-N

$$R_2NH \ (150) \xrightarrow[\text{phase-transfer catalyst, base}]{R'\text{-}X} R_2N-R' \ (151)$$

Simple amines (150) are not amenable to deprotonation under conventional base-mediated phase-transfer conditions (pKa > 23), consequently the monoalkylation of substrates of this type has not received much attention (V. Gomez-Parra, F. Sanchez, T. Torres, *Synthesis* 1985, 282). In contrast, there has been extensive investigation into the alkylation of more acidic N-H compounds (150) such as, nitrogen heterocycles (see for example: A. Chadro *et al.*, *Synth. Commun.* 1991, 21, 535; M. Bourak, R. Gallo, *Heterocycles* 1990, **31**, 447; G. Bram *et al.*, *Synthesis* 1985, 543; R.S. Davidson *et al.*, *Tetrahedron Lett.* 1983, **24**, 5907; D. Landini, F. Rolla, *Synthesis* 1976, 389; A. Barco, S. Benetti, G. Pollini, *Synthesis* 1976, 124), anilines (S. Voskresensky, M. Makosza, *Synth. Commun.* 2000, **30**, 3523; S.A. Brown, C.J. Rizzo, *Synth. Commun.* 1996, **26**, 4065; U.R. Kalkote *et al.*, *Synth. Commun.* 1991, **21**, 1889; E.V. Dehmlow *et al.*, *Tetrahedron Lett.* 1985, **26**, 297), amides (K. Osowska-Pacewicka, A. Zwierzak, Synthesis 1996, 333; T. Gajda, A. Zwierzak, *Synthesis* 1981, 1005; P. Mattingly, M. Miller, *J. Org. Chem.* 1981, **46**, 1557; R. Brehme, *Synthesis* 1976, 113), and sulfoximines (C.R. Johnson, M.O. Lavergne, *J. Org. Chem.* 1993, **58**, 1922).

Recently a method for the monoalkylation of *N*-sulfonylamino acid esters (e.g. 152) has been described (D. Albanese *et al.*, *Eur. J. Org. Chem.* 2000, 1443). High yields, and low levels of racemisation were reported using this chemistry.

O_2N H N CO_2Me S O_2 (152) —[Me-I; $BnNEt_3Cl$ (10mol%), solid K_2CO_3, CH_3CN, RT, 15h]→ O_2N Me N CO_2Me S O_2 (153) 91%

The monoalkylation of trifluoroacetamide (154) has also been utilised in the synthesis of α-amino acids (e.g. 155) (D. Albanese, D. Landini, M. Penso, *J. Org. Chem.* 1992, **57**, 1603; D. Landini, M. Penso, *J. Org. Chem.* 1991, **56**, 420).

$$F_3C\text{-}C(O)NH_2\ (154) \xrightarrow[\substack{BnNEt_3Cl\ (5mol\%) \\ solid\ K_2CO_3 \\ CH_3CN,\ reflux,\ 2h}]{CH_3CH(Br)CO_2Et} F_3C\text{-}C(O)\text{-}NH\text{-}CH(CH_3)CO_2Et\ (155)\quad 72\%$$

The selective mono *N*-alkylation of Fischer carbene complexes (e.g. 156) has been reported (L. Zhao, H. Matsuyama, M. Iyoda, *Chem. Lett.*, 1996, 827). Dialkylation of these substrates is also possible under more forcing conditions.

$$(OC)_5Cr{=}C(NH_2)CH_3\ (156) \xrightarrow[\substack{n\text{-}Bu_4NBr\ (10mol\%) \\ 50\%\ aq.\ NaOH, \\ CH_2Cl_2,\ RT}]{CH_3\text{-}I} (OC)_5Cr{=}C(NHMe)CH_3\ (157)\quad 40\%$$

(c) Other C-X bond forming reactions

$$Na_2S\ (158)\ \text{or}\ R-SH\ (159) \xrightarrow[\substack{phase\text{-}transfer\ catalyst \\ base}]{R\text{-}X} R\text{-}S\text{-}R\ (160)$$

Dialkylsulfides have been prepared *via* the phase-transfer alkylation of both sodium sulfide (B. Jursic, J. Branko, *J. Chem. Res. (S)* 1989, 104; M. Lancaster, D.J.H. Smith, *Synthesis* 1982, 582; D. Landini, F. Rolla, *Synthesis* 1974, 565) and alkylthiols (A.W. Herriott, D. Picker, *Synthesis* 1975, 447). *S*-Alkylation of sulfinates has also been employed in the synthesis of sulfones (J.K. Crandall, C. Pradat, *J. Org. Chem.* 1985, **50**, 1327).

Carbon-sulfur bond formation has also been achieved *via* reaction of methyl methanethiolsulfonate with stabilised carbanion species (B. Wladislaw *et al.*, *Synthesis*, 1997, 420). Recently this reagent, in conjunction with chiral phase-transfer catalyst (162), was employed in the diastereoselective preparation of ketosulfoxide (163) (B. Wladislaw *et al.*, *Tetrahedron*, 1999, **55**, 12023):

(161) → (163), 80% d.e.; catalyst (162); PhH, CH_2Cl_2, solid K_2CO_3, $MeSO_2SMe$, RT, 2h

Alkylation of phosphites to give alkyl phosphonates (e.g. 165) under phase-transfer conditions has been reported (see for example: W. Ye, X. Liao, *Synthesis* 1985, 986):

$(i\text{-PrO})_2P(O)\text{-}H$ (164) + $Cl\text{-}CH_2\text{-}CO_2i\text{-}Pr$ → $(i\text{-PrO})_2P(O)CH_2CO_2i\text{-}Pr$ (165), 78%; $BnNEt_3Cl$ (6mol%), 50% aq. NaOH, CH_2Cl_2, 0°C, 90min

Carbon-bromine bond formation *via* reaction of arylalkynes (e.g. 166) with carbon tetrabromide has been described recently (E. Abele *et al.*, *J. Chem. Res. (S)*, 1998, 618):

(166) → (167), 66%; CBr_4, 18-crown-6 (5mol%), solid KOH, PhMe, RT, 4h

Halogen exchange processes are also often conducted using quaternary ammonium halides (e.g. n-Bu_4NBr), however these usually involve homogeneous reaction systems and so are outside the scope of this review. For a discussion of issues relating to heterogeneous halogen exchange, see (S. Dermeik, Y. Sasson, *J. Org. Chem.* 1985, **50**, 879).

5. Oxidation

Phase-transfer oxidation reactions employing a wide range of stochiometric oxidants (including molecular oxygen, peroxides, hypochlorites, permanganates and periodates) have been reported. Methods for the oxidation of many functional groups (e.g. alcohols, aldehydes, ketones, diols, alkenes, alkynes, alkanes, amines, amides, sulfides, acetals, aromatics) have been developed, however for the purposes of this review we will limit discussion to the oxidation of alcohols and alkenes since these are the functionalities that have recieved most attention. A number of reviews which cover additional material have been published recently (S. Schrader, E.V. Dehmlow, *Org. Prep. Proc. Int. Ed.* 2000, **32**, 123; R.A. Sheldon, *Chemtech* 1991, 21, 566), and further information on phase-transfer oxidations is available in various phase-transfer texts cited at the start of ths review.

(a) Alcohol Oxidation

Permanganate is readily extracted into organic solvents by a wide variety of phase-transfer catalysts and this has led to the development of an effective method for the oxidation of primary alcohols (e.g. 168) to carboxylic acids (see for example: A. Mahmood, G.E. Robinson, L. Powell, *Org. Process Res. Dev.* 1999, **3**, 363; A. Fatiadi, *Synthesis* 1987, 85; A. Herriot, D. Picker, *Tetrahedron Lett.* 1974, 1511).

(168) → (169) 77%

$NaMnO_4$, Et_4NHSO_4 (0.8mol%), H_2O, PhMe, 70°C, 6.5h

The oxidation of alcohols using hypochlorite ion has also been reported. With this oxidant it has been shown that aliphatic primary alcohols react slowly, secondary alcohols are rapidly oxidised to ketones, and primary benzylic alcohols (e.g. 170) give aldehydes (see for example: G.A. Mirafzal, A.M. Lozeva, *Tetrahedron Lett.* 1998, **39**, 7263; J.-S. Do, T.-C. Chou, *Ind. Eng. Chem.* 1990, **29**, 1095; G.A. Lee, H.H. Freedman, *Israel J. Chem.* 1985, **26**, 229). It has also been reported that use of ethyl acetate as the organic phase leads to substantially increased rate of reaction.

$$\text{PhCH}_2\text{OH}\ (170) \xrightarrow[\text{n-Bu}_4\text{NBr (17mol\%)},\ \text{EtOAc}]{5\%\ \text{aq.NaOCl}} \text{PhCHO}\ (17)\quad 93\%$$

The phase-transfer oxidation of alcohols (e.g. 171) to aldehydes and ketones using hydrogen peroxide in conjunction with molybdate and tungstate co-catalysts has received significant attention in recent years (see for example: K. Sato *et al., Bull Chem. Soc. Jpn.* 1999, **72**, 2287; K. Sato *et al., J. Am. Chem. Soc.* 1997, **119**, 12386; Y. Ishii *et al., J. Org. Chem.* 1988, **53**, 3587). This process appears to be a highly effective method for transformations of this type and is usually carried out without the addition of an organic solvent:

$$\text{n-C}_7\text{H}_{15}\text{CH}_2\text{OH}\ (171) \xrightarrow[(\text{n-C}_8\text{H}_{17})_3\text{NMeHSO}_4\ \text{(cat.)},\ 30\%\ \text{aq. H}_2\text{O}_2,\ 90^\circ\text{C, 4h}]{\text{Na}_2\text{WO}_4\ \text{(cat.)}} \text{n-C}_7\text{H}_{15}-\text{CHO}\ (172)\quad 87\%$$

(c) Oxidation of alkenes

Permanganate oxidation of alkenes under phase-transfer conditions has been shown to give either oxidative cleavage or dihydroxylation depending upon the pH of the reaction medium and the work-up employed (see for example: V. Bhushan, R. Rathore, S. Chandrasekaran, *Synthesis* 1984, 431; T. Ogino *Tetrahedron Lett.* 1980, 177; T. Ogino, M. Mochizuki, *Chem. Lett.* 1979, 443). This has resulted in the development of procedures for the preparation of both carboxylic acids (D.G. Lee, S.E. Lamb, V.S. Chang, *Org. Synth. Col. Vol. VII* 1990, 397; A.P. Krapcho, J.R. Larson, J.M. Eldridge, *J. Org. Chem.* 1977, **42**, 3749; C.M. Starks, *J. Am. Chem. Soc.* 1971, **93**, 195);

$$\text{CH}_3(\text{CH}_2)_{17}\text{CH=CH}_2\ (173) \xrightarrow[\text{Andogen 464 (cat.)},\ \text{CH}_2\text{Cl}_2,\ \text{H}_2\text{SO}_4,\ \text{AcOH}]{\text{KMnO}_4} \text{CH}_3(\text{CH}_2)_{17}-\text{CO}_2\text{H}\ (174)\quad 75\%$$

and diols (W.P. Weber, J.P. Sheperd, *Tetrahedron Lett.* 1972, 4907):

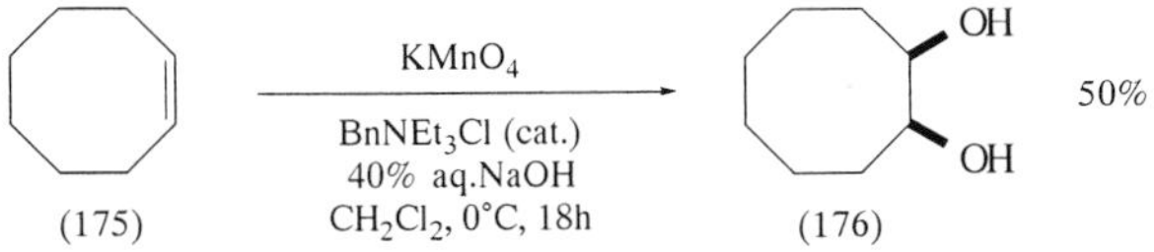

A wide variety of phase-transfer epoxidations of alkenes have also been developed. For example, the epoxidation of unactivated alkenes (e.g. cholesterol, 177) has been reported using magnesium monoperphthalate (P. Brougham *et al.*, *Synthesis*, 1987, 1015):

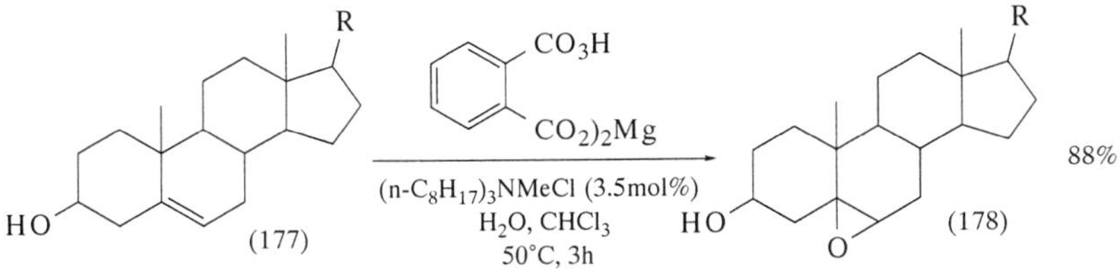

The use hydrogen peroxide in conjunction with molybdate and tungstate co-catalysts has also been applied to alkene epoxidation (see for example: L. Salles, J.M. Bregeault, R. Thouvenot, *C.R. Acad. Sci. II C*, 2000, **3**, 183; D.C. Duncan *et al.*, *J. Am. Chem. Soc.* 1995, **117**, 681; C. Aubry *et al.*, *Inorg. Chem.* 1991, **30**, 4409; L.J. Csanyi, K. Jaky, *J. Mol. Catal.* 1990, **61**, 75; Y. Ishii *et al.*, *J. Org. Chem.* 1988, **53**, 3587; C. Venturello, E. Alneri, M. Ricci, *J. Org. Chem.* 1983, **51**, 3831). A range of different phase-transfer catalysts have been used for this process, however quaternary ammonium salts incorporating long-chain alkane substituents are preferred, as illustrated by the epoxidation of styrene (179) (G.D. Yadav, A.A. Pujari, *Org. Process Res. Dev.* 2000, **4**, 88):

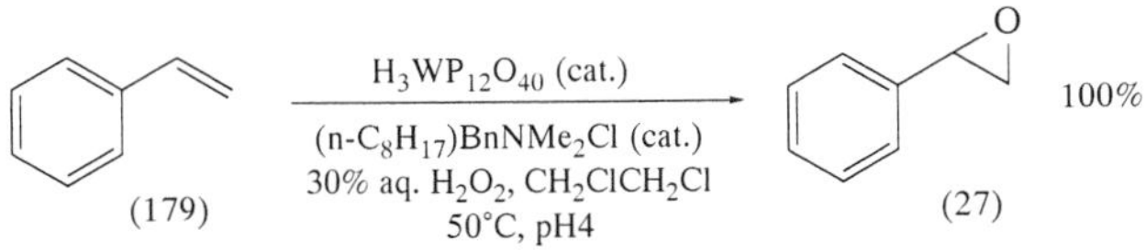

A similar oxidation process employing cetylpyridinium peroxotungstophosphate (PCWP), has also been utilised in the oxidative cleavage of silylenol ethers (e.g. 180) (S. Sakaguchi *et al.*, *J. Org. Chem.* 1999, **64**, 5954).

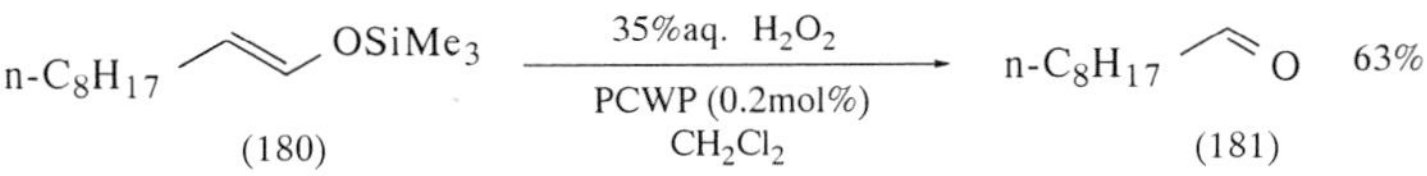

n-C_8H_{17} OSiMe_3 (180) $\xrightarrow[\text{PCWP (0.2mol\%)},\ CH_2Cl_2]{35\%\text{aq. } H_2O_2}$ n-C_8H_{17} O (181) 63%

The phase-transfer catalysed epoxidation of electron deficient alkenes has also been widely investigated. Early work in this area established that basic hydrogen peroxide, basic *tert*-butylperoxide and sodium hypochlorite could all be used for oxidations of this type, and that moderate levels of enantioselectivity could be obtained with chiral catalysts such as (47) (see for example: H. Pluim, H. Wynberg, *J. Org. Chem.* 1980, **45**, 2498; H. Wynberg, B. Marsman, *J. Org. Chem.* 1980, **45**, 158; R. Helder *et al.*, *Tetrahedron Lett.* 1976, 1831). More recently a number of chiral quaternary ammonium catalysts (e.g. 182, 183) have been reported to be highly effective for the asymmetric epoxidation of a range of chalcone substrates (e.g. 72) . With catalysts such as (182), it has been shown that the oxidation can be carried out at room temperature using sodium hypochlorite, giving the epoxides (e.g. 184) in high enantiomeric excess (B. Lygo, P.G. Wainwright, *Tetrahedron* 1999, **55**, 6289; B. Lygo, P.G. Wainwright, *Tetrahedron Lett.* 1998, **39**, 1599). By employing the same catalyst in conjunction with concentrated potassium hypochlorite, it has been reported that the reactions can be conducted at lower temperatures, leading to even higher levels of enantioselectivity (E.J. Corey, F.-Y. Zhang, *Org. Lett.* 1999, **1**, 1287). The use of catalyst (183) and basic hydrogen peroxide has also been shown to give highly enantioselective epoxidation with substrates such as (72) (S. Arai, H. Tsuge, T. Shioiri, *Tetrahedron Lett.* 1998, **39**, 7563). Intriguingly this process gives the same enantiomer of the product (184) to that obtained with catalyst (182).

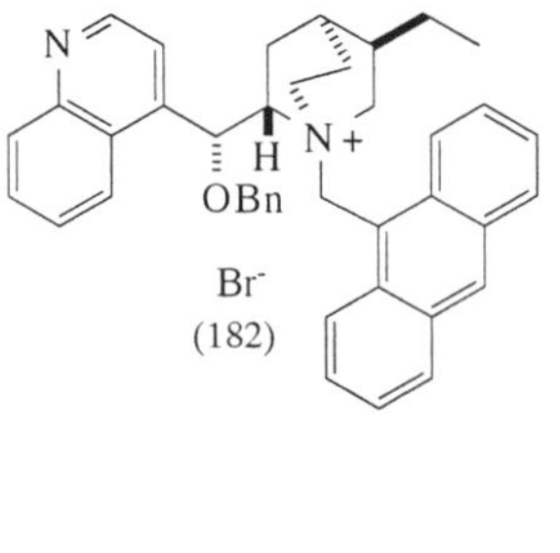

N H N+ OBn Br^- (182)

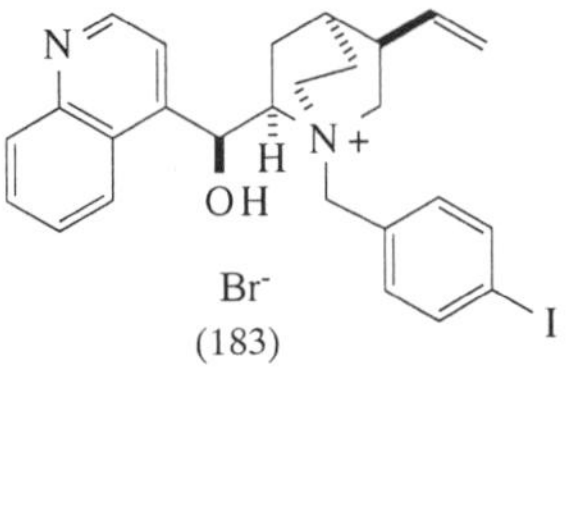

N H N+ OH Br^- I (183)

Ph–CH=CH–C(O)–Ph (72) → catalyst, oxidant, solvent → (184)

10mol% (182), 11%aq. NaOCl, PhMe, RT, 18h — 90% (86%e.e.)
10mol% (182), 8M aq. KOCl, PhMe, -40°C, 12h — 96% (93%e.e.)
5mol% (183), 30% aq. H_2O_2, LiOH, n-Bu_2O, 4°C, 37h — 97% (84%e.e.)

To date investigations into the application of asymmetric phase-transfer catalysts in the epoxidation of other types of enone have generally resulted in lower levels of enantioselectivity (see for example, S. Arai *et al.*, *Synlett* 1998, 1201), however recently a highly enantioselective epoxidation of enone (185) using stochiometric amounts of chiral quaternary ammonium salt (47) was reported (L. Alcaraz *et al.*, *J. Org. Chem.* 1998, **63**, 3526; G. Macdonald *et al.*, *Tetrahedron Lett.* 1998, **39**, 5433):

(185) → catalyst (47) (100mol%), solid NaOH (cat.), t-BuOOH, PhMe, RT, 6days → (186)

32% (82% based on recovered 185) 89%e.e.

Another asymmetric alkene oxidation of note involves the oxidation of enolates with molecular oxygen. Using chiral catalyst (41), this oxidation process has been shown to give hydroxyketones (e.g. 188) in up to 79% enantiomeric excess (M. Masui, A. Ando, T. Shioiri, *Tetrahedron Lett.* 1988, **29**, 2835):

(187) → catalyst (41) (5mol%), 50% aq.NaOH, PhMe, $(EtO)_3P$, O_2, RT, 5h → (188)

95% (79%e.e.)

6. Reduction

A number of reducing agents have been utilised under phase-transfer conditions including, borohydrides, aluminium hydrides, hydrogen and dithionites. Generally the chemoselectivity of these reducing systems tends to be similar to that obtained with homogeneous systems, and this is illustrated below.

(a) Reduction of Carbonyl Compounds

Solid lithium aluminium hydride in conjunction with a phase-transfer catalyst, generates a powerful reducing agent that has been shown to be capable of reacting with most types of carbonyl functionality (e.g. 189) (see for example: E.V. Dehmlow, R. Cyrankiewicz, *J. Chem. Res. (S)* 1990, 24; V. Gevorgyan, E. Lukevics, *J. Chem. Soc., Chem. Commun.* 1985, 1234).

O
NEt$_2$ (189) — solid LiAlH$_4$ / BnNEt$_3$Cl (5mol%) / PhH, 80°C, 1h → NEt$_3$ (190) 86%

Aqueous borohydride in conjunction with simple quaternary ammonium salts provides a milder reducing system (V. Yadav, G.D. Yadav, J.R. Vyas, *Chim. Oggi* 2000, **18**, 39; M.M. Cook, M.E. Halpern, *Chim. Oggi* 1998, **16**, 44) and has been reported to be effective for the selective reduction of aldehydes in the presence of ketones (J.R. Blanton, *Synth. Commun.* 1997, **27**, 2093; C.S Rao, A.A. Deshmukh, B.J. Patel, *Indian J. Chem. B* 1986, **25**, 626; G. Lamaty M.H. Riviere, J.P. Roque, *Bull. Soc. Chim. Fr.* 1983, 33):

CHO (191) + O (70) — aq. NaBH$_4$ / n-Bu$_4$NBr(cat.) / PhH → OH (192) 95% + OH (193) 7%

It has been reported that the reactivity of this latter system can be increased by employing quaternary ammonium salts containing a β-hydroxy group. With this modification ketones were rapidly reduced to the corresponding alcohols (see for example: T.C. Pochapsky, P.M. Stone, S.S. Pochapsky, *J. Am. Chem. Soc.* 1991, **113**, 1460; S. Colonna, R. Fornasier, *Synthesis* 1979, 531). The application of chiral phase-transfer catalysts in this process was also investigated, however only low levels of enantioselectivity were obtained.

(b) Reduction of Alkenes

Phase-transfer catalysed hydrogenation involving a combination of rhodium trichloride and $(n\text{-}C_8H_{17})_3NMeCl$ (Aliquat 336) has been reported to be effective for the reduction of a range of alkenes (e.g. 194) at low hydrogen pressures (see for example: K.S. Weddle, J.D. Aiken, R.G. Finke, *J. Am. Chem. Soc.* 1998, **120**, 5653; J. Blum *et al.*, *J. Org. Chem.* 1987, **52**, 2804; I. Amer *et al.*, *Tetrahedron Lett.* 1987, **28**, 1321; J. Azran *et al.*, *J. Mol. Catal.* 1986, **34**, 229).

(194) → (195), 96%
$(n\text{-}C_8H_{17})_3NMeCl$ (5.5mol%), $RhCl_3$ (0.5mol%), H_2 (1atm), H_2O, CH_2ClCH_2Cl, 30°C, 4h

Closely related hydrogenations of arenes are also possible (J.D. Aiken III, Y. Lin, R.G. Finke, *J. Mol. Cat.* 1996, **114**, 29; K.R. Januszkiewicz, H. Alper, *Organometallics* 1983, **2**, 1055).

(196) → (197), 91% (92%d.e.)
$Na_2S_2O_4$, $NaHCO_3$, $(n\text{-}C_8H_{17})_3NMeCl$ (30mol%), H_2O, PhH, 100°C, 2h

Selective reduction of conjugated alkenes (e.g. 196) using sodium dithionite has also been reported (F. Camps, J. Coll, J. Guitart, *Tetrahedron* 1986, **42**, 4603).

(c) Other Reduction Processes

The formation and reduction of alkylazides has been reported to be an effective method for the generation of alkylamines (e.g. 200) (F. Rolla, *J. Org. Chem.* 1982, **47**, 4327). In this case, the starting azide (199) was also prepared using phase-transfer catalysis (W.P. Reeve, M.L. Nahr, *Synthesis* 1976, 823).

Br, CO_2t-Bu (198) → 25% aq.NaN_3, $(n\text{-}C_8H_{17})_3NMeCl$ (5mol%), RT, 24h → N_3, CO_2t-Bu (199) 94%

(199) → aq.$NaBH_4$, $(n\text{-}C_{12}H_{25})P(n\text{-}Bu)_3Br$ (10mol%), PhMe, RT → H_2N, CO_2t-Bu (200) 72%

A variety of phase-transfer hydrogenolysis reactions have also been described (see for example: S. Calet, H. Alper, *Organomet.* 1987, **6**, 1625; A. Zoran, Y. Sasson, J. Blum, *J. Mol. Catal.* 1984, **27**, 349; R. Bar, Y. Sasson, J. Blum, *J. Mol. Catal.* 1982, **16**, 175; J.E. Hallgren, G.M. Lucas, *J. Organomet. Chem.* 1981, **212**, 135).

7. Concluding Remarks

The use of phase-transfer reactions in synthetic chemistry has developed substantially over the last 30 years, and in this review we have attempted to illustrate the vast range of chemical transformations have now been described. Favourable environmental aspects, and the simplicity of the techniques involved, make these processes highly attractive alternatives to many homogeneous reaction systems. Recent advances in the development of enantioselective phase-transfer reactions have further emphasised the utility of this technology and this is an area of research that is likely to develop rapidly in the future.

Second Edition of Rodd's Chemistry of Carbon Compounds,
Volume V, Topical Volumes
Asymmetric Catalysis, edited by M.Sainsbury

Chapter 5

ADDITIONS TO CARBONYL COMPOUNDS

ALISON S. FRANKLIN

The development of methods for the enantioselective synthesis of chiral alcohols and amines, based on nucleophilic additions to aldehydes and ketones or imines, hydrazones, oximes and nitrones, has been an area of intense interest within the synthetic organic chemistry community over the last few decades. Three distinct strategies for the stereoselective generation of the new sp^3 centre have been adopted: addition of an achiral nucleophile to a chiral electrophile; the coupling of a chiral nucleophile and an achiral electrophile; and the use of a chiral species to promote the reaction.

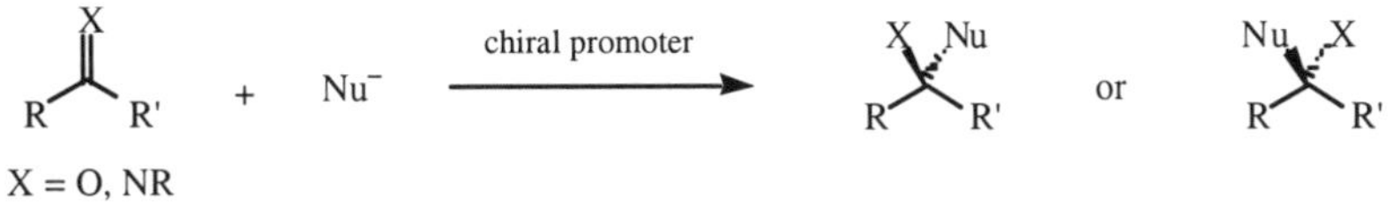

X = O, NR

The third approach is particularly appropriate as many of these additions are accelerated by Lewis acids or indeed have negligible rates in their absence. However, the development of efficient systems which not only provide high yields and good levels of asymmetric induction, but which also utilise sub-stoichiometric quantities of the chiral catalyst, is not trivial. This review will focus on methods published up to the end of 2000 which have been reported to employ no more than 50 mol% of a chiral species to promote the enantioselective addition of achiral nucleophiles to prochiral C=O and C=N functionalities.
Several general reviews, many of which are updated annually and whose coverage includes reactions within the scope of this chapter, have appeared (Catalytic applications of transition metals: L. Haughton and J.M.J. Williams, *J. Chem. Soc., Perkin Trans. 1*, 2000, 3335; Catalytic asymmetric processes: H. Tye, *J. Chem. Soc., Perkin Trans. 1*, 2000, 275). The use of enzymes as chiral catalysts in preparative biotransformations (S.M. Roberts, *J. Chem. Soc., Perkin Trans. 1*, 2000, 611) and developments in the area of solid supported catalysts (Y.R. de

Miguel, *J. Chem. Soc., Perkin Trans. 1*, 2000, 4213; S.J. Shuttleworth, S.M. Allin and P.K. Sharma, *Synthesis*, 1997, 1217) are also regularly surveyed and these strategies will not be discussed here.
Material has been organised based on the nucleophile employed, with approaches providing alcohols considered prior to those for the synthesis of amines. Catalytic and non-catalytic approaches for the 1,2-addition of organometallic reagents to C=N bonds have been reviewed recently (D. Enders and U. Reinhold, *Tetrahedron: Asymmetry*, 1997, **8**, 1895), as have methods for catalytic enantioselective additions to imines (S. Kobayashi and H. Ishitani, *Chem. Rev.*, 1999, **99**, 1069). A brief overview of the use of chiral Lewis acids in catalytic asymmetric ene and hydrocyanation reactions has appeared (K. Narasaka, *Synthesis*, 1991, 1) and the use of enantioselective carbonyl addition reactions for the catalytic asymmetric synthesis of fluoroorganic compounds has been surveyed (K. Iseki, *Tetrahedron*, 1998, **54**, 13887).

1. Additions of Alkyl and Aryl Organometallics

(a) Organolithium reagents

A number of chiral amine and amino alcohol-derived ligands have been employed in the enantioselective addition reactions of organolithium species. However, the majority of these methods require 2.6 equivalents of the chiral ligand, with severe reductions in enantioselectivity being observed when sub-stoichiometric quantities are employed. Some success has been achieved with *N*-(4-methoxyphenyl)imines and chiral amines (1) and (2).

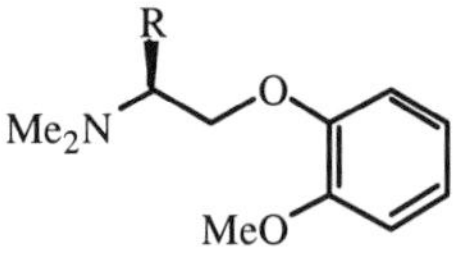

(1) 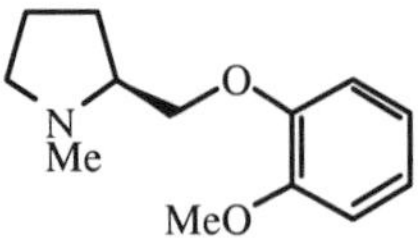 (2)

As little as 5 mol% of (1, R = Bn) promotes the addition of alkyl and vinyllithium species to aryl and α,β-unsaturated imines in 33-66% ee and 70-99% yield (K. Tomioka *et al.*, *Tetrahedron Lett.*, 1991, **31**, 3095; I.

Inoue *et al.*, *Tetrahedron*, 1994, **50**, 4429; I. Inoue *et al.*, *Tetrahedron: Asymmetry*, 1995, **6**, 2527), although higher selectivities were obtained using a stoichiometric quantity of the ligand. Similarly, 25 mol% of proline-derived amine (2) provides moderate selectivity (6-20% ee, 33-96% yield) in the addition of alkyl and phenyllithiums to phenyl and α,β-unsaturated imines (C.A. Jones *et al.*, *Tetrahedron Lett.*, 1995, **36**, 7885; C.A. Jones *et al.*, *J. Chem. Soc., Perkin Trans. 1*, 1997, 2891). The enantiomeric amine product can be obtained using (1, R = Pr^i). The highest selectivities reported to date have been achieved using bis(oxazoline) (3), which gives chiral amines in 51-82% ee and 81-98% yield from the addition of alkyl and vinyllithiums to aryl, aliphatic and α,β-unsaturated imines (S.E. Denmark, N. Nakajima and O. J.-C. Nicaise, *J. Am. Chem. Soc.*, 1994, **116**, 8797). Again, the stoichiometric process generally provides higher enantiomeric excesses.

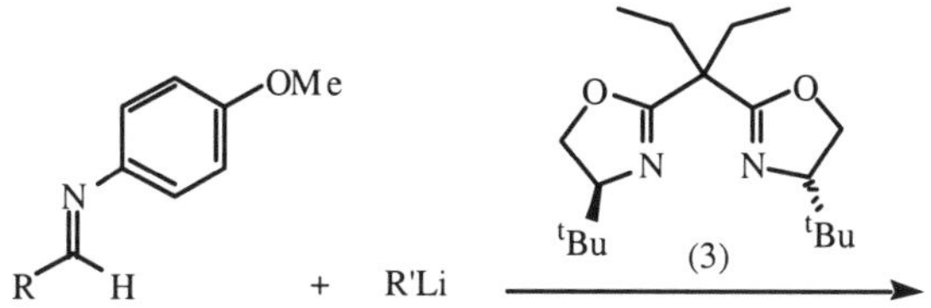

(b) Organozinc additions

The addition of dialkylzinc reagents to aldehydes in the presence of chiral ligands is a highly effective reaction. The chiral catalyst-mediated enantioselective addition of diethylzinc to benzaldehyde has become a benchmark reaction for new ligands since Oguni and Omi reported their first successes in this area (N. Oguni *et al.*, *Chem. Lett.*, 1983, 841; N. Oguni and T. Omi, *Tetrahedron Lett.*, 1984, **25**, 2823). A vast array of amino alcohols, diols and Ti(IV) complexes have been shown to promote the addition which has also been shown to be subject to chirality amplification when catalysts of <100% ee are employed (N. Oguni, Y. Matsuda and T. Kaneko, *J. Am. Chem. Soc.*, 1988, **110**, 7877; R. Noyori and M. Kitamura, *Angew. Chem. Int. Ed. Engl.*, 1991, **30**, 49; C. Girard and H.B. Kagan, *Angew. Chem. Int. Ed.*, 1998, **37**, 2922). Comprehensive reviews of the plethora of catalysts available for the addition of dialkylzincs, and the more reactive aryl, vinyl and alkynyl species, to carbonyl compounds have been published (L. Pu and H.-B. Yu, *Chem. Rev.*, 2001, **101**, 757; K. Soai and S. Niwa, *Chem. Rev.*, 1992, **92**, 833). Methods have also been developed for the generation and subsequent enantioselective addition reactions of more highly functionalised organozinc reagents (P. Knochel and R.D. Singer, *Chem. Rev.*, 1993, **93**, 2117).

RCHO + R'_2Zn —chiral catalyst→ RCH(OH)R' (*)

The catalytic asymmetric Reformatsky reaction of ethyl bromoacetate has met with much more limited success. 1,2-Amino alcohols have been shown to promote addition of the zinc enolate to aldehydes, although low levels of asymmetric induction are frequently observed with less than one equivalent of the ligand (15-46% ee, 41-100% yield) (K. Soai and Y. Kawase, *Tetrahedron: Asymmetry*, 1991, **2**, 781; D. Pini, A. Mastantuono and P. Salvadori, *Tetrahedron: Asymmetry*, 1994, **5**, 1875; A. Mi *et al.*, *Tetrahedron: Asymmetry*, 1995, **6**, 2641; A. Mi *et al.*, *Synth. Commun.*, 1997, **27**, 1469). The most successful ligand to date is diaminodiol (4) which provides 23-78% ee (21-90% yield) when used in 50 mol% for the addition of ester and Weinreb amide-derived Reformatsky reagents to a range of aldehydes and to acetophenone (J.M. Andrés *et al.*, *Tetrahedron*, 1997, **53**, 3787; J.M. Andrés, R. Pedrosa and A. Pérez-Encabo, *Tetrahedron*, 2000, **56**, 1217). The *N*-Me ephedrine catalysed addition of a difluoro Reformatsky reagent to benzaldehyde has also been reported (45-47% yield, 54-79% ee) (M. Braun, A. Vonderhagen and D. Waldmüller, *Liebigs Ann.*, 1995, 1447).

OH Me N Me N OH Ph Me Me Ph

(4)

The zinc-mediated addition of terminal alkynes to aryl aldehydes has also been described, propargylic alcohols are formed in 65-90% yield and 62-

85% ee in the presence of 10 mol% (5) or (6) (Z. Li *et al.*, *Synthesis*, 1999, 1453).

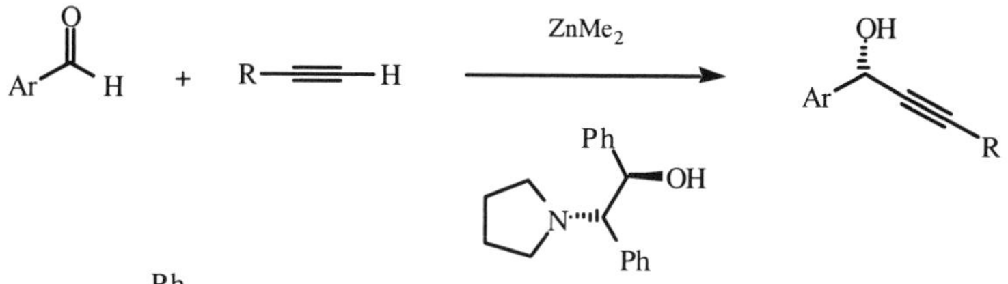

(c) Organoaluminium, boron, tin and silicon species

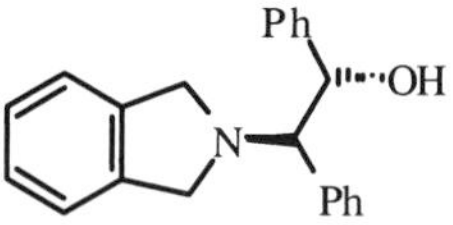

A limited number of successful additions of alkylaluminium reagents to aldehydes catalysed by chiral diol-titanium(IV) complexes have been reported. Reaction of trialkylaluminium reagents with aryl aldehydes proceeds in excellent yields (80-100%) with moderate to good enantioselectivity (43-86% ee) in the presence of 20 mol% binol and $Ti(OPr^i)_4$ (A.S.C. Chan, F.-Y. Zhang and C.-W. Yip, *J. Am. Chem. Soc.*, 1997, **119**, 4080). Complementary asymmetric induction and quantitative yields (90-96% ee) were obtained when (*S*)-5,5',6,6',7,7',8,8'-octahydro-1,1'-bi-2-naphthol was employed in place of (*R*)-binol. The fluorotitanium complex derived from TiF_4 (14 mol%) and chiral diol (7) (15 mol%) mediates the addition of trimethylaluminium to aryl, heteroaryl and unsaturated aldehydes (41-93% yield, 54-85% ee) (B.L. Pagenkopf and E.M. Carreira, *Tetrahedron Lett.*, 1998, **39**, 9593) whilst $Ti(OPr^i)_4$ and 20 mol% α,α,α',α'-tetraaryl-1,3-dioxolane-4,5-dimethanol (TADDOL) promotes the reaction of triethylaluminium with aryl and unsaturated aldehydes (88-100% yield, 43-99% ee) (J.-F. Lu, J.-S. You and H.-M. Gau, *Tetrahedron: Asymmetry*, 2000, **11**, 2531).

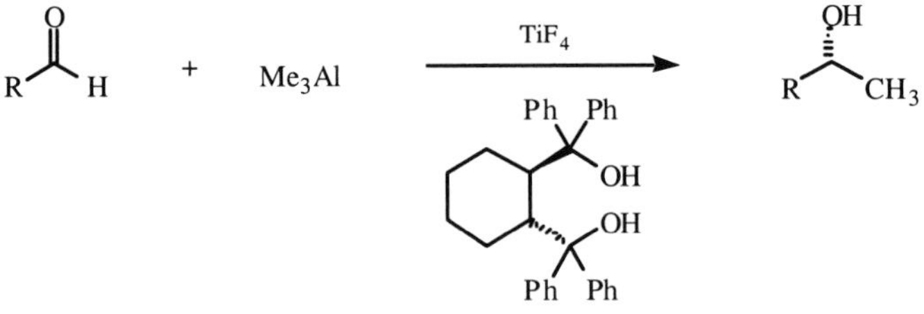

$$\mathrm{RCHO} + \mathrm{Me_3Al} \xrightarrow[\text{(7)}]{\mathrm{TiF_4}} \mathrm{R{-}CH(OH){-}CH_3}$$

(7)

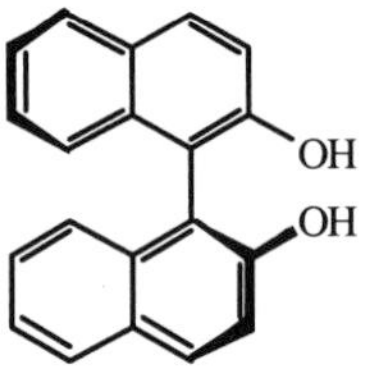

binol

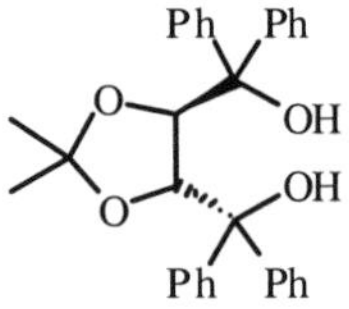

TADDOL

The addition of organoboronic acids to aldehydes is catalysed by Rh(I) complexes. The addition of phenylboronic acid to 1-naphthaldehyde proceeds in 78% yield and 41% ee in the presence of phosphine (8) (M. Sakai, M. Ueda and N. Miyaura, *Angew. Chem. Int. Ed.*, 1998, **37**, 3279). Enantioselective aryl transfer (82-96% ee) from $ArSnMe_3$ to *N*-sulfonyl imines is also promoted by 3 mol % of a rhodium complex of (8) in 31-90% yield (T. Hayashi and M. Ishigedani, *J. Am. Chem. Soc.*, 2000, **122**, 976).

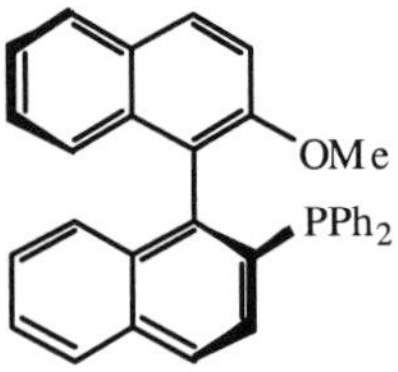

(8)

Corey has employed a tin to boron transmetallation and chiral oxazaborolidine (9) for the asymmetric alkynylation (72-80% yield,

85-97% ee) of alkyl and aryl aldehydes (E.J. Corey and K.A. Cimprich, *J. Am. Chem. Soc.*, 1994, **116**, 3151).

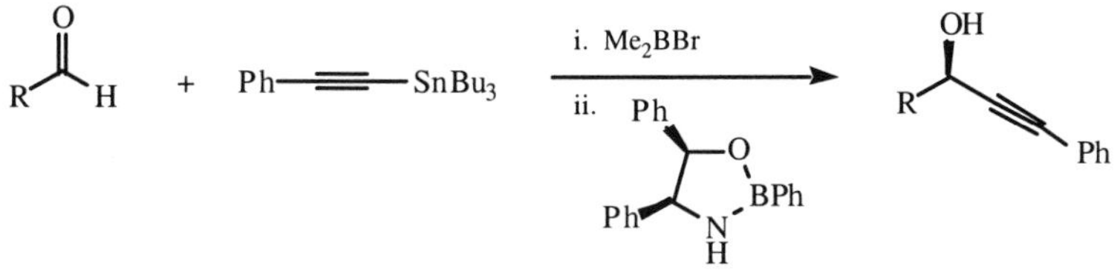

(9)

Trifluoromethyl transfer from TMS-OTf to aldehydes and ketones is catalysed by 10 mol% *N*-(*p*-trifluoromethylbenzyl)cinchoninium fluoride (10, R = (4-$CF_3C_6H_4$)CH_2) (87->99% yield, 15-51% ee) (K. Iseki, T. Nagai and Y. Kobayashi, *Tetrahedron Lett.*, 1994, **35**, 3137), whilst triaminosulfonium salt (11) catalyses the reaction with aryl, unsaturated and alkyl aldehydes (71-99% yield, 10-52% ee) (Y. Kuroki and K. Iseki, *Tetrahedron Lett.*, 1999, **40**, 8231).

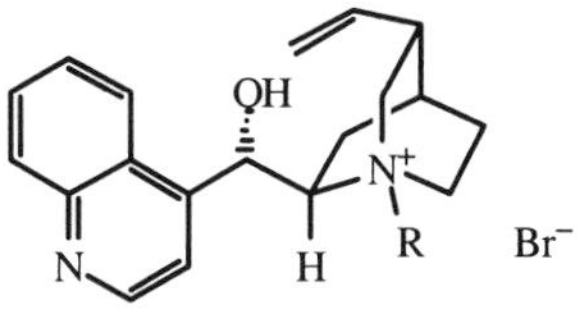

(10) 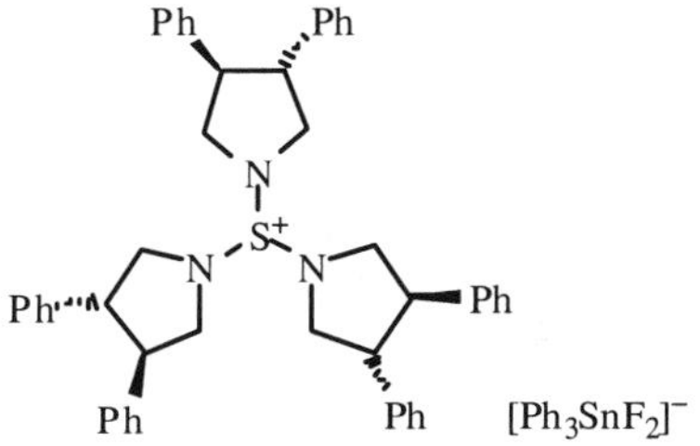 (11)

2. Addition of Allyl Metal Species

(a) Allyl tin reagents

The enantioselective addition of allyltributyltin (12, R = H) to aldehydes has been extensively studied. In 1993, Keck *et al.* and Tagliavini, Umani-Ronchi and co-workers reported that a Ti(IV).binol complex could be employed as a catalyst in this reaction. A combination of 20 mol% binol, $TiCl_2(OPr^i)_2$ and molecular sieves provides good yields (75-96%) of homoallylic alcohols from aliphatic, α,β-unsaturated, aryl and heteroaryl aldehydes in 80-98% ee (A.L. Costa *et al.*, *J. Am. Chem. Soc.*, 1993, **115**, 7001). Similarly, the use of 10 mol% $Ti(OPr^i)_4$ and molecular sieves in the presence of either 10 mol% binol or trifluoromethanesulfonic acid and 20 mol% binol also leads to high yields (78-98%) and 77-96% ee (G.E. Keck, K.H. Tarbet and L.S. Geraci, *J. Am. Chem. Soc.*, 1993, **115**, 8467). An improved reaction protocol, which employs preformed $Ti(OPr^i)_2$.binol (10 mol%) and provides products in 59-94% yield and 83-95% ee was reported shortly afterwards (G.E. Keck and L.S. Geraci, *Tetrahedron Lett.*, 1993, **34**, 7827). Chiral amplification is observed in a modification of this procedure when binol of 50% ee is employed, the use of racemic binol "poisoned" with D-(–)-diisopropyl tartrate also led to good levels of asymmetric induction (82-92% ee) (J.W. Faller, D.W.I. Sams and X. Liu, *J. Am. Chem. Soc.*, 1996, **118**, 1217).

Bu_3Sn ... R + R'CHO —Lewis acid→ R'CH(OH)CH₂C(R)=CH₂

(12)

The addition of methallyltributyltin (12, R = Me) using 20 mol% binol and 10 mol% $Ti(OPr^i)_4$ proceeds in 80-99% yield and 86-99% ee (G.E. Keck, D. Krishnamurthy and M.C. Grier, *J. Org. Chem.*, 1993, **58**, 6543), whilst β-ketoesters are accessible by oxidative cleavage of the alkene following the reaction of (12, R = CH_2CO_2Me) with a range of aldehydes (60-100% yield, 93-99% ee) (G.E. Keck and T. Yu, *Org. Lett.* 1999, **1**, 289). Further improvements in the enantioselectivity of the reaction of 2-substituted reagents (12, R ≠ H) were obtained when a longer period of catalyst premixing was employed (19-96% yield, 92-99% ee) (S. Weigand and R. Brückner, *Chem. Eur. J.*, 1996, **2**, 1077).

The less reactive allenyltin species requires 50 mol% binol and provides both allenyl and propargyl alcohols (1:4 to 1:2.3) in lower yields (25-64%) but with 82->99%ee (G.E. Keck, D. Krishnamurthy and X. Chen, *Tetrahedron Lett.*, 1994, **35**, 8323).

The titanium complexes of a number of closely related biaryl diols, (13), (14) and 7,7'-bisbenzyloxy-1,1'-bi-2-naphthol, have also been employed to promote the enantioselective addition of allyltributyltin to aldehydes

(B.H. Lipshutz *et al.*, *Tetrahedron Lett.*, 1997, **38**, 753; E. Brenna, L. Scaramelli and S. Serra, *Synlett*, 2000, 357; M. Bandin *et al.*, *Eur. J. Org. Chem.*, 2000, 491) whilst a combination of 10 mol% binol.Ti$(OPr^i)_4$ and CuCl and one equivalent of stannane (15) leads to the reaction of aryl and unsaturated aldehydes with allyl bromide in moderate yields (42-52% yield, 18-63% ee) (K.K. Majumdar, *Tetrahedron: Asymmetry*, 1997, **8**, 2079).

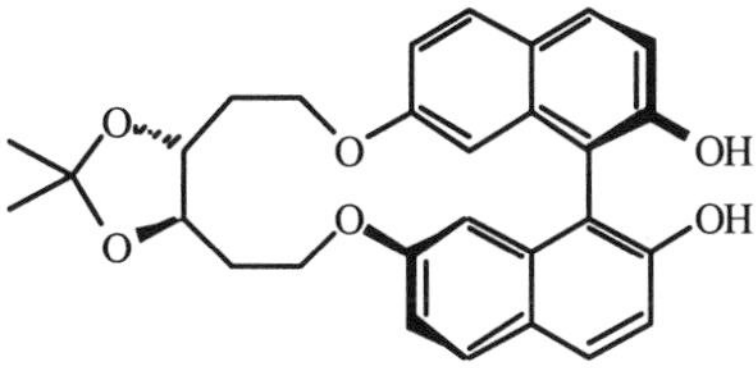

CO_2Me Et OH OH O O O O OH OH O S Sn

(13) 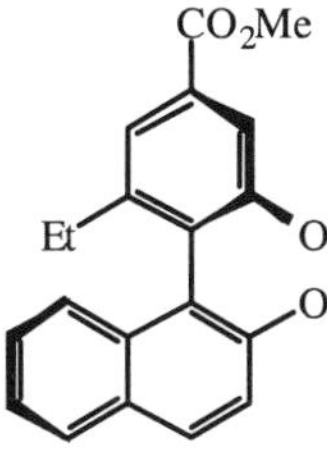 (14) 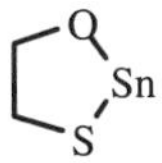 (15)

Use of 20 mol % Zr$(OPr^i)_2$.binol in the presence of molecular sieves also provides chiral homoallylic alcohols (34-84% yield, 87-93% ee) from allyltin additions, the reaction rate was increased relative to the titanium mediated process and a non-linear effect was again observed (P. Bedeschi *et al.*, *Tetrahedron Lett.*, 1995, **36**, 7897). The addition of 0.5-10 mol% of achiral calixarene (16) to the complex derived from $ZrCl_4(THF)_2$ and binol (2-10 mol%) leads to activation in the absence of molecular sieves giving products in 38-78% yield and 77-96% ee (S. Casolari *et al.*, *Chem. Commun.*, 1997, 2123). Zirconium catalyst (17) has also been shown to provide strong activation of carbonyl substrates (76-91% yield, 93-94% ee) (H. Hanawa *et al.*, *Tetrahedron Lett.*, 2000, **41**, 5543) and the use of tetraallyltin in combination with binol and $TiCl_2(OPr^i)_4$ (20 mol%) provides tertiary alcohols from its addition to aryl methyl ketones (75-91% yield, 44-65% ee) (S. Casolari, D. D'Addario and E. Tagliavini, *Org. Lett.*, 1999, **1**, 1061).

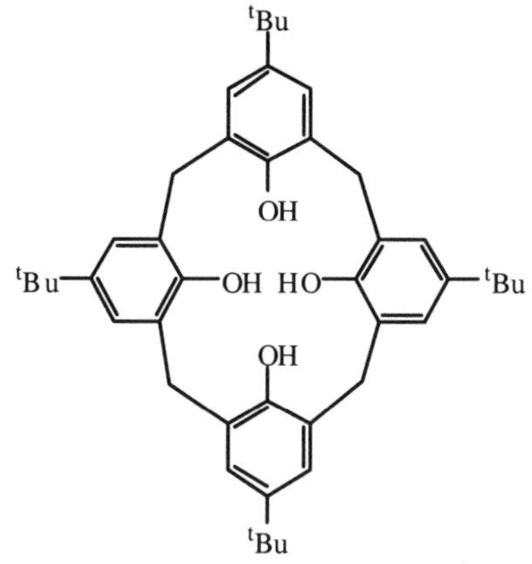

(16)

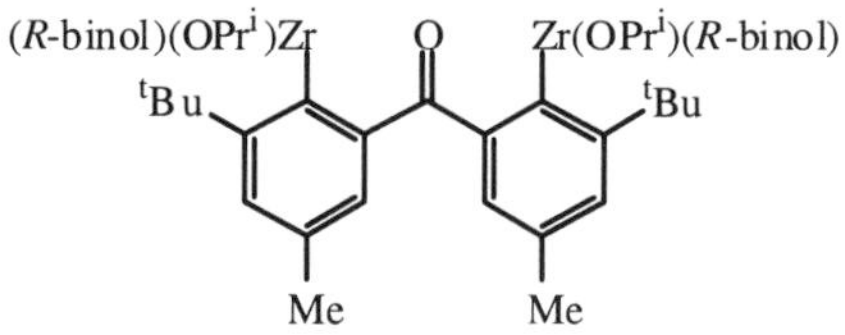

(17)

Trimethyl borate (50 mol%) in combination with $Ti(OPr^i)_2$.binol (5-10 mol%) has also been shown to enhance the rate of reaction of allyltributyltin with alkyl and aryl aldehydes (71-91% yield, 90-96% ee) and to promote the addition of allenyltributyltin, giving the corresponding propargylic alcohols (3.5:1 to 49:1 propargyl:allenyl) in 44-72% yield and 92->97% ee (C.-M. Yu *et al.*, *Synlett*, 1997, 889). These workers have also developed the use of bifunctional reagents, of the general form R_2MSR', which act in stoichiometric quantities to accelerate the reaction. Initial studies involved the use of Me_3SiSPr^i (C.-M. Yu *et al.*, *Tetrahedron Lett.*, 1996, **37**, 7095) but Et_2BSPr^i proved to be the most effective additive, resulting in 77-93 % yield and 93-97% ee when used in combination with 10 mol% $Ti(OPr^i)_4$ and 20 mol% binol (C.-M. Yu *et al.*, *Chem. Commun.*, 1997, 761). The same reaction system also provides access to propargylic alcohols (52-86% yield, 91-95% ee) (C.-M. Yu *et al.*, *Chem. Commun.*, 1997, 763), 1,1-disubstituted allenes (41-97% yield, 81-97% ee) (C.-M. Yu *et al.*, *Angew. Chem. Int. Ed.*, 1998, **37**, 2392) and dienylic products (53-83% yield, 88-97% ee) (C.-M. Yu *et al.*, *Chem. Commun.*, 1998, 2749; *idem, J. Chem. Soc., Perkin Trans. 1*, 1999, 3557). The increased reactivity observed is attributed to the formation of $^nBu_3SnSPr^i$ and the Lewis acidity of the boron.
The chiral acyloxyborane (CAB) complex (18) also acts as a Lewis acid catalyst (50 mol%) for the *syn* selective addition of an (*E*)-crotylstannane

(70-74% yield, 78-92% ds, 70-99% ee) (J.A. Marshall and M.R. Palovich, *J. Org. Chem.*, 1998, **63**, 4381).

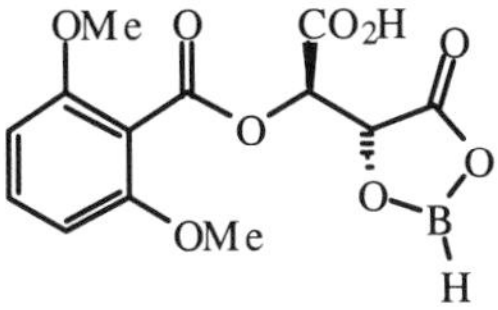

(18)

An alternative approach to the catalytic enantioselective allylation of aldehydes by allyltributyltin employs a silver(I) complex. The use of 5 mol% of a binap.AgOTf complex leads to high levels of asymmetric induction (88-97% ee, 47-94% yield) in additions to a range of aldehydes (A. Yanagisawa *et al.*, *J. Am. Chem. Soc.*, 1996, **118**, 4723). Binap.AgOTf also promotes the reaction of methallyl and crotylstannanes with aryl and unsaturated aldehydes (45-96% yield, 88-98% ee), the latter reactions providing the *anti* product with 85% ds (A. Yanagisawa *et al.*, *Synlett*, 1997, 88). A combination of 20 mol% AgOTf and diphenyl-thiophosphoramide (19) provides homoallylic alcohols in 65-80% yield but with lower enantioselectivity (54-63% ee) (M. Shi and W.-S. Sui, *Tetrahedron: Asymmetry*, 2000, **11**, 773).

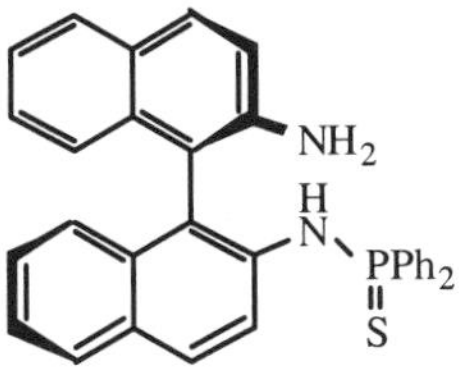

(19)

Bis(oxazoline)-containing catalysts have also been shown to induce asymmetry in allylation reactions, but selectivity is generally lower than that observed with binaphthyl ligands. Only 32-46% ee (45-85% yield) is observed in reactions promoted by 10 mol% $Zn(OTf)_2$ and bis(oxazoline) (20) (P.G. Cozzi *et al.*, *Tetrahedron Lett.*, 1997, **38**, 145) whilst 5 mol% of rhodium(III) complex (21) gives good yields (84-98%) and variable selectivity (43-80% ee) (Y. Motoyama, H. Narusawa and H. Nishiyama, *Chem. Commun.*, 1999, 131).

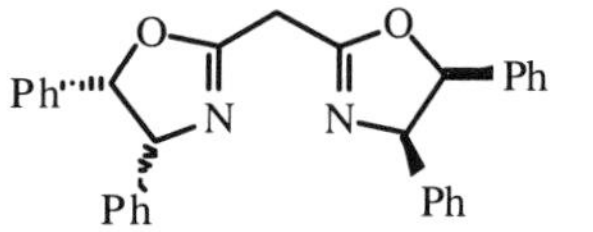

(20)

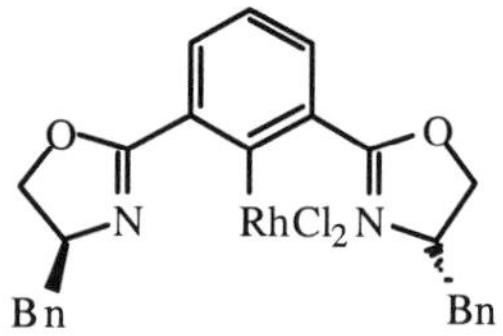

(21)

A limited number of reports of the catalytic enantioselective addition of allyltributyltin to imines have been published. Bis(π-allylpalladium) complex (22) catalyses the reaction with *N*-benzylimines providing amines in 64-83% yield and 40-80% ee (H. Nakamura, K. Nakamura and Y. Yamamoto, *J. Am. Chem. Soc.*, 1998, **120**, 4242), whilst an ene-like mechanism is proposed for the addition of allylstannanes to ethyl glyoxylimine in the presence of (*p*-tolyl)binap.copper(I) complexes (74-95% yield, 68-98% ee) (X. Fang *et al.*, *J. Org. Chem.*, 1999, **64**, 4844).

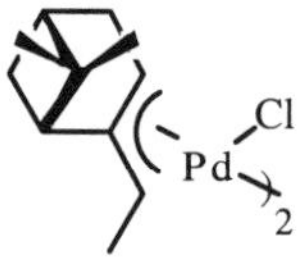

(22)

(b) Allyl silicon reagents

Allylsilanes are inherently much less reactive than the corresponding allylstannanes, leading to a corresponding reduction in the number of methods which have been developed for their catalytic enantioselective carbonyl addition reactions.
In 1991, allyltrimethylsilanes bearing 2- and 3-substituents were reported to react with aryl, unsaturated and aliphatic aldehydes in the presence of 20 mol% tartrate-derived CAB complex (23, R = H) in a *syn* selective fashion (21-81% yield, 80-97% ds, 55-96% ee) (K. Furuta, M. Mouri and H. Yamamoto, *Synlett*, 1991, 561). Rate improvements were observed when (23, R = 3,5-$(CF_3)_2C_6H_3$) was employed, although some loss of

enantioselectivity also occurred (55-99% yield, 92-94% ds, 48-91% ee) (K. Ishihara *et al.*, *J. Am. Chem. Soc.*, 1993, **115**, 11490).

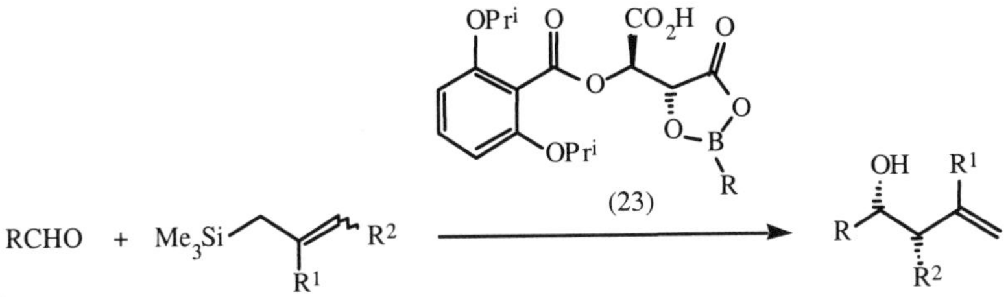

The successful reaction of allyltrimethylsilane catalysed by 10 mol% of a highly reactive complex derived from TiF_4 and binol (69-93% yield, 60-94% ee) has been reported more recently (D.R. Gauthier Jr. and E.M. Carreira, *Angew. Chem. Int. Ed. Engl.*, 1996, **35**, 2363). Addition of substituted allylsilanes and stannanes to alkyl glyoxylates also proceeds in the presence of binol and $TiCl_2(OPr^i)_2$ or $TiBr_2(OPr^i)_2$ to give α-hydroxy esters in 35-80% yield and 2-86% ee (S. Aoki *et al.*, *Tetrahedron*, 1993, **49**, 1783). Moderate to good *syn* selectivity (53-84% ds) was observed with 3-substituted reagents.

The reaction of allyl and crotyltrimethoxysilane with aromatic and unsaturated aldehydes is catalysed by 6 mol% (*p*-tolyl)binap.silver(I) fluoride (67-99% yield, 78-96% ee) (A. Yanagisawa *et al.*, *Angew. Chem. Int. Ed.*, 1999, **38**, 3701). High *anti* selectivity (92-96% ds) is observed regardless of the initial geometry of the crotylsilane.

The Lewis base-catalysed addition of allyltrichlorosilane to aldehydes provides an alternative strategy for the synthesis of homoallylic alcohols. Denmark has investigated the use of his chiral phosphoramides in this reaction with (24) providing 40-79% yield and 53-59% ee when used in 10-50 mol%, although better results (81% yield, 60% ee) were obtained when one equivalent of (24) was employed (S.E. Denmark *et al.*, *J.Org. Chem.*, 1994, **59**, 6161). The reaction of allyl and crotyltrichlorosilanes with aryl aldehydes is catalysed by 10-20 mol% of phosphoramide (25) (K. Iseki *et al.*, *Tetrahedron Lett.*, 1996, **37**, 5149). The reaction is stereospecific with (*Z*)-crotyltrichlorosilane providing *syn* products and *anti* products being obtained from the corresponding (*E*)-alkene (95->99% ds, 47-95% yield, 72-88% ee).

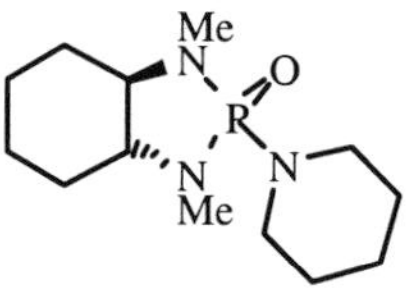

(24) 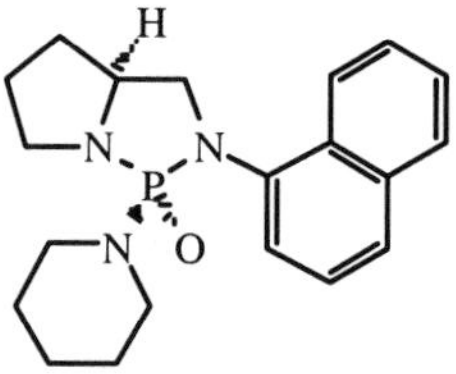 (25)

Chiral formamide (26) is also an efficient catalyst for the reaction with aliphatic aldehydes (53-97% yield, 68-98% ee), when used in combination with stoichiometric quantities of HMPA (K. Iseki *et al.*, *Tetrahedron Lett.*, 1998, **39**, 2767; *idem, Tetrahedron*, 1999, **55**, 977). (*E*)-Crotyltrichlorosilane provides *anti* products with >99% ds but reactions of the (*Z*)-crotylsilane are essentially non selective, providing the *syn* product in only 3% ee at –20 °C and a 3:2 mixture of diastereomers with the *anti* product predominating in 98% ee at –78 °C.

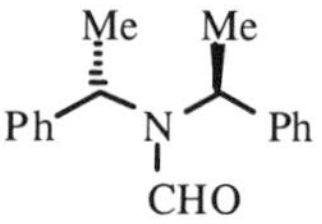

(26)

Formamide (26) also catalyses the formation of allenic alcohols (93-99% *vs.* propargylic) from the addition of propargyl and allenyltrichlorosilane to aliphatic aldehydes (37-71% yield, 56-95% ee), although extended reaction times (14 days) are required (K. Iseki, Y. Kuroki and Y. Kobayashi, *Tetrahedron: Asymmetry*, 1998, **9**, 2889).
Biquinoline dioxide (27) has been shown to promote the addition of allyltrichlorosilane to aryl and unsaturated aldehydes (68-91% yield, 71-92% ee) in the presence of excess Pr^i_2EtN (M. Nakajima *et al.*, *J. Am. Chem. Soc.*, 1998, **120**, 6419). Benzaldehyde also reacts with substituted trichlorosilanes under these conditions (52-70% yield, 49-86% ee). A method for the *in situ* generation of silanes from a range of allylic chlorides and bromides has also been developed, with subsequent addition of (27) and benzaldehyde providing products in 52-95% yield and 43-91% ee (M. Nakajima, M. Saito and S. Hashimoto, *Chem. Pharm. Bull.*, 2000, **48**, 306).

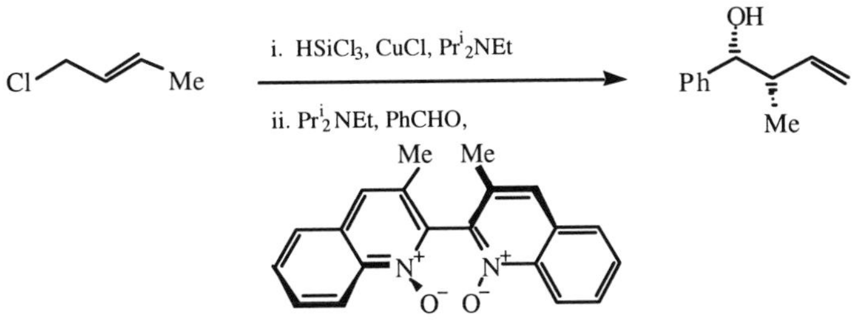

Cl⁄⁄Me → Ph(OH)(Me)

i. $HSiCl_3$, CuCl, Pr^i_2NEt

ii. Pr^i_2NEt, PhCHO,

(27)

In addition to promoting the reaction with tin species, bis(π-allyl palladium) complex (22) catalyses the addition of allyltrimethylsilane to aryl *N*-benzylimines providing amines in 60-95% yield and 64-84% ee (K. Nakamura, H. Nakamura and Y. Yamamoto, *J. Org. Chem.*, 1999, **64**, 2614).

(c) Allyl chromium reagents

The catalytic asymmetric addition of an *in situ* generated allylchromium species to aldehydes was reported in 1999 (M. Bandini *et al.*, *Angew. Chem. Int. Ed.*, 1999, **38**, 3357). A combination of $CrCl_3$, Mn and chiral ligand (28) produces a Cr(II) complex which is employed in 10 mol% to promote the addition of allyl halides to aromatic and aliphatic aldehydes in 40-85% yield and 42-89% ee. The reaction of crotyl bromide and other 3-substituted halides with aromatic aldehydes provides *syn* products (60-75% ds, 25-72% yield, 58-90% ee) when a 2:1 ratio of (28):Cr is employed, whereas a slight preference for the *anti* diastereomer (67% ds) is observed with a 1:1 (28):Cr complex (M. Bandini, P.G. Cozzi and A. Umani-Ronchi, *Angew. Chem. Int. Ed.*, 2000, **39**, 2327).

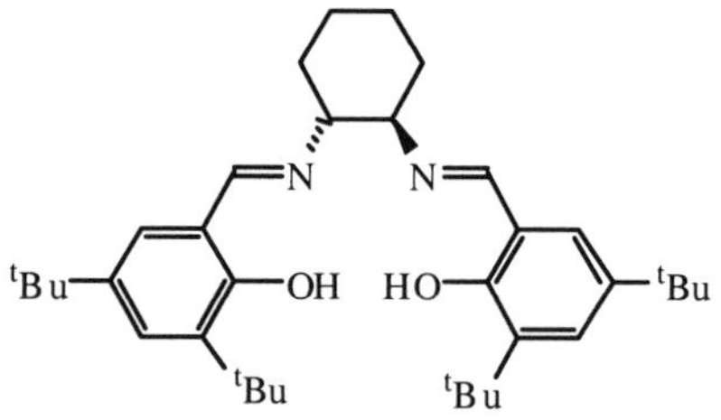

(28)

3. Additions of enolates and their equivalents

(a) The aldol reaction

In recent years, the aldol reaction has become one of the most powerful and versatile methods available for the synthesis of carbon-carbon bonds in a stereocontrolled fashion. Two main strategies have been employed to control the absolute stereochemistry of addition of achiral substrates: under appropriate conditions, an enolate bearing chiral ligands can be generated; alternatively a chiral Lewis acid can be used to promote a Mukaiyama aldol reaction. In general, it is the latter approach which provides more opportunities for the development of catalytic processes. A number of reviews of this area have appeared in recent years, Arya and Qin have surveyed advances in asymmetric enolate methodology (P. Arya and H. Qin, *Tetrahedron*, 2000, **56**, 917) whilst Nelson and Mahrwald have detailed the catalytic enantioselective reactions of preformed enolate equivalents and Lewis acid-mediated aldol reactions respectively (S.G. Nelson, *Tetrahedron: Asymmetry*, 1998, **9**, 357; R. Mahrwald, *Chem. Rev.*, 1999, **99**, 1095). New catalytic concepts for asymmetric aldol reactions have also been considered (H. Gröger, E.M. Vogl and M. Shibasaki, *Chem. Eur. J.*, 1998, **4**, 1137) and catalytic enantioselective Mukaiyama aldol reactions have been highlighted (T. Bach, *Angew. Chem. Int. Ed. Engl.*, 1994, **33**, 417). The use of enzymes, particularly aldolases, and other biochemical catalysts such as catalytic antibodies has also made an important contribution in this area. Machajewski and Wong have described both chemical and biochemical strategies (T.D. Machajewski and C.-H. Wong, *Angew. Chem. Int. Ed.*, 2000, **39**, 1353) whilst other reviews have focussed on chemoenzymatic processes (H.J.M.

Gijsen *et al.*, *Chem. Rev.*, 1996, **96**, 443; C.-H. Wong *et al.*, *Angew. Chem. Int. Ed. Engl.*, 1995, **34**, 412).

(i) Additions of preformed enolates and enolate equivalents

The Lewis acid-mediated addition of silyl enol ethers or silylketene acetals to aldehydes or acetals, the Mukaiyama aldol reaction, has been widely studied since it first came to prominence as a useful asymmetric process in the 1980s. Early workers investigated the use of chiral substrates, with the development of chiral Lewis acid catalysts receiving increased attention in more recent years.
A major contribution has been made by Teruaki Mukaiyama's research group who have studied the use of chiral diamines such as (29) in combination with tin(II) triflate and a butyltin halide or acetate for the reaction of aldehydes and silylketene acetals (T. Mukaiyama, *Aldrichimica Acta*, 1996, **29**, 59). Initial studies required stoichiometric quantities of the diamine but variation of the reaction conditions, in particularly the solvent employed, allowed the quantity of (29) to be reduced to 10-20 mol% without compromising yields and selectivities (S. Kobayashi *et al.*, *Tetrahedron*, 1993, **49**, 1761; S. Kobayashi, Y. Fujishita and T. Mukaiyama, *Chem. Lett.*, 1990, 1455; T. Mukaiyama *et al.*, *Chem. Lett.*, 1990, 129; T. Mukaiyama, H. Uchiro and S. Kobayashi, *Chem. Lett.*, 1990, 1147).

RCHO + OTMS / Me / SEt
$Sn(OTf)_2$
N Me, H N
(29)
OH O / R / SEt / Me
65-80% yield
89-100 % ds
89->98% ee

The scope of the reaction was later extended to include alkynals (86-95% ds, 63-82% yield, 86-91% ee), unsubstituted silylthioketene acetals (48-81% yield, 77-93% ee) and the silylketene acetal of phenyl 2-benzyloxy acetate (90->98% ds, 68-87% yield, 80-96% ee) (T. Mukaiyama *et al.*, *Chem. Lett.*, 1991, 989; S. Kobayashi *et al.*, *Tetrahedron: Asymmetry*, 1991, **2**, 635; S. Kobayashi and T. Kawasuji, *Synlett*, 1993, 911).

CAB complex (23) catalyses the reaction of silylenol ethers and silylketene acetals with a variety of aldehydes. Use of 20 mol% (23, R = H) leads to the generation of *syn* β-hydroxy ketones (64->95% ds, 49-97% yield, 76-94% ee) and esters (82->95% ds, 57-99% yield, 77->95% ee) respectively (K. Furuta, T. Maruyama and H. Yamamoto, *J. Am. Chem. Soc.*, 1991, **113**, 1041; *idem, Synlett*, 1991, 439). The use of (23, R = 3,5-$(CF_3)_2C_6H_3$) resulted in increased reaction rates, whilst improved enantioselectivities were observed in the presence of (23, R = 2-$PhOC_6H_4$) (K. Ishihara *et al.*, *Bull. Chem. Soc. Jpn.*, 1993, **66**, 3483).

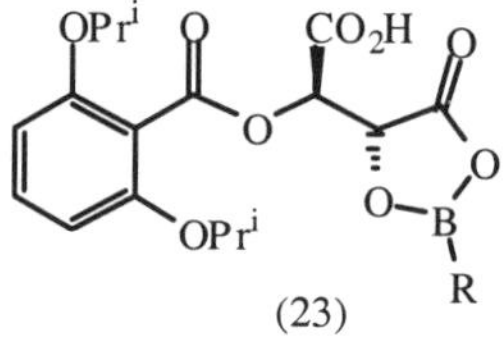

OPr^i O CO_2H O

O O

OPr^i O–B

R

(23)

Chiral Lewis acids prepared from borane-tetrahydrofuran complex and α,α-disubstituted glycine arenesulfonamides (30) and (31) have been shown to promote the reaction of 2,2-dimethylsilyl enol ethers (59-89% yield, 84-99% ee) with aryl and aliphatic aldehydes (E.R. Parmee *et al.*, *J. Am. Chem. Soc.*, 1991, **113**, 9365). The reaction was later extended to unsubstituted and monosubstituted silylketene acetals (E.R. Parmee *et al.*, *Tetrahedron Lett.*, 1992, **33**, 1729). Propionates provide *anti* β-hydroxy esters (77-94% ds, 60->98% ee) with 20 mol% (30), whilst (31) is the preferred ligand for the reaction of unsubstituted silylketene acetals and silylthioketene acetals with aliphatic aldehydes (81-93% ee).
More recently, (31) has been employed in the synthesis of α,α-dihalo-β-hydroxy esters (80->99% yield, 13-98% ee) from aliphatic, aromatic and unsaturated aldehydes and difluoro- and bromofluorosilylketene acetals (K. Iseki *et al.*, *Tetrahedron Lett.*, 1997, **38**, 1447; *idem*, *Tetrahedron,* 1997, **53**, 10271; K. Iseki, Y. Kuroki and Y. Kobayashi, *Synlett*, 1998, 437; *idem*, *Tetrahedron*, 1999, **55**, 2225). Interestingly, it has been shown that the absolute configuration of the product is dependent on the reaction temperature, with the (*R*)-alcohol favoured at –78 °C and the (*S*)-alcohol predominating at –20 °C.

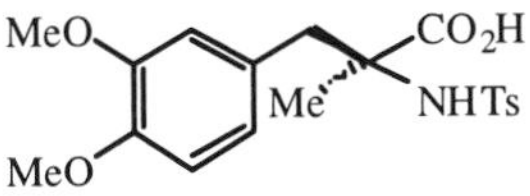

(30) 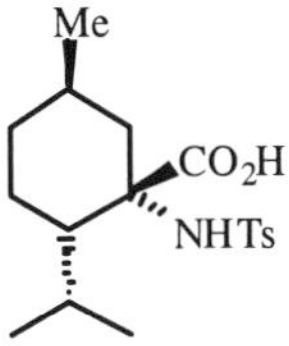 (31)

Tryptophan-derived oxazaborolidine (32, R = ^{n}Bu) (20 mol%) provides chiral β-hydroxy ketones from terminal trimethylsilyl enol ethers and a range of aldehydes (56-100% yield, 86-93% ee) whilst 40 mol% is required for the successful reaction of benzaldehyde and the silyl enol ether of cyclopentanone (71% yield, 94% *syn*, 92% ee) (E.J. Corey, C.L. Cywin and T.D. Roper, *Tetrahedron Lett.*, 1992, **33**, 6907). The scope of the reaction has since been extended to include other substituted silyl enol ethers by employing 10 mol% of (32, R = 3,5-$(CF_3)_2C_6H_3$) (91-99% yield, 78->99% ds, 89->99% ee), although silylketene acetals are not good substrates for either catalyst (K. Ishihara, S. Kondo and H. Yamamoto, *Synlett*, 1999, 1283).

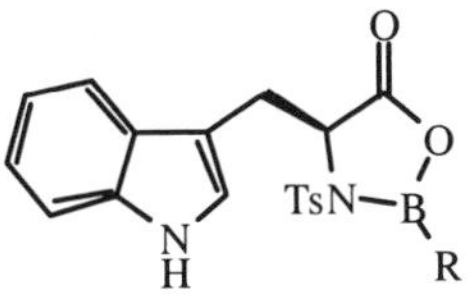

(32) 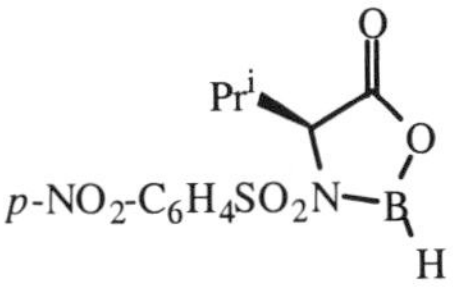 (33)

The oxazaborolidine (33) formed from *N*-(*p*-nitrophenylsulfonyl)valine is also an efficient catalyst for the reaction of silylketene acetals (60-97% yield, 60-76% ds, 60->98% ee), provided that the reaction is conducted in nitroethane (S. Kiyooka, Y. Kaneko and K. Kume, *Tetrahedron Lett.*, 1992, **33**, 4927; S. Kiyooka *et al.*, *Tetrahedron Lett.*, 1997, **38**, 3553). Although titanium(IV) chloride is one of the most commonly used Lewis acid catalysts in Mukaiyama aldol reactions, the development of chiral titanium complexes has lagged behind that of tin and boron-based systems. Binol-derived titanium oxo species (34) has been employed for the addition of the *tert*-butyldimethylsilylketene acetal of *tert*-butyl thioacetate to aryl and unsaturated aldehydes (91-98% yield, 36-85% ee) (T. Mukaiyama *et al.*, *Chem. Lett.*, 1990, 1015) whilst binol.$TiCl_2$

catalyses the reaction of thioacetate and thiopropionate-derived silylketene acetals with glyoxylates (54-73% yield, 73-99% ds, 77->99% ee), fluoral (38-64% yield, 52-56% ds, 55-96% ee) and other β-heteroatom-bearing aldehydes (47-85% yield, 52-92% ds, 80-98% ee) (K. Mikami and S. Matsukawa, *J. Am. Chem. Soc.*, 1993, **115**, 7039; K. Mikami *et al.*, *Synlett*, 1995, 1057; K. Mikami and S. Matsukawa, *J. Am. Chem. Soc.*, 1994, **116**, 4077). The latter reactions are believed to proceed *via* an ene type mechanism.

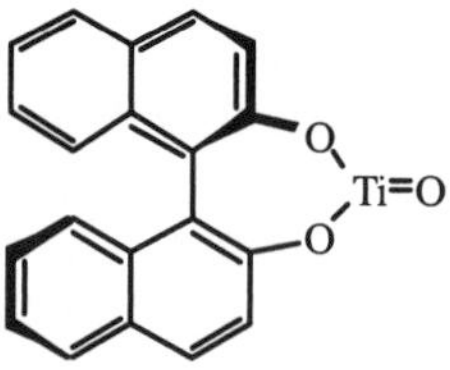

(34)

Carreira and co-workers have shown that titanium complex (35) can be used in 0.5-5 mol% as a catalyst for the reaction of aldehydes and unsubstituted silylketene acetals (72-98% yield, 88-97% ee) (E.M. Carreira, R.A. Singer and W. Lee, *J. Am. Chem. Soc.*, 1994, **116**, 8837). Commercially-available enolate equivalent 2-methoxypropene can also be employed in these reactions, providing direct access to β-hydroxymethyl ketones (77-99% yield, 66-98% ee) (E.M. Carreira, W. Lee and R.A. Singer, *J. Am. Chem. Soc.*, 1995, **117**, 3649), whilst acetoacetate-derived silyldienol ether (36) also gives high yields (79-97%) and selectivities (80-94% ee) (R.A. Singer and E.M. Carreira, *J. Am. Chem. Soc.*, 1995, **117**, 12360). A method for the *in situ* generation of (35) has also been developed (R.A. Singer and E.M. Carreira, *Tetrahedron Lett.*, 1997, **38**, 927), comparable or even superior results (81-99% yield, 91-99% ee) were obtained from the subsequent aldol reactions.

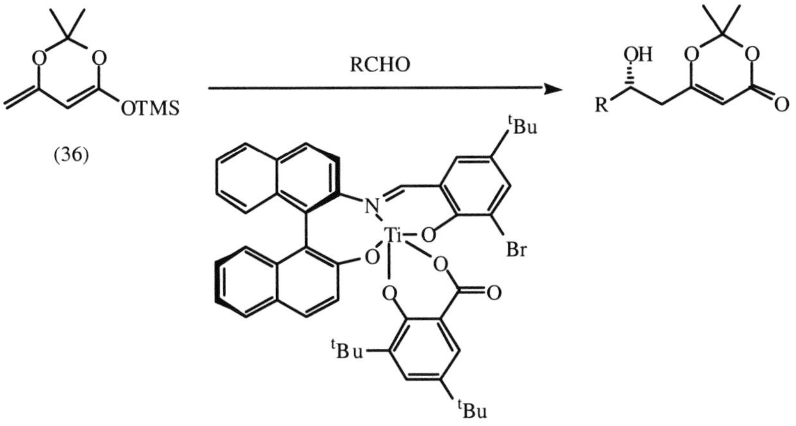

(36)

(35)

Keck and Krishnamurthy found that the binol.Ti(OPr^i)$_4$ conditions which had previously been employed for allylation reactions needed to be re-examined before success could be achieved in asymmetric aldol additions. Concentration effects were found to be particularly important, with reactions between unsubstituted thioketene acetals and a range of aldehydes providing high yields (70-90%) and selectivity (89->98% ee) under optimal conditions (G.E. Keck and D. Krishnamurthy, *J. Am. Chem. Soc.*, 1995, **117**, 2363). The Mukaiyama aldol reactions of (36) and related silyl dienol ethers are catalysed by Ti(OPr^i)$_2$.binol (8-50 mol%), providing δ-hydroxy-β-ketoester derivatives in 15-93% yield, 75-82% ds and 33->99% ee (M. Sato *et al.*, *Heterocycles*, 1995, **41**, 1435; A. Soriente *et al.*, *Tetrahedron: Asymmetry*, 2000, **11**, 2255; M. De Rosa, A. Soriente and A. Scettri, *Tetrahedron: Asymmetry*, 2000, **11**, 3187). A combination of Ti(OPr^i)$_4$ and binol also promotes the *syn* selective reaction of 2-(trimethylsilyloxy)furan with aryl and aliphatic aldehydes (55-99% yield, 24-75% ds, 52->96% ee) (M. Szlosek *et al.*, *J. Org. Chem.*, 1998, **63**, 5169; M. Szlosek and B. Figadère, *Angew. Chem. Int. Ed.*, 2000, **39**, 1799), whilst aliphatic aldehydes and silylthioketene acetal (37) combine in the presence of 5 mol% TiCl$_2$(OPr^i)$_2$ and binol (83-95% yield, 97-98% ee) (S. Matsukawa and K. Mikami, *Tetrahedron: Asymmetry*, 1995, **6**, 2571). The addition of one equivalent of perfluorophenol increased the enantioselectivity of the reaction between the corresponding trimethylsilylenol ether and decanal.

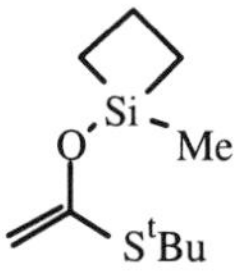

(37)

Up to 10 mol% of (diiodobinol).zirconium complex (38) has been used to promote the highly *anti* selective reaction of propionate derived silyl-ketene acetals (58-98% yield, 86-95% ds, 81-97% ee) (H. Ishitani *et al.*, *J. Am. Chem. Soc.*, 2000, **122**, 5403).

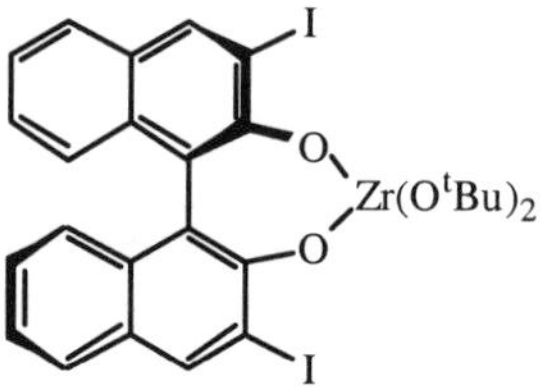

(38)

Evans *et al.* have developed bis(oxazolinyl) complexes (39) and (40) as highly effective Lewis acids for enantioselective organic synthesis (J.S. Johnson and D.A. Evans, *Acc. Chem. Res.*, 2000, **33**, 325). Their use in asymmetric aldol reactions is generally limited to those additions which employ α-benzyloxyacetaldehyde or α-ketoesters and α-diketones, although other workers have achieved broader success in protic (aqueous) media (37-98% yield, 62-85% ds *syn*, 32-85% ee) (S. Kobayashi, S. Nagayama and T.Busujima, *Chem. Lett.*, 1999, 71; *idem*, *Tetrahedron*, 1999, **55**, 8739) and by employing the preformed tributylstannyl enolate of cyclohexanone (66-74% yield, 64-67% ds *syn*, 84-88% ee) (A. Yanagisawa *et al.*, *Synlett*, 1998, 958).

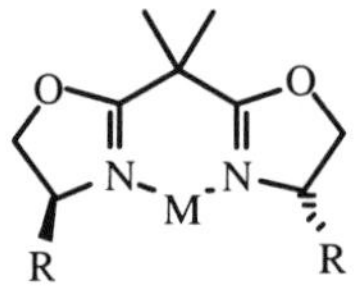

(39) 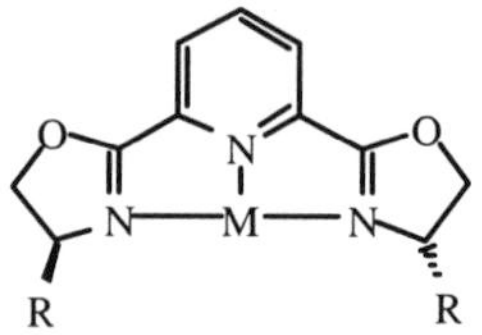 (40)

Evans and co-workers have shown that a wide variety of cyclic and acyclic silylenol ethers and silylketene acetals react with α-benzyloxy-acetaldehyde in a highly enantioselective fashion in the presence of 0.5-10 mol% (40, R = Ph, M = $Cu(SbF_6)_2$) to give (*S*)-alcohols in 48-100% yield and 85-99% ee (D.A. Evans, J.A. Murry and M.C. Kozlowski, *J. Am. Chem. Soc.*, 1996, **118**, 5814; D.A. Evans *et al.*, *J. Am. Chem. Soc.*, 1999, **121**, 669). Both (*E*)- and (*Z*)-2-methylthioketene acetals lead to selective formation of the *syn* diastereomer (86-97% ds) whilst the enantiomeric products are obtained using (39, R = ^{t}Bu, M = $Cu(OTf)_2$). Appropriate choice of the catalyst also provides selective access to either diastereomeric product from the reaction of silyl enol ethers and silyl thioketene acetals with pyruvates and α-diketones (D.A. Evans *et al.*, *J. Am. Chem. Soc.*, 1997, **119**, 7893; D.A. Evans, D.W.C. MacMillan and K.R. Campos, *J. Am. Chem. Soc.*, 1997, **119**, 10859; D.A. Evans *et al.*, *J. Am. Chem. Soc.*, 1999, **121**, 686). *Syn* (2*S*,3*S*) products are favoured (77-100% yield, 90-98% ds, 93-99% ee) when (39, R = ^{t}Bu, M = $Cu(OTf)_2$) is employed whilst tin(II) complexes yield *anti* products. Use of (39, R = Bn, M = $Sn(OTf)_2$) is optimal for the reaction of glyoxylates (72-90% yield, 90-96% ds, 92-98% ee) whilst (40, R = Ph, M = $Sn(OTf)_2$) provides the (2*S*,3*R*) product (76-94% yield, 95-99% ds, 92-99% ee) from pyruvates. Diethyl ketomalonate reacts with silyl enol ethers in the presence of (39, R = Ph, M = $Cu(OTf)_2$) in high yields (80-95%) and 58-93% ee (F. Reichel *et al.*, *Chem. Commun.*, 1999, 1505).
Aryl aldehydes combine with a dimethylsubstituted silylketene acetal in low to good yields (36-98%) but with moderate enantioselectivity (27-49% ee) when chiral disodium amide (40) and $Ln(OTf)_3$ are employed as catalysts (K. Uotsu, H. Sasai and M. Shibasaki, *Tetrahedron: Asymmetry*, 1995, **6**, 71). Chiral triaryl carbenium ion (41) also provides another, but as yet only moderately selective, strategy for catalysis (≤50 %ee) (C.-T. Chen *et al.*, *J. Am. Chem. Soc.*, 1997, **119**, 11341).

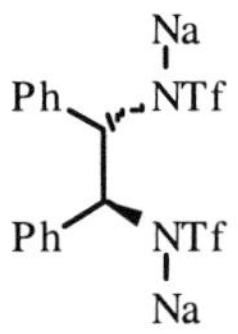

(41)

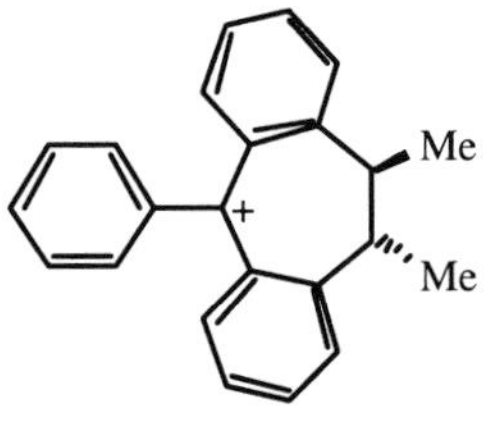

(42)

An alternative strategy for Lewis acid catalysis of the Mukaiyama aldol reaction is based on activation of the nucleophilic component. This is readily achieved using a (*p*-tolyl)binap.CuF_2 complex and dienolate equivalent (36), the reaction proceeds in 48-98% yield and 65-95% ee in the presence of only 2 mol% of the catalyst (J. Krüger and E.M. Carreira, *J. Am. Chem. Soc.*, 1998, **120**, 837). Mechanistic studies indicated that a chiral copper(I) enolate participates in the catalytic cycle (B.L. Pagenkopf, J. Krüger and A. Stojanovic, *Angew. Chem. Int. Ed.*, 1998, **37**, 3124). A comparison of aldehyde activation (with $Ti(OPr^i)_2$binol) and enolate activation (with CuF_2-(*p*-tolyl)binap) indicated that comparable results could be obtained (18-45% yield, 60-75% ee *vs.* 35-80% yield, 48-77% ee) (G. Bluet and J.-M. Campagne, *Tetrahedron Lett.*, 1999, **40**, 5507). Phase transfer catalysis has also been investigated in the context of the reaction of benzaldehyde with silylenol ethers and for the addition of aliphatic aldehydes to silylketene acetal (43). *N*-Benzylcinchoninium fluoride (10, R = Bn) promotes the reaction of cyclic and acyclic silyl enol ethers with benzaldehyde (55-76% yield, 70-82% ds, 25-70% ee) (A. Ando *et al.*, *Tetrahedron Lett.*, 1993, **34**, 1507; T. Shioiri, A. Bohsako and A. Ando, *Heterocycles*, 1996, **42**, 93), whilst α-amino-β-hydroxy esters can be prepared in 48-81% yield (50-93% ds, 70-95% ee) from the reactions of (43) in the presence of 10 mol% of cinchonidinium salt (44) (M. Horikawa, J. Busch-Petersen and E.J. Corey, *Tetrahedron Lett.*, 1999, **40**, 3843).

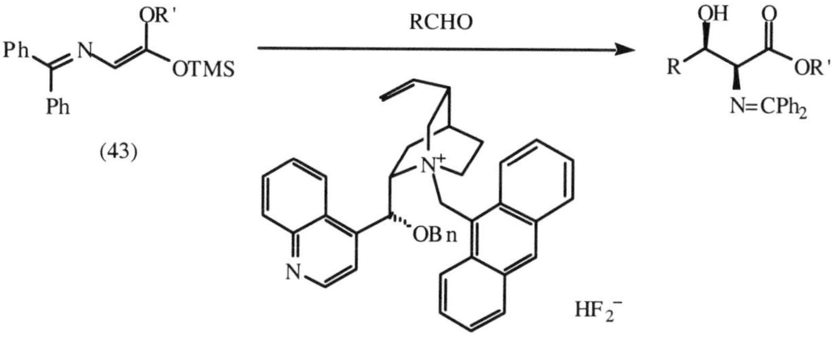

RCHO

(43)

HF_2^-

(44)

Chiral Rh(I) catalysts bearing diphosphine ligands were employed in an enantioselective aldol reaction as early as 1987, however, very low levels of asymmetric induction were observed (5-12% ee) (M. Reetz and A.E. Vougiokas, *Tetrahedron Lett.*, 1987, **28**, 793). Palladium complexes have proved to be much more effective with 1-5mol% binap.$PdCl_2$ or (*p*-tolyl)-binap.$PdCl_2$ in the presence of AgOTf and molecular sieves catalysing the reaction of silyl enol ethers and aryl and aliphatic aldehydes in wet DMF (56-96% yield, 70-89% ee) (M. Sodeoka, K. Ohrai and M. Shibasaki, *J. Org. Chem.*, 1995, **60**, 2648; M. Sodeoka *et al.*, *Synlett*, 1997, 463). Platinum complex (45) (5 mol%) has been employed in the reaction of a dimethylsubstituted silylketene acetal with aryl, unsaturated and primary aliphatic aldehydes (92-99% yield, 46-95% ee) (O. Fujimura, *J. Am. Chem. Soc.*, 1998, **120**, 10032).

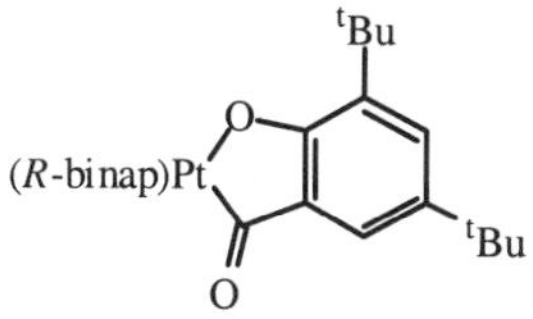

(45)

The more recent development of chiral Lewis basic catalysts for the addition of trichlorosilylenol ethers to aldehydes has provided a complementary strategy to the traditional Mukaiyama aldol reaction for the synthesis of β-hydroxy ketones. Initial success was obtained using 10 mol% of phosphoramide (46) to promote the addition reactions of the trichlorosilyl enolates of methyl acetate and cyclohexanone (78-95% yield, 65-99% ds, 38-93% ee), with later studies exploring the scope of the reaction with the trichlorosilylenol ethers of cyclic and acyclic ketones (79-98% yield, 84->99% ds, 50-97% ee) (S.E. Denmark *et al.*, *J. Am. Chem. Soc.*, 1996, **118**, 7404; S.E. Denmark, K.-T. Wong and R.A. Stavenger, *J. Am. Chem. Soc.*, 1997, **119**, 2333; *idem.*, *J. Org. Chem.*, 1998, **63**, 918; *idem*, *Tetrahedron*, 1998, **54**, 10389; S.E. Denmark *et al.*, *J. Am. Chem. Soc.*, 1999, **121**, 4982). The addition was designed to proceed through a hexacoordinate siliconate species, in place of the open transition state invoked for Mukaiyama aldol reactions, and is effective for a wide variety of aldehydes, although those bearing primary alkyl substituents undergo enolisation rather than reaction.

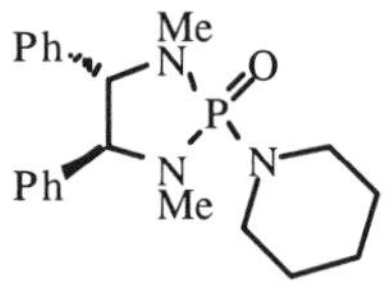

(46) 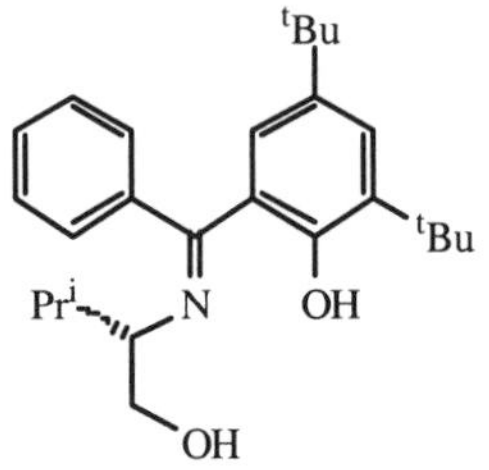 (47)

The addition of preformed tributylstannyl enolates to aldehydes is catalysed by binap.AgOTf (33-98% yield, 83->99% ds, 41-96% ee) (A. Yanagisawa *et al.*, *J. Am. Chem. Soc.*, 1997, **119**, 9319), whilst enol trichloroacetates react with benzaldehyde in the presence of 10 mol% Bu_3SnOMe and binap,AgOTf (59-94% yield, 78-92% ds, 84-95% ee) (A. Yanagisawa *et al.*, *J. Am. Chem. Soc.*, 1999, **121**, 892). The latter paper also details the use of diketene as an enolate equivalent with 20 mol% Bu_3SnOMe and 22 mol% (*p*-tolyl)-binap.AgOTf (59% yield, 84% ee) whilst 20 mol% of chiral imine (47), in combination with one equivalent of $Ti(OPr^i)_4$, promotes the reaction of diketene with aryl, unsaturated and aliphatic aldehydes (59-69% yield, 67-87% ee) (N. Oguni, K. Tanaka and H. Ishida, *Synlett*, 1998, 601).

(ii) Addition of in situ generated enolates

Although major advances have been made in the catalytic enantioselective aldol reactions of preformed enolates, effective reagents for the one-pot enolisation of ketones and their subsequent nucleophilic addition have proved to be much more elusive.
In 1997, Shibasaki and co-workers reported that their heterobimetallic-binol complexes catalysed the addition of methyl ketones to aliphatic aldehydes, 20 mol% $LaLi_3(binol)_3$ gave 28-90% yield and 44-94% ee (Y.M.A. Yamada *et al.*, *Angew. Chem. Int. Ed. Engl.*, 1997, **36**, 1871). The range of ketones which could be employed was later expanded by addition of hexamethyldisilazane (7.2 mol%) to reactions mediated by 8 mol% $LaLi_3(binol)_3$ (50-91% yield, 30-93% ee) (N. Yoshikawa *et al.*, *J. Am. Chem. Soc.*, 1999, **121**, 4168). A combination of $Ba(OPr^i)_2$ and the monomethylether of binol (5 mol%) also promotes the aldol reaction of unmodified ketones and aliphatic aldehydes (77-99% yield, 50-70% ee) (Y.M.A. Yamada and M. Shibasaki, *Tetrahedron Lett.*, 1998, **39**, 5561). The reactions of acetone and hydroxyacetone with aryl and α-substituted aliphatic aldehydes are catalysed by amino acids, in particular proline (38-97% yield, 60->95% ds *anti*, 60->99% ee) (B. List, R.A. Lerner and C.F. Barbas III, *J. Am. Chem. Soc.*, 2000, **122**, 2395; W. Notz and B. List, *J. Am. Chem. Soc.*, 2000, **122**, 7386), whilst Trost and Ito have developed a three component protocol, consisting of 5 mol% (48), 10 mol% Et_2Zn and 15 mol% triphenylphosphine sulfide, for the addition of aryl methyl ketones to aliphatic aldehydes (24-79% yield, 68-99% ee) (B.M. Trost and H. Ito, *J. Am. Chem. Soc.*, 2000, **122**, 12003).

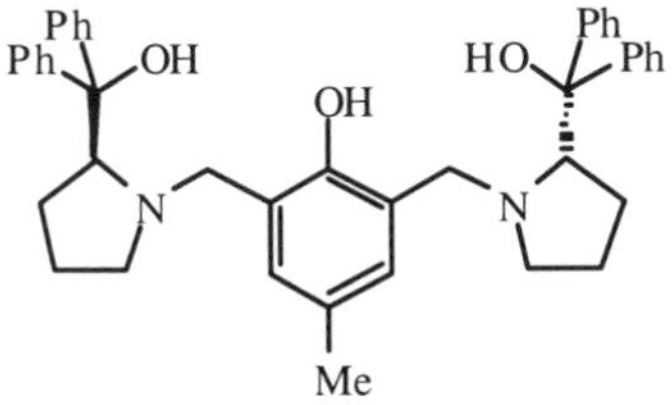

Ph Ph OH OH HO Ph Ph N N Me

(48)

The aldol reactions of 2-cyanopropionates are catalysed by a rhodium(I) complex of chiral diferrocenyl bisphosphine ligand (*S*,*S*)-(*R*,*R*)-PhTRAP. (2*S*,3*S*)-α-Cyano-β-hydroxy esters are formed in good yields (61-96%), with moderate to high stereoselectivity (45-84% ds, 31-93% ee) (R. Kuwano, H. Miyazaki and Y. Ito, *Chem. Commun.*, 1998, 71; *idem, J. Organomet. Chem.*, 2000, **603**, 18). A catalytic reductive aldol process using 6.5 mol% binap., 2.5 mol% $[Rh(COD)Cl]_2$ and Et_2MeSiH, in combination with phenyl acrylate provides *syn* α-methyl-β-hydroxy esters from aryl and aliphatic aldehydes (48-82% yield, 64-80% ds, 45-88% ee) (S.J. Taylor, M.O. Duffey and J.P. Morken, *J. Am. Chem. Soc.*, 2000, **122**, 4528).

(b) Additions to imines

The catalytic asymmetric addition of enolates, and their equivalents, to imines has received far less attention than the corresponding aldol reactions, with only a handful of successful approaches being reported. Aryl *N*-(2-hydroxyphenyl)imines (49) and binaphthyldiol-Zr(IV) complexes have been extensively studied. 6,6'-Dibromobinol and $Zr(O^tBu)_4$ combine in a 2:1 ratio to give a species which catalyses (5-10 mol%) the addition of silylketene acetals in the presence of 5-30 mol% *N*-methylimidazole (56-100% yield, 80->98% ee) (H. Ishitani, M. Ueno and S. Kobayashi, *J. Am. Chem. Soc.*, 1997, **119**, 7153). The use of 10 mol% of bis(binol) (50), $Zr(O^tBu)_4$ and *N*-methylimidazole results in the opposite sense of absolute asymmetric induction (75-98% yield, 82-95% ee), as does the reaction of a silylthioketene acetal mediated by $Zr(O^tBu)_4$ (20 mol%), 6,6'-dibromo-3,3'-(4-nitrophenyl)binol (40 mol%) and *N*-methylimidazole (60 mol%) (80% yield, 84% ee) (H. Ishitani, T. Kitazawa and S. Kobayashi, *Tetrahedron Lett.*, 1999, **40**, 2161; S. Kobayashi *et al.*, *J. Org. Chem.*, 1999, **64**, 4220). Selective synthesis of both *syn* and *anti* α-hydroxy-β-amino esters can be achieved by appropriate choice of the protecting group in (51). *Syn* products are formed in the reactions of (51, P = TBDMS) (65-100% yield, 82->99% ds, 91-98% ee) whilst the *anti* diastereomer is favoured using (51, P = Bn) (68-100% yield, 57-92% ds, 76-96% ee) (S. Kobayashi, H. Ishitani and M. Ueno, *J. Am. Chem. Soc.*, 1998, **120**, 431).

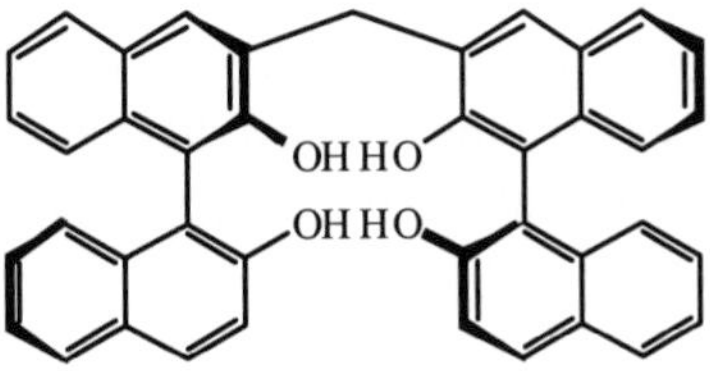

OH HO
OH HO

(50)

HO
N
Ar H

\+ PO OR OTMS

$Zr(O^tBu)_4$,
R-6,6'-dibromo-binol

Me N Me N

OH NH O Ar OR OTBDMS

or

OH NH O Ar OR OBn

(49) (51)

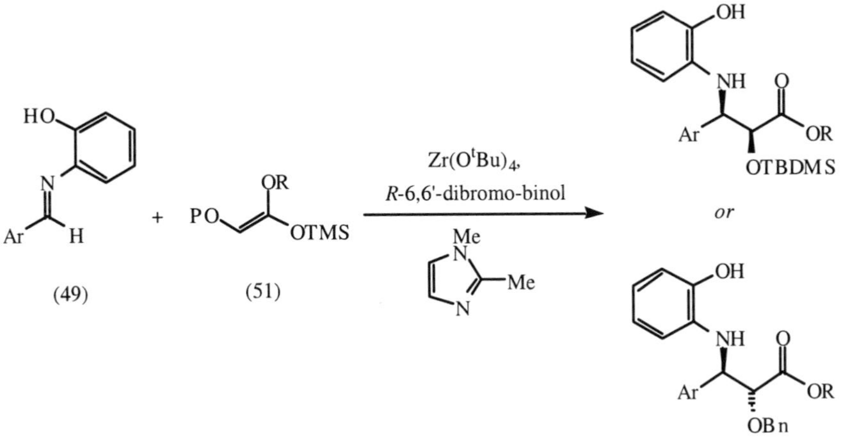

Aliphatic aldehyde-derived products are accessible *via* hydrazone (52) and the 3,3'-dibromobinol complex (39-66% yield, 86-96% ee) (S. Kobayashi, Y. Hasegawa and H. Ishitani, *Chem. Lett.*, 1998, 1131), whilst the reaction of 2-(trimethylsilyloxy)furans and (49) is catalysed by $Ti(OPr^i)_4$ and binol (50-80% yield, 66-96% ds, 28-54% ee) (S.F. Martin and O.D. Lopez, *Tetrahedron Lett.*, 1999, **40**, 8949).

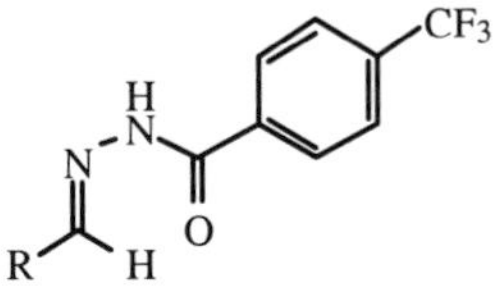

CF_3
H N N O
R H

(52)

The reaction of *N*-tosyl α-imino esters and silyl enol ethers is catalysed by 10 mol% of a $CuClO_4$-(*p*-tolyl)binap complex (65-93% yield, 75-96% ds

anti, 46->99% ee) (D. Ferraris *et al.*, *J. Am. Chem. Soc.*, 1998, **120**, 4548; *idem*, *J. Org. Chem.*, 1998, **63**, 6090). Binuclear hydroxyl-bridged palladium-binap. catalysts (3-5 mol%) have been employed for the analogous reaction of *N*-(4-methoxyphenyl)glyoxylimines (62-95% yield, 53-90% ee) (E. Hagiwara, A. Fujii and M. Sodeoka, *J. Am. Chem. Soc.*, 1998, **120**, 2474).
The direct Mannich reaction of unmodified aryl alkyl ketones with iminium equivalent (53) occurs in the presence of 30 mol% $LiAl(binol)_2$ and $La(OTf)_3$ and molecular sieves to give α-chiral β-amino ketones (61-76% yield, 31-44% ee) (S. Yamasaki, T. Iida and M. Shibasaki, *Tetrahedron Lett.*, 1999, **40**, 307; *idem*, *Tetrahedron*, 1999, **55**, 8857).
A three-component asymmetric condensation reaction between 4-methoxyaniline, acetone and aryl and aliphatic aldehydes in the presence of 35 mol% proline (35-90% yield, 70-96% ee) has also been reported (B. List, *J. Am. Chem. Soc.*, 2000, **122**, 9336). *Syn* selectivity (67->95% ds) is observed with other dialkyl ketones (50-96% yield, 65-99% ee).

MeO NEt$_2$

(53)

Enantioselective lactam formation has been achieved in the reaction of lithium ester enolates with *N*-(4-methoxyphenyl)imines upon addition of 20 mol% of chiral ligands (54), (55) and (39, R = Pr^i) (75-99% yield, 65-90% ee) (H. Fujieda *et al.*, *J. Am. Chem. Soc.*, 1997, **119**, 2060; K. Tomioka *et al.*, *Chem. Commun.*, 1999, 715; T. Kamabara and K. Tomioka, *Chem. Pharm. Bull.*, 1999, **47**, 720).

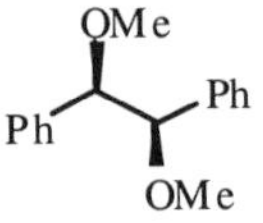

OMe Ph Ph OMe

(54)

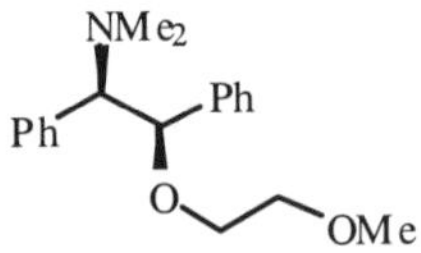

NMe$_2$ Ph Ph O OMe

(55)

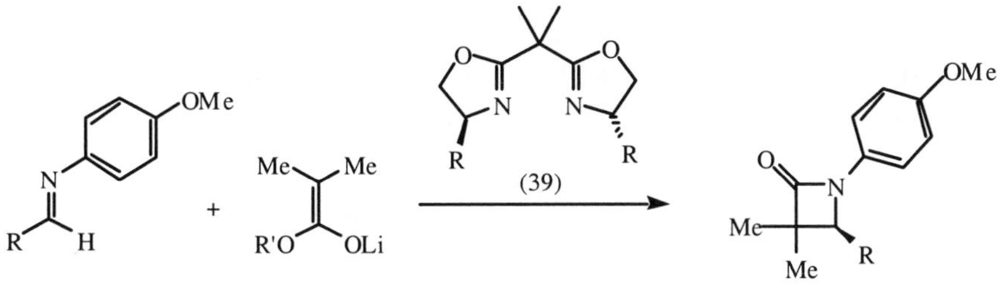

(39)

(c) Darzen's and Baylis-Hillman reactions

The formation of chiral epoxides by asymmetric addition of α-halo anions to aldehydes, with subsequent ring closure provides an alternative to alkene oxidation approaches. Limited success has been achieved, with early experiments involving the addition of benzyl bromide to aryl aldehydes in the presence of potassium hydroxide and 20-50 mol% of camphor-derived alcohols (56) and (57) (89-100% yield, 36-60% ee) (N. Furukawa, Y. Sugihara and H. Fujihara, *J. Org. Chem.*, 1989, **54**, 4222; A.-H. Li *et al.*, *J. Org. Chem.*, 1996, **61**, 489). α-Chloro ketones react with aryl and aliphatic aldehydes in the presence of 10 mol% *N*-(*p*-trifluoromethylbenzyl)cinchoninium fluoride (10, R = $(4\text{-}CF_3C_6H_4)CH_2$) and lithium hydroxide (32-99% yield, 35-86% ee) (S. Arai and T. Shioiri, *Tetrahedron Lett.*, 1998, **39**, 2145; S. Arai *et al.*, *Chem. Commun.*, 1999, 49; *idem*, *Tetrahedron*, 1999, **55**, 6375) whilst only aryl aldehydes provide α,β-epoxy ketones when sodium hydroxide and glucose-fused azacrown (58) are employed (29-74% yield, 11-62% ee) (P. Bakó *et al.*, *Tetrahedron: Asymmetry*, 1999, **10**, 4539).

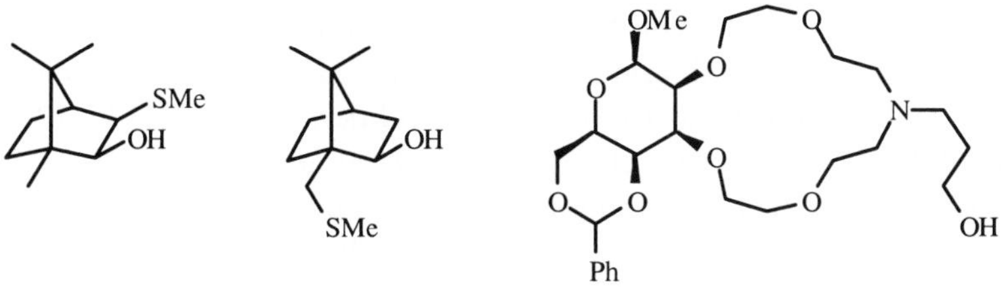

(56) (57) (58)

Quinine-derived phase transfer catalyst (59) (10 mol%) and potassium hydroxide promote the addition of phenylsulfonylchloromethane to aryl aldehydes (69-94% yield, 64-81% ee) (S. Arai, T. Ishida and T. Shioiri, *Tetrahedron Lett.*, 1998, **39**, 8299).

ArCHO + $PhSO_2CH_2Cl$ → (KOH, (59)) Ar–epoxide–SO_2Ph

(59)

The Baylis-Hillman reaction provides α-methylene-β-hydroxy esters and ketones from the corresponding α,β-unsaturated carbonyl species. The reaction is frequently very slow but both high pressure and phosphine additives have been shown to provide useful rate improvements. Asymmetric approaches have generally focussed on the use of chiral amines in place of the DABCO employed in the racemic version (S.E. Drewes and G.H.P. Roos, *Tetrahedron*, 1988, **42**, 4653; P. Langer, *Angew. Chem. Int. Ed.*, 2000, **39**, 3049).
Quinidine (60) provides moderate levels of asymmetric induction (18-45% ee, 40-50% yield) in reactions performed at 3-10 kbar (I.E. Markó, P.R. Giles and N.J. Hindley, *Tetrahedron*, 1997, **53**, 1015) whilst 10 mol% of oxygen-bridged quinidine derivative (61) promotes the highly enantioselective reaction (91-99% ee, 31-58% yield) of 1,1,1,3,3,3-hexafluoroisopropyl acrylate with a range of aldehydes (Y. Iwabuchi *et al.*, *J. Am. Chem. Soc.*, 1999, **121**, 10219). Both methyl vinyl ketone and methyl acrylate react with aryl aldehydes at 10 kbar in the presence of 1 mol% hydroquinone and 15 mol% DABCO derivative (62) (45-72% yield, 10-47% ee) (T. Oishi, H. Oguri and M. Hirama, *Tetrahedron: Asymmetry*, 1995, **6**, 1241) and proline-derived bicyclic amine (63) catalyses the addition of vinyl ketones to aryl aldehydes in the presence of sodium tetrafluoroborate (17-93% yield, 21-72% ee) (A.G.M. Barrett, A.S. Cook and A. Kamimura, *Chem. Commun.*, 1998, 2533).

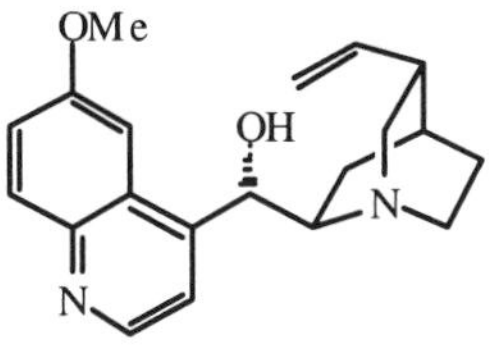

(60) 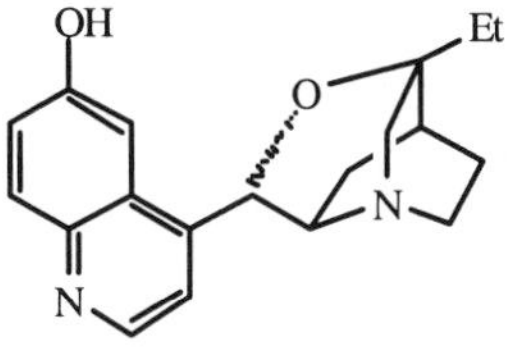 (61) 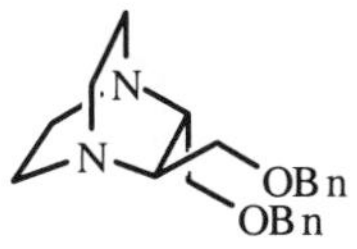 (62)

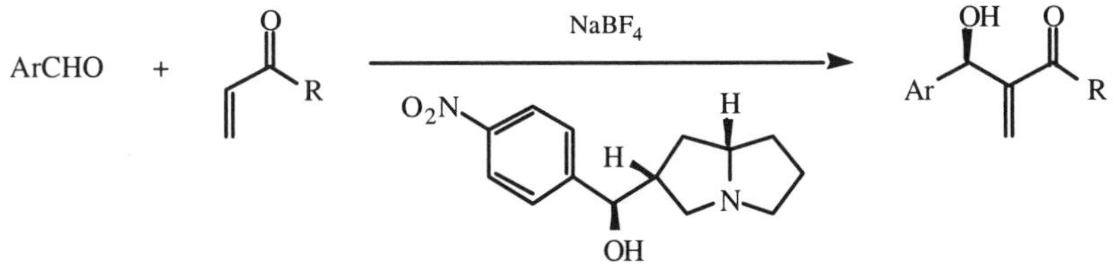

$NaBF_4$

ArCHO + →

(63)

Binol and tributyl phosphine have been shown to act as mild cooperative catalysts for the reaction of acrylates and cyclic enones with aryl and aliphatic aldehydes, 16 mol% Ca(binol) and 10 mol% Bu_3P provide moderate results (62% yield, 56% ee) from the addition of 2-cyclopentenone to 3-phenylpropionaldehyde (Y.M.A. Yamada and S. Ikegami, *Tetrahedron Lett.*, 2000, **41**, 2165). Binap. promotes the reaction of pyrimidinyl aldehydes and methyl acrylate (18-24% yield, 37-44% ee) (T. Hayase *et al.*, *Chem. Commun.*, 1998, 1271) whilst Barrett and Kamimura have developed an alternative approach to Baylis-Hillman products based on the use of $PhSSiMe_3$ or $PhSeSiMe_3$ in place of DABCO. Yamamoto's CAB catalyst (23) was employed in the carbon-carbon bond forming step (9-59% yield, 63->97% ee) but subsequent oxidative elimination (46-88% yield, 50-96% ee) of the phenylsulfanyl or phenylselenyl group was required to regenerate the alkene (A.G.M. Barrett and A. Kamimura, *J. Chem. Soc., Chem. Commun.*, 1995, 1755).

RCHO + [enone R'] —PhSeSiMe3, (23)→ [β-hydroxy ketone, PhSe] —H2O2→ [product]

(23)

4. Ene reactions

The Lewis acid-catalysed carbonyl-ene reaction, between an aldehyde and an alkene bearing an allylic hydrogen atom, provides homoallylic alcohols. Developments in asymmetric approaches were highlighted in 1995 (D.J. Berrisford and C. Bolm, *Angew. Chem. Int. Ed. Engl.*, 1995, **34**, 1717) and several overviews of this area have also appeared (K. Mikami and M. Shimizu, *Chem. Rev.*, 1992, **92**, 1021; K. Mikami, M. Terada and S. Narisawa, *Synlett*, 1995, 347; K. Mikami, *Pure Appl. Chem.*, 1996, **68**, 639).

Effective catalytic asymmetric ene reactions generally require the use of reactive aldehydes or activated alkenes. Chloral and highly electron-deficient benzaldehydes are good substrates, with the first catalytic asymmetric ene reaction being observed in the presence of 20 mol% of binaphthylaluminium Lewis acid (64) (35-88% yield, 49-88% ee) (K. Maruoka *et al.*, *Tetrahedron Lett.*, 1988, **29**, 3967). Fluoral and chloral undergo asymmetric ene reactions with trisubstituted alkenes in the presence of 10 mol% $TiCl_2$.binol or $TiBr_2$.binol and molecular sieves (35-94% yield, 91-98% ds *syn*, 11-96% ee) (K. Mikami *et al.*, *Tetrahedron: Asymmetry*, 1994, **5**, 1087; *idem, Tetrahedron*, 1996, **52**, 85), whilst 2-methylpropene and chloral combine in 55% yield and 22% ee in the presence of 20 mol% boron trifluoride-(-)-menthylethyl etherate (I.M. Akhmedov *et al.*, *Synth. Commun.*, 1994, **24**, 137).

Cl_3CCHO + [alkene, R] → [product: Cl_3C–CH(OH)–CH$_2$–C(=CH$_2$)–R] (catalyst 64: binaphthol bearing two $SiPh_3$ groups, with O–Al–Me)

(64)

The ene reactions of 1,1-disubstituted alkenes and glyoxylate esters are promoted by 1-10 mol% TiX_2.binol (X = Cl, Br) in the presence of molecular sieves (72-98% yield, 83-98% ee) (K. Mikami, M. Terada and T. Nakai, *J. Am. Chem. Soc.*, 1989, **111**, 1940; *idem, J. Am. Chem. Soc.*, 1990, **112**, 3949). Lower yields (49-69%, 63->99% ee) were obtained using 2,3-benzo-exocyclic alkenes (F.T. van der Meer and B.L. Feringa, *Tetrahedron Lett.*, 1992, **33**, 6695), whilst the increased nucleophilicity of vinyl sulfides and selenides allows the reaction to proceed with only 0.5 mol% $TiCl_2$.binol (83-95% yield, 69->99% ee) (M. Terada, S. Matsukawa and K. Mikami, *J. Chem. Soc., Chem. Commun.*, 1993, 327). Methallyltriphenylsilane also undergoes an ene reaction with methyl glyoxylate (45% yield, 95%ee) (K. Mikami and S. Matsukawa, *Tetrahedron Lett.*, 1994, **35**, 3133) and 4-oxobutenoates and 4-oxo-butynoates are highly effective replacements for glyoxylates (60-93% yield, 72-97% ee) in reactions with 1,1-disubstituted exocyclic alkenes (K. Mikami, A. Yoshida and Y. Matsumoto, *Tetrahedron Lett.*, 1996, **37**, 8515).

A μ-oxo titanium species has been proposed as the active catalyst for these ene reactions (D. Kitamoto, H. Imma and T. Nakai, *Tetrahedron Lett.*, 1995, **36**, 1861; M. Terada *et al.*, *Chem. Commun.*, 1997, 281), which have also been shown to be subject to asymmetric amplification when $TiCl_2$.binol of 13-67% ee is employed (K. Mikami and M. Terada, *Tetrahedron*, 1992, **48**, 5671). The use of racemic binol "poisoned" with D-(−)-diisopropyl tartrate gives comparable enantioselectivity to the enantiopure binol catalyst for the reactions of chloral with exocyclic alkenes (J.W. Faller and X. Liu, *Tetrahedron Lett.*, 1996, **37**, 3449) whilst the addition of (*R*,*R*)-TADDOL or a biphenol to $Ti(OPr^i)_4$ and (*R*)-binol leads to the self-assembly of a catalyst which performs well in asymmetric glyoxylate ene reactions (50-62% yield, 91-97% ee) (K. Mikami *et al.*, *Angew. Chem. Int. Ed. Engl.*, 1997, **36**, 2768; M. Chavarot *et al.*, *Tetrahedron: Asymmetry*, 1998, **9**, 3889). Catalysts prepared from 6,6'-dibromobinol and Ti(IV) or Yb(III) have also been reported (63-89% yield, 94-97% ds *syn*, 61-89% ee and 78% yield, 38% ee respectively), as has the use of chiral diol (65) in combination with $Ti(OPr^i)_4$ and $TiCl_4$ (38-56% yield, 48-52% ee) (M. Terada, Y. Motoyama and K. Mikami, *Tetrahedron Lett.*, 1994, **35**, 6693; C. Qian and T. Huang, *Tetrahedron Lett.*, 1997, **38**, 6721; G. Manickam and G. Sundararajan, *Tetrahedron: Asymmetry*, 1999, **10**, 2913).

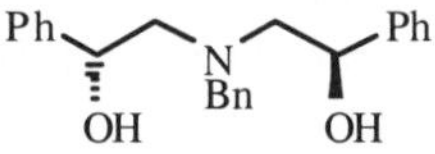

Ph N Ph
Bn
OH OH

(65)

Evans has shown that his bis(oxazolinyl) Lewis acids are highly effective in the ene reactions of ethyl glyoxylate and a wide range of unactivated alkenes, including substrates such as hexene which cannot be employed with other catalysts. (*S*)-Homoallylic alcohols are formed using (39, R = ^tBu, M = $Cu(SbF_6)_2$) whilst (39, R = Ph, M = $Cu(OTf)_2$) provides (*R*)-products (62-99% yield, 76-97% ee) (D.A. Evans *et al.*, *J. Am. Chem. Soc.*, 1998, **120**, 5824). A detailed investigation of the scope and selectivity of the reaction has been undertaken and pyruvate esters have also been shown to be suitable substrates (76-95% yields, ≥98% ee) (D.A. Evans *et al.*, *J. Am. Chem. Soc.*, 2000, **122**, 7936). The corresponding Lewis acid derived from (40, R = Ph) and $Yb(OTf)_3$ is a less efficient catalyst (60-85% yield, 16-49% ee) (C. Qian and L. Wang, *Tetrahedron: Asymmetry*, 2000, **11**, 2347).

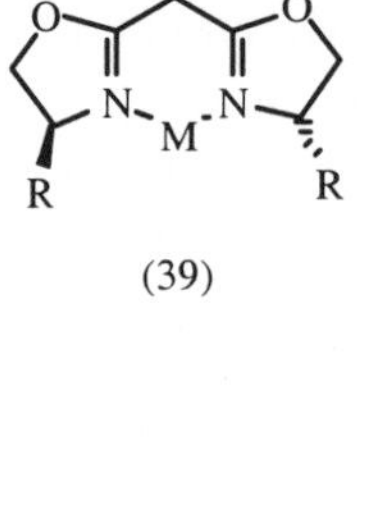

O O
N M N
R R

(39)

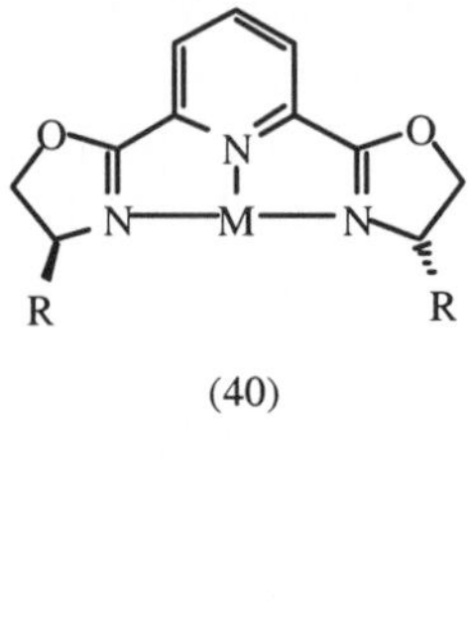

O N O
N M N
R R

(40)

The glyoxylate ene reaction is also promoted by 10 mol% Pd(*p*-tolyl)-binap)$(MeCN)_2(SbF_6)_2$ (83-97% yield, 69-85% ds, 73-88% ee) (J. Hao, M. Hatano and K. Mikami, *Org. Lett.*, 2000, **2**, 4059).
The only successful catalytic asymmetric ene reactions of imines reported to date have employed *N*-tosyl α-iminoesters and copper(I)-(*p*-tolyl-binap) complexes. High selectivities (82-99% ee, 62-94% yield) are obtained in the presence of as little as 0.1 mol% of the catalyst (W.J.I. Drury *et al.*, *J. Am. Chem. Soc.*, 1998, **120**, 11006; S. Yao, X. Fang and K.A. Jørgensen, *Chem. Commun.*, 1998, 2547).

5. Cyanide as nucleophile

(i) Additions to aldehydes and ketones

Cyanohydrins are of interest as synthetic intermediates and as key substrates in biological systems. A vast array of catalysts for the enantioselective addition of HCN and TMS-CN to aldehydes have been developed (R.J.H. Gregory, *Chem. Rev.*, 1999, **99**, 3649). These reactions are also subject to efficient asymmetric catalysis by oxynitrilase enzymes (F. Effenberger, *Angew. Chem. Int. Ed. Engl.*, 1994, **33**, 1555; H. Griengl *et al.*, *Chem. Commun.*, 1997, 1933).
A key breakthrough was made in 1981, when it was shown that 20 mol% cyclic dipeptide (66) catalysed the hydrocyanation of benzaldehyde in up to 90% ee, however, it was found that the product racemised under the reaction conditions giving only 12% ee at 90% conversion (J. Oku and S. Inoue, *J. Chem. Soc., Chem. Commun.*, 1981, 229; J. Oku, N. Ito and S. Inoue, *Makromol. Chem.*, 1982, **183**, 579). Variation of the reaction time, solvent, aldehyde, cyanide source and dipeptide employed provided variable results (25-100% yield, 0-≥98% ee) (S. Asada, Y. Kobayashi and S. Inoue, *Makromol. Chem.*, 1985, **186**, 1755; Y. Kobayashi *et al.*, *Bull. Chem. Soc. Jpn.*, 1986, **59**, 893; K. Tanaka, A. Mori and S. Inoue, *J. Org. Chem.*, 1990, **55**, 181; H.J. Kim and W.R. Jackson, *Tetrahedron: Asymmetry*, 1994, **5**, 1541; Y. Kobayashi *et al.*, *Chem. Lett.*, 1986, 931; J. Oku, N. Ito and S. Inoue, *Makromol. Chem.*, 1979, **180**, 1089; A. Mori *et al.*, *Chem. Lett.*, 1989, 2119), whilst closely related imidazolidinedione (67) was also shown to promote an enantioselective addition (67-93% yield, 16-41 % ee) (H. Danda, *Bull. Chem. Soc. Jpn.*, 1991, **64**, 3743).

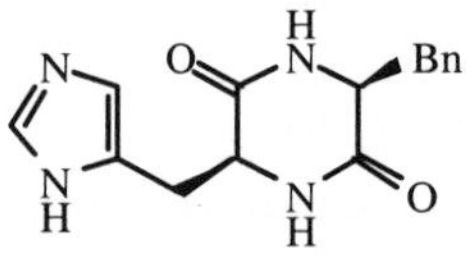

(66) 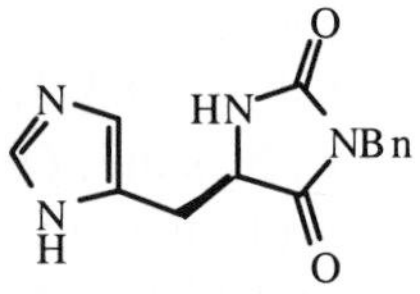 (67)

The first Lewis acid-catalysed trimethylsilylcyanation was reported in 1986 using chiral boranes (68, X = OMe, Cl), although very low selectivity was observed (12-16% ee, 45-55% yield) (M.T. Reetz, F. Kunisch and P. Heitmann, *Tetrahedron Lett.*, 1986, **27**, 4721). Later studies employed 20 mol% $Ti(OPr^i)_4$ and L-(+)-diisopropyl tartrate (79-88% yield, 60-83% ee), a combination of $BiCl_3$, L-(+)-diethyl tartrate and butyllithium (87-100% yield, 20-72% ee) and a Zr(IV)-TADDOL complex (51% yield, 72% ee) (M. Hayashi, T. Matsuda and N. Oguni, *J. Chem. Soc., Chem. Commun.*, 1990, 1364; *idem, J. Chem. Soc., Perkin Trans. 1*, 1992, 3135; M. Wada *et al.*, *Tetrahedron: Asymmetry*, 1997, **8**, 3939; T. Ooi *et al.*, *Synlett*, 2000, 1133).

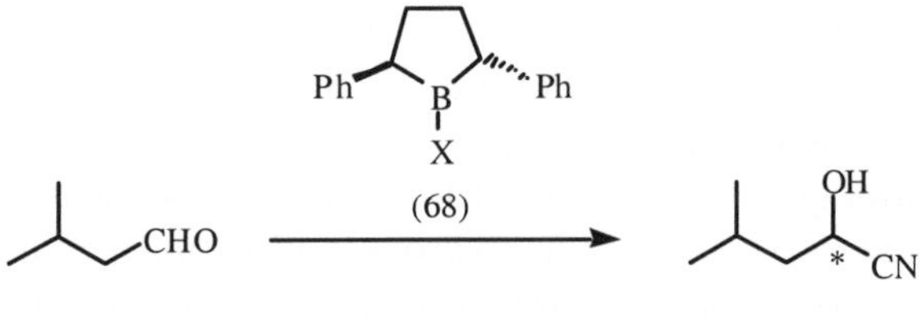

(68)

Binol has also been shown to provide some degree of asymmetric induction in these reactions. $Ti(OPr^i)_4$ and binol (20 mol%) promote the reaction of TMS-CN with aliphatic aldehydes (>90% yield, 33-75% ee), whilst 1 mol% of the mono lithium salt of binol gives 38-99% yield and 0-59% ee with aryl, unsaturated and aliphatic aldehydes and 10 mol% $La_2(3,3'\text{-}(CH_2CH_2OMe)_2\text{binol})_3$ provides cyanohydrins in 56-82% yield with 48-73% ee (M. Mori, H. Imma and T. Nakai, *Tetrahedron Lett.*, 1997, **38**, 6229; I.P. Holmes and H.B. Kagan, *Tetrahedron Lett.*, 2000, **41**, 7453; C. Qian, C. Zhu and T. Huang, *J. Chem. Soc., Perkin Trans. 1*, 1998, 2131).

Imines (69) and (70, R = Ph, 3-indolyl) derived from 2-hydroxy-1-naphthaldehyde and chiral amines have also been reported as ligands for asymmetric cyanation reactions. A combination of 10-20 mol% (69) and trimethylaluminium promotes the trimethylsilylcyanation of aryl and aliphatic aldehydes (48-95% yield, 37-69% ee) whilst (70) and $Ti(OEt)_4$ catalyse hydrocyanation reactions (20-99% yield, 37-91% ee) (A. Mori *et al.*, *Synlett*, 1991, 563; H. Ohno *et al.*, *J. Org. Chem.*, 1992, **57**, 6778; A. Mori *et al.*, *Tetrahedron Lett.*, 1991, **32**, 4333; H. Nitta *et al.*, *J. Am. Chem. Soc.*, 1992, **114**, 7969; H. Abe *et al.*, *Chem. Lett.*, 1992, 2443).

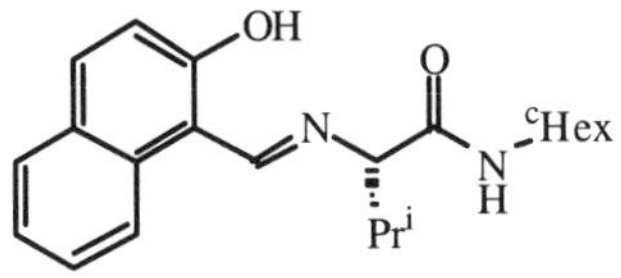

(69) 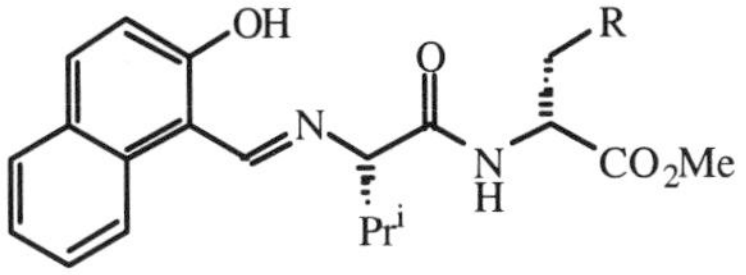 (70)

Valinol salicylimine (71) and $Ti(OPr^i)_4$ (20 mol%) mediate the reaction of TMS-CN and aromatic, unsaturated and aliphatic aldehydes (48-85% yield, 20-96% ee), poorest results are obtained with electron-deficient aromatic and aliphatic aldehydes (M. Hayashi *et al.*, *J. Chem. Soc., Chem. Commun.*, 1991, 1752; *idem, J. Org. Chem.*, 1993, **58**, 1515; *idem, Tetrahedron*, 1994, **50**, 4385). The complex formed from related imine (72) and $Ti(OPr^i)_4$ has also been reported to catalyse the reaction of aryl and unsaturated aldehydes with TMS-CN (31-89% yield, 48-91% ee), although improved results were obtained in the benzaldehyde addition when one equivalent of the Lewis acid was employed (Y. Jiang *et al.*, *Tetrahedron: Asymmetry*, 1995, **6**, 405; *idem, Tetrahedron: Asymmetry*, 1995, **6**, 2915).

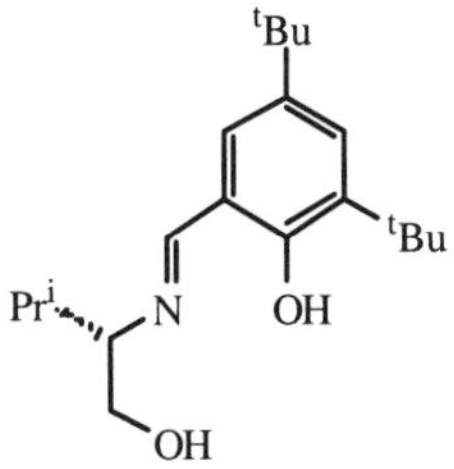

(71) 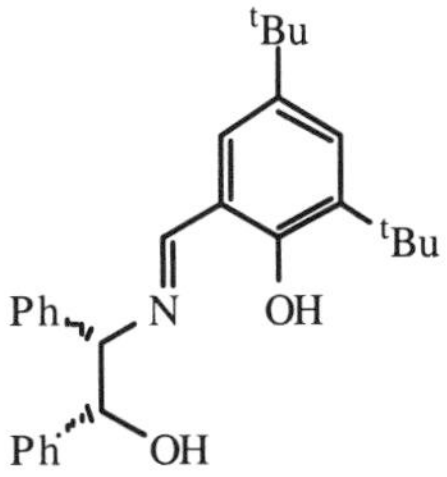 (72)

The salen complexes generated by reaction of $Ti(OPr^i)_4$ with salicylimines (28) and (73) are also catalysts for the trimethylsilyl-cyanation of aryl, unsaturated and aliphatic aldehydes (58-85% yield, 22-87% ee and 43-

100% yield, 30-92% ee), again aliphatic and electron-deficient aromatic aldehydes are the poorest substrates (W. Pan *et al.*, *Synlett*, 1996, 337; Y. Jiang *et al.*, *Tetrahedron*, 1997, **53**, 14327; Y. Belokon' *et al.*, *J. Chem. Soc., Perkin Trans. 1*, 1997, 1293; Y.N. Belokon' *et al.*, *J. Am. Chem. Soc.*, 1999, **121**, 3968). The $Ti(OPr^i)_4$-(28) complex (0.1-0.5 mol%) also catalyses the addition of TMS-CN to alkyl aryl ketones (64-100% yield, 32-72% ee) (Y.N. Belokon' *et al.*, *Tetrahedron Lett.*, 1999, **40**, 8147).

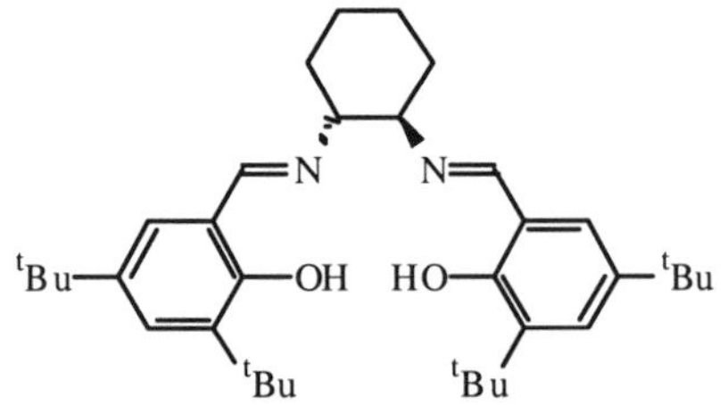

(28)

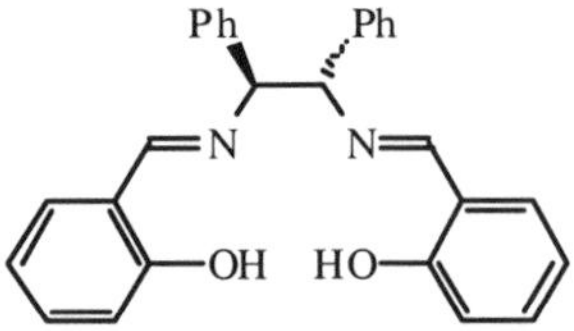

(73)

An oxovanadium complex of (28) (0.1 mol%) provides improved selectivity (68-95% ee) in the aldehyde additions, although reduced reactivity is observed (Y.N. Belokon', M. North and T. Parsons, *Org. Lett.*, 2000, **2**, 1617), whilst cyclophane complex (74) promotes the addition of TMS-CN to benzaldehyde (90% yield, 84% ee) (Y. Belokon', *Tetrahedron: Asymmetry*, 1997, **8**, 3245). The mono lithium salt of (28) is also an active catalyst (1 mol%) for TMS-CN additions (64-99% yield, 0-97% ee) (I.P. Holmes and H.B. Kagan, *Tetrahedron Lett.*, 2000, **41**, 7457).

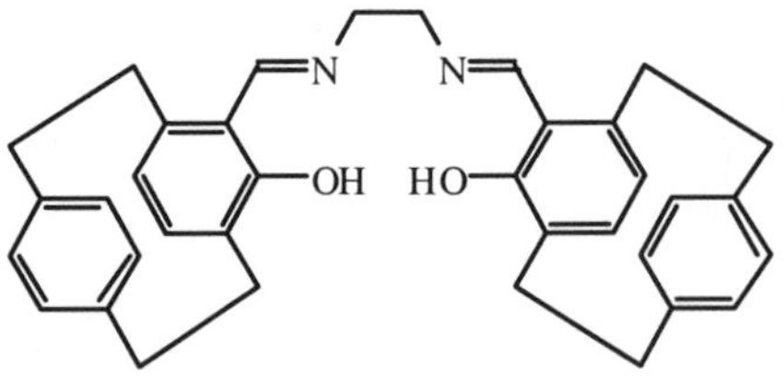

(74)

Camphor-derived bis(amide) (75) forms a chiral complex with $Ti(OPr^i)_4$, which has been employed for the addition of TMS-CN to aryl, unsaturated and aliphatic aldehydes with uniformly high selectivity (87-98% ee, 51-

96% yield) (C.-D. Hwang, D.-R. Hwang and B.-J. Uang, *J. Org. Chem.*, 1998, **63**, 6762).

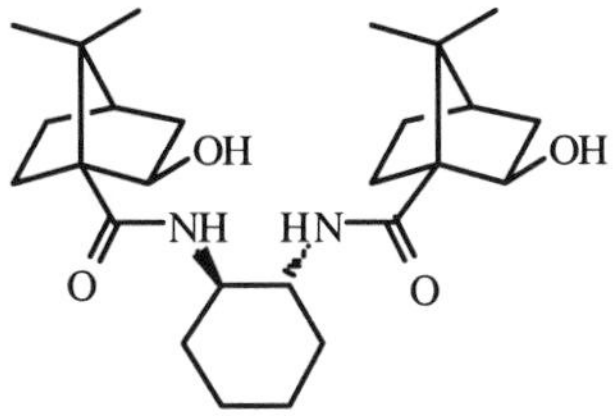

(75)

Bis(oxazolinyl) ligands have also been employed in asymmetric cyanation reactions. A combination of 12 mol% (76) and 20 mol% (77) provides cyanide and carbonyl binding sites to give trimethylsilylcyanohydrins in 24-94% yield and 52-95% ee, aliphatic aldehydes are better substrates than benzaldehyde or unsaturated aldehydes under these conditions (E.J. Corey and Z. Wang, *Tetrahedron Lett.*, 1993, **34**, 4001). The $AlCl_3$ and $YbCl_3$ complexes of (40, R = Pr^i) also catalyse the addition of TMS-CN to aldehydes (84-100% yield, 44-96% ee and 60-96% yield, 45-75% ee) (I. Iovel *et al.*, *Tetrahedron: Asymmetry*, 1997, **8**, 1279; H.C. Aspinall, N. Greeves and P.M. Smith, *Tetrahedron Lett.*, 1999, **40**, 1763).

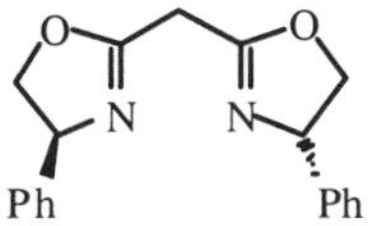

(76) 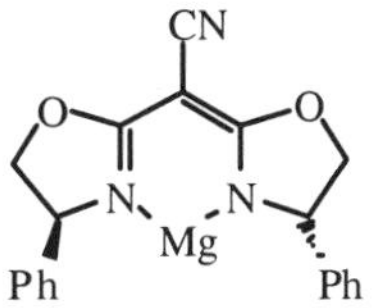 (77) 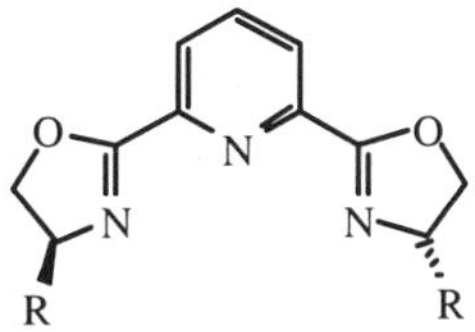 (40)

Chiral phosphine oxides (78) to (82) have also been used as ligands in asymmetric cyanation reactions. A samarium(III) complex of (78) gives 29-90% ee with aryl aldehydes (W.-B. Yang and J.-M. Fang, *J. Org. Chem.*, 1998, **63**, 1356) whilst $Ti(OPr^i)_4$ and (79) provide variable selectivity (15-98% ee, 70-95% yield) (J.-M. Brunel, O. Legrand and G. Buono, *Tetrahedron: Asymmetry*, 1999, **10**, 1979). The chloroaluminium complex (80) contains both Lewis acidic and Lewis basic centres and catalyses TMS-CN additions in the presence of 36 mol% tributylphosphine oxide (86-100% yield, 83-98% ee) (Y. Hamashima *et al.*, *J. Am. Chem. Soc.*, 1999, **121**, 2641). Carbohydrate-derived ligands (81) and (82) form complexes with Et_2AlCl and $Ti(OPr^i)_4$ respectively and can be used without a phosphine oxide additive for additions to aldehydes (82-98% yield, 70-80% ee) and aryl alkyl ketones (72-92% yield, 69-95% ee) (M. Kanai, Y. Hamashima and M. Shibasaki, *Tetrahedron Lett.*, 2000, **41**, 2405; Y. Hamashima, M. Kanai and M. Shibasaki, *J. Am. Chem. Soc.*, 2000, **122**, 7412).

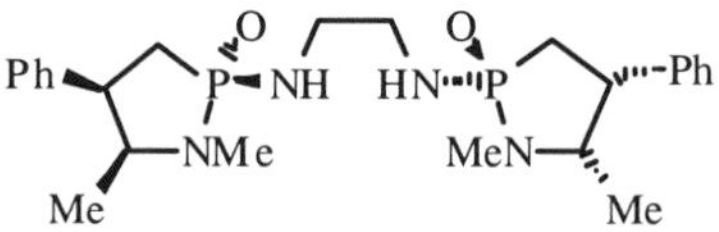

(78) 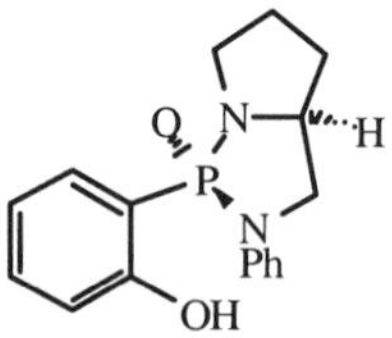 (79)

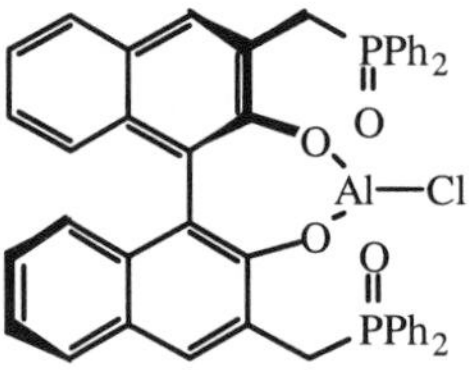

(80) 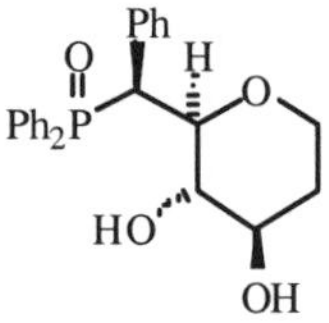 (81) 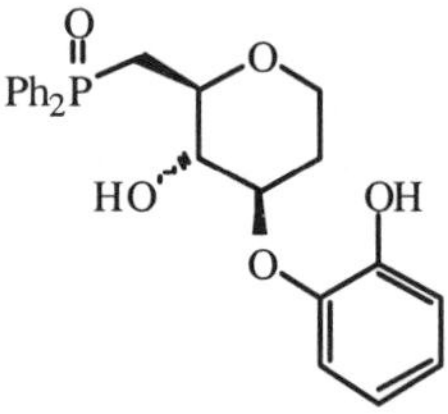 (82)

The complex formed between $Y_5O(OPr^i)_{13}$ and a 1,3-diferrocenyl-1,3-diketone (0.2-1 mol%) provides trimethylsilylcyanohydrins in high yields (>95%) but with variable selectivity (10-91% ee), again electron deficient aldehydes are poor substrates, as are aliphatic aldehydes which also give products with the opposite absolute configuration to those obtained from aryl aldehydes (A. Abiko and G. Wang, *J. Org. Chem.*, 1996, **61**, 2264; *idem, Tetrahedron*, 1998, **54**, 11405).

(ii) Additions to imines

The Strecker reaction provides a synthetic route to α-amino acids, based on hydrolysis of the α-aminonitrile species generated by addition of cyanide to imines. Asymmetric variants have employed a range of catalysts (L. Yet, *Angew. Chem. Int. Ed.*, 2001, **40**, 875).
Cyclic dipeptide (83), derived from phenylalanine and a lower homologue of arginine, provides variable selectivity (<10->99% ee, 71-97% yield) in the hydrocyanation of *N*-benzhydrylimines (M.S. Iyer *et al.*, *J. Am. Chem. Soc.*, 1996, **118**, 4910) with bicyclic guanidine (84) giving more consistent results (80-99% yield, 50-88% ee) (E.J. Corey and M.J. Grogan, *Org. Lett.*, 1999, **1**, 157).

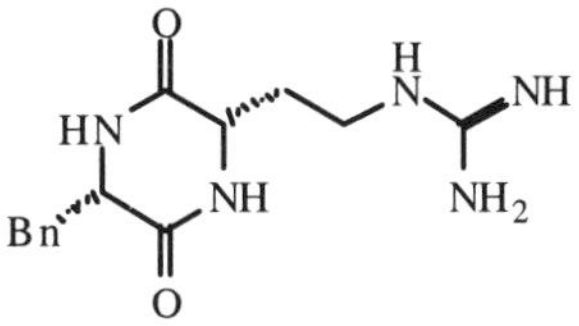

(83) 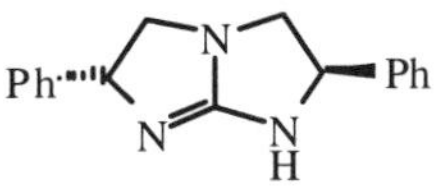 (84)

The chloroaluminium complex of salen ligand (28) has been employed in the reactions of HCN with aryl and aliphatic imines (69-99% yield, 37-95% ee), subsequent addition of trifluoroacetic anhydride leads to isolation of the trifluoroacetamido products (M.S. Sigman and E.N. Jacobsen, *J. Am. Chem. Soc.*, 1998, **120**, 5315). This reaction protocol was also used in additions mediated by (85), which provided improved selectivity (77-97% ee, 65-99% yield) (M.S. Sigman, P. Vachal and E.N. Jacobsen, *Angew. Chem. Int. Ed.*, 2000, **39**, 1279). A combination of related peptido salicylimine (86) and $Ti(OPr^i)_4$ (10 mol%) promotes the reaction of *N*-benzhydrylimines with TMS-CN (61-97% yield, 76-97% ee) (C.A. Krueger *et al.*, *J. Am. Chem. Soc.*, 1999, **121**, 4284; J.R. Porter *et al.*, *J. Am. Chem. Soc.*, 2000, **122**, 2657) whilst $Ti(OPr^i)_4$ and binol catalyse the addition to *N*-benzylbenzaldimine (60% yield, 30% ee) (M. Mori, H. Imma and T. Nakai, *Tetrahedron Lett.*, 1997, **38**, 6229).

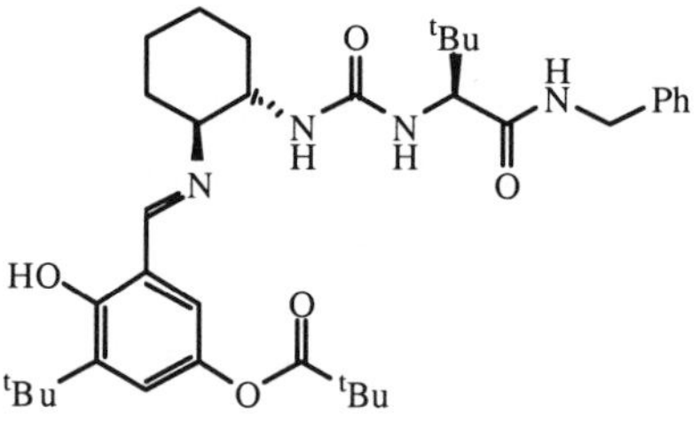

(85) 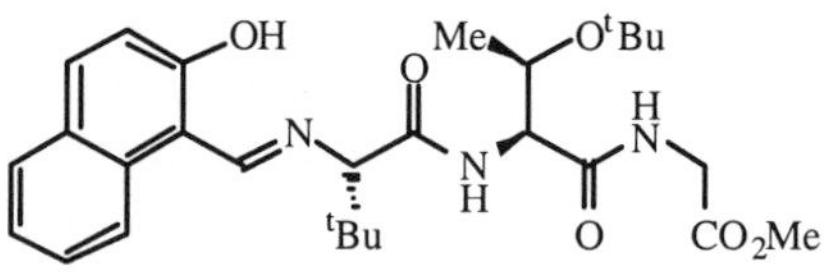 (86)

Bifunctional catalyst (80) has been employed for the trimethylsilyl-cyanation of aryl, unsaturated and aliphatic *N*-fluorenylimines (66-97% yield, 70-96% ee) (M. Takamura *et al., Angew. Chem. Int. Ed.*, 2000, **39**, 1650) whilst Bu_3SnCN was used as the cyanide source for additions to 2-hydroxyphenylaldimines mediated by 5 mol% of a combination of $Zr(O^tBu)_4$, 6,6'-(dibromo)binol, 3,3'-(dibromo)binol and *N*-methyl-imidazole (55-98% yield, 74-92% ee) (H. Ishitani, S. Komiyama and S. Kobayashi, *Angew. Chem. Int. Ed.*, 1998, **37**, 3186; H. Ishitani *et al., J. Am. Chem. Soc.*, 2000, **122**, 762). The latter catalyst system is also effective for the three-component Strecker reaction of HCN, 2-amino-phenols and aldehydes (76-100% yield, 84-94% ee).
Jacobsen has also employed catalyst (85) for the hydrocyanation of ketoimines (45-100% yield, 41-95% ee) (P. Vachal and E.N. Jacobsen, *Org. Lett.*, 2000, **2**, 867) whilst 10 mol% $Ti(OPr^i)_2$.binol and 20 mol% TMEDA promote the reaction of TMS-CN and the *N*-benzylimine of acetophenone (80% yield, 56% ee) (J.J. Byrne *et al., Tetrahedron Lett.*, 2000, **41**, 873).

6. Nitroalkane anions

Although the addition of nitroalkane anions to aromatic aldehydes is frequently accompanied by spontaneous dehydration to give the corresponding nitroalkene, nitro alcohols can be isolated from the Henry reaction of aliphatic aldehydes, resulting in broad synthetic utility (F.A. Luzzio, *Tetrahedron*, 2001, **57**, 915).
Asymmetric catalysis of this process has been achieved by Shibasaki and co-workers using their heterobimetallic-binol complexes (M. Shibasaki, H. Sasai and T. Arai, *Angew. Chem. Int. Ed. Engl.*, 1997, **36**, 1236). Initial studies provided chiral alcohols (79-91% yield, 73-90% ee) from

the reaction of nitromethane and aldehydes in the presence of $La_3(O^tBu)_9$, binol and LiCl (H. Sasai *et al.*, *J. Am. Chem. Soc.*, 1992, **114**, 4418). Improved results were later obtained using $LaLi_3(6,6'\text{-}(Et_3SiC{\equiv}C)_2binol)_3$ and $SmLi_3(6,6'\text{-}(Et_3SiC{\equiv}C)_2binol)_3$. Diastereoselective and enantioselective nitroalkane additions occurred in the presence of 3.3 mol% of the lanthanum catalyst (70-97% yield, 89-93% ds *syn*, 93-97% ee) whilst 5-8 mol% of the samarium species provided optimum results (52-58% yield, 74-95% ee) for the addition of nitromethane to α,α-difluoro aldehydes (H. Sasai *et al.*, *J. Org. Chem.*, 1995, **60**, 7388; K. Iseki *et al.*, *Tetrahedron Lett.*, 1996, **37**, 9081). Chiral guanidine (87) has also been employed in catalytic asymmetric Henry reactions (R. Chinchilla, C. Nájera and P. Sánchez-Aguilló, *Tetrahedron: Asymmetry*, 1994, **5**, 1393) although only moderate results were obtained (31-33% yield, 33-54% ee).

Me NEt2 Me
Ph N N(H) Ph
(87)

$$RCHO + CH_3NO_2 \xrightarrow{(87)} R\text{-}CH(OH)\text{-}CH_2NO_2$$

The addition of nitromethane to *N*-phosphinoylimines is promoted by 20 mol% $YbK.(binol)_3$ (41-93% yield, 69-91% ee) (K. Yamada *et al.*, *Angew. Chem. Int. Ed.*, 1999, **38**, 3504). The reaction is restricted to non-enolisable aldimines as the *N*-phosphinoylimines could not prepared from enolisable aldehydes.

7. Miscellaneous nucleophiles

Hydrophosphonylation reactions are also catalysed by heterobimetallic-binol complexes. $LiAl(binol)_2$ (10 mol%) promotes the addition of dimethylphosphite to aryl and unsaturated aldehydes (53-95% yield, 55-90% ee), with $LaLi_3(binol)_2$ extending the scope of the reaction to include aliphatic aldehydes (63-93% yield, 36-95% ee) (T. Arai *et al.*, *J. Org. Chem.*, 1996, **61**, 2926; H. Sasai *et al.*, *Tetrahedron Lett.*, 1997, **38**, 2717). *N*-Benzhydrylimines react in the presence of 5-20 mol% $LaK_3(binol)_3$ to give the corresponding α-aminophosphonates (62-97% yield, 49-96% ee) (H. Sasai *et al.*, *J. Org. Chem.*, 1995, **60**, 6656).

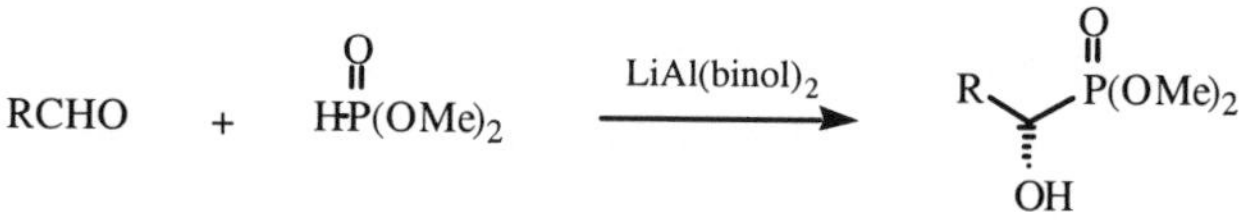

$$RCHO + HP(O)(OMe)_2 \xrightarrow{LiAl(binol)_2} RCH(OH)P(O)(OMe)_2$$

The benzoin reaction is subject to catalysis by thiazolium and triazolium salts, as well as enzymes, and a variety of chiral salts have been employed for the parent reaction. A combination of triethylamine and (88) gives 18-45% yield and 16-30% ee, whilst polycyclic thiazolium (89) gives improved yields, but lower selectivity (31-100% yield, 1-26% ee) (C.A. Dvorak and V.H. Rawal, *Tetrahedron Lett.*, 1998, **39**, 2925; A.U. Gerhard and F.J. Leeper, *Tetrahedron Lett.*, 1997, **38**, 3615). Triazolium salts (90, R = Ph, Me) and (91) provide the opposite sense of absolute asymmetric induction to (89) and higher enantioselectivity (11-47% yield, 40-83% ee) but poor results are obtained with menthyl derivative (92) (20% yield, 35% ee) (R.L. Knight and F.J. Leeper, *J. Chem. Soc., Perkin Trans. 1*, 1998, 1891; W. Tagaki, Y. Tamura and Y. Yano, *Bull. Chem. Soc. Jpn.*, 1980, **53**, 478).

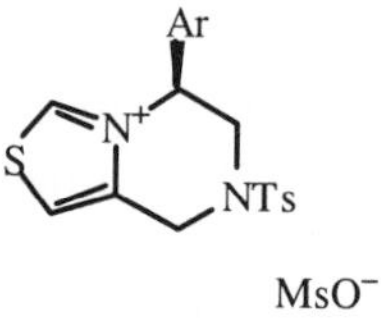

Ar, NTs, MsO^-

(88)

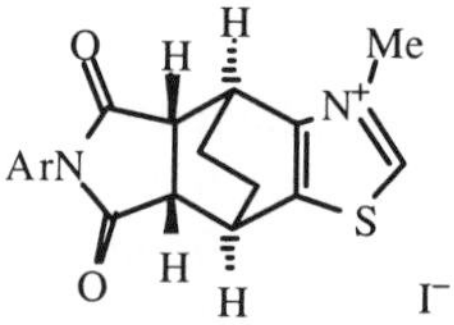

ArN, Me, I^-

(89)

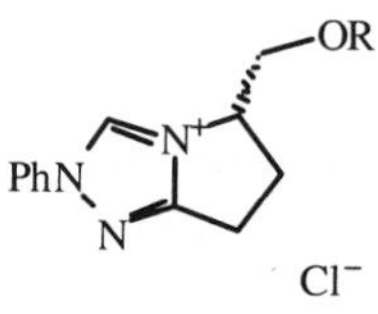

PhN, OR, Cl^-

(90)

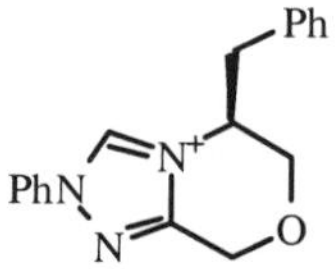

PhN, Ph

(91)

Me, ClO_4^-

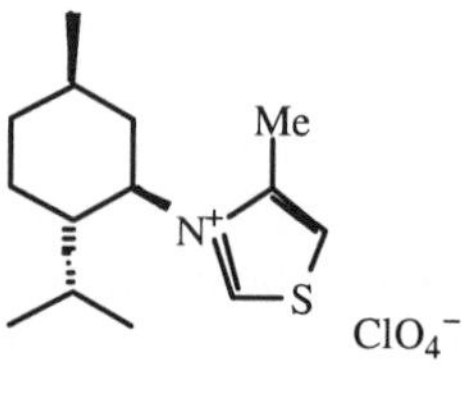

(92)

Much broader reactivity was observed with 1.25 mol% of triazolium (93) in the presence of potassium carbonate, with a number of aryl aldehydes reacting in 22-72% yield and 20-86% ee (D. Enders, K. Breuer and J.H. Teles, *Helv. Chim. Acta*, 1996, **79**, 1217).

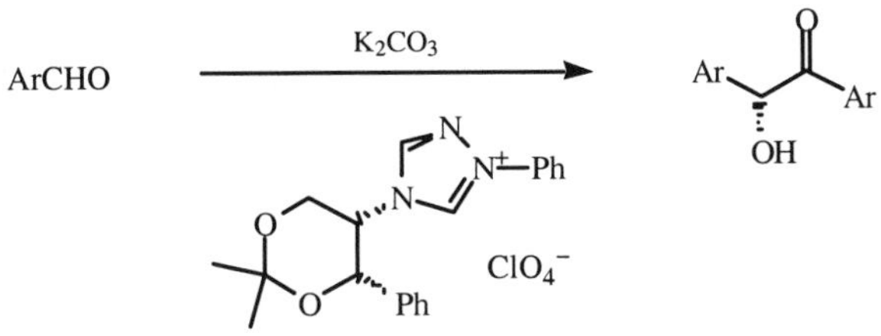

(93)

Limited success has been achieved in the asymmetric Friedel-Crafts reactions of activated aromatic substrates. Camphor-derived dibornacyclopentadienyl zirconium catalyst (94) has been employed for the addition of ethyl pyruvate to 1-hydroxynaphthalene (56% yield, 84% ee) whilst 5-15 mol% $Ti(OPr^i)_4$ and 6,6'-(dibromo)binol promote the *para*-selective (75-89%) addition of fluoral to phenylalkyl ethers (89-90% yield, 54-90% ee) (G. Erker and A.A.H. van der Zeijden, *Angew. Chem. Int. Ed. Engl.*, 1990, **29**, 512; A. Ishii, V.A. Soloshonik and K. Mikami, *J. Org. Chem.*, 2000, **65**, 1597).

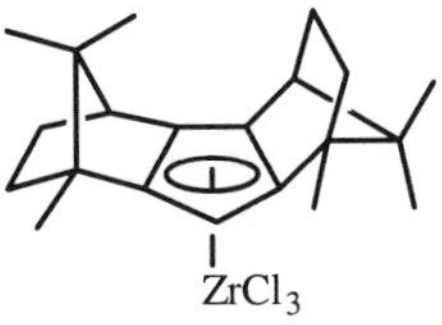

(94)

Fu and co-workers have developed the earlier work of Pracejus to form α-chiral esters (80-97% yield, 68-80% ee) by addition of methanol to ketenes in the presence of 10 mol% of azaferrocene catalyst (95) (B.L. Hodous, J.C. Ruble and G.C. Fu, *J. Am. Chem. Soc.*, 1999, **121**, 2637).

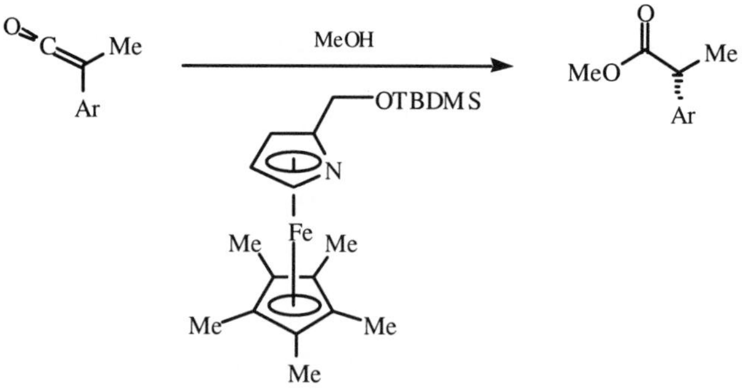

MeOH

OTBDMS

(95)

8. Overview

In common with the diverse reactivity of carbonyl substrates, a vast array of strategies for the corresponding catalytic asymmetric processes have been explored. The ready complexation and concomitant activation of carbonyl functionalities has lead to many of these approaches focussing on the use of chiral Lewis acids, although *in situ* modification of the nucleophile has also provided highly enantioselective additions. It is interesting, however, to note that a limited number of ligand frameworks dominate the field with 1,1'-binaphthyl species and, more recently, bis(oxazolinyl) complexes showing the greatest versatility.

Second Edition of Rodd's Chemistry of Carbon Compounds,
Volume V, Topical Volumes
Asymmetric Catalysis, edited by M.Sainsbury

Chapter 6

CONJUGATE ADDITION REACTIONS

Nicholas C. O. Tomkinson

1. Introduction

The conjugate addition reaction is a fundamental process in organic chemistry. The reaction involves the addition of nucleophiles to acceptor substituted double and triple bonds and results in the formation of a new carbon-carbon, carbon-nitrogen, carbon-oxygen or carbon-sulfur bond. The reaction often leads to the generation of one, two or even three new stereogenic centres and so considerable effort has been made to develop catalytic asymmetric methods, particularly under the influence of an external chiral ligand or chiral catalyst. Indeed, over the past ten years there has been a plethora of such new methods, and this chapter adumbrates some of the major recent advances in this area. It also provides an overview of the arsenal of catalysts available to carry out this important class of synthetic reaction.
There are several strategies that may be considered for the stereoselective conjugate addition reaction. Diastereoselective conjugate addition reactions to chiral Michael acceptors, along with the addition of stoichiometric chirally modified nucleophiles to prochiral Michael acceptors has been well documented elsewhere (B. E. Rossiter and N. M. Swingle, *Chem. Rev.* 1992, **92**, 771). The catalytic asymmetric conjugate addition is a far more challenging problem that is a very active area of research. Progress in the subject has been extensively reviewed recently (M. Kanai and M. Shibasaki, in *Catalytic Asymmetric Synthesis*; 2nd edn., I. Ojima Ed. Wiley-VCH: Weinheim, 2000, 569; M. Yamaguchi, in *Comprehensive Asymmetric Catalysis*, E. N. Jacobsen, A. Pfaltz and H. Yamamoto Eds. Springer: Berlin, 1999, **3**, 1121; K. Tomioka and Y. Nagaoka, in *Comprehensive Asymmetric Catalysis*, E. N. Jacobsen, A. Pfaltz and H. Yamamoto Eds. Springer: Berlin, 1999, **3**, 1105; A. Alexakis, in *Transition Metal Catalyzed Reactions*, S. I. Murahashi and S. G. Davies Eds. IUPAC Blackwell Science: Oxford, 1999, 303; N. Kraus, *Angew. Chem., Int. Ed. Engl.*,1998, **37**, 283; A. Alexakis, in *Transition Metals For Organic Synthesis*, M. Beller and C. Bolm Eds. Wiley-VCH:

Weinheim, 1998, **1**, 504; N. Kraus and A. Gerold, *Angew. Chem., Int. Ed. Engl.*,1997, **36**, 186). In this chapter only catalytic asymmetric approaches to conjugate addition reactions will be covered. Enantioselective additions using stoichiometric amounts of chiral reagent will not be presented.
In considering the enantioselective conjugate addition reaction it is necessary to consider the stereochemical consequences of the reaction and for this purpose the transformation can be subdivided into two classes: Firstly, the addition of a prochiral donor to an acceptor, where the asymmetric centre is formed on the donor atom. Secondly, the addition of a donor to a prochiral acceptor, where the two enantiofaces of the acceptor must be discriminated by the donor atom and the new chiral centre is formed on the acceptor.
On addition to a prochiral acceptor atom there are two diastereofaces of the double bond, known as Si and Re and the reagent needs to discriminate between them. Approach of a chiral reagent selectively to either of these faces will result in a stereoselective transformation. With cyclic substrates the double bond is held in the s-*trans* conformation due to the geometrical constraints of the ring, however, with the conjugate addition to acyclic substrates the situation is further complicated by the possibility of the double bond adopting either an *s-cis* or *s-trans* conformation with respect to the acceptor substituent (e.g. a carbonyl group). On addition of a nucleophile to an α,β-unsaturated ester, the acceptor can adopt sixteen possible conformations when complexed to a chiral metal species. This fact shows the enormity of the task presented in developing asymmetric catalysts for this class of substrate, and it pays great tribute to the ingenuity of chemists who are attempting to solve this significant problem!

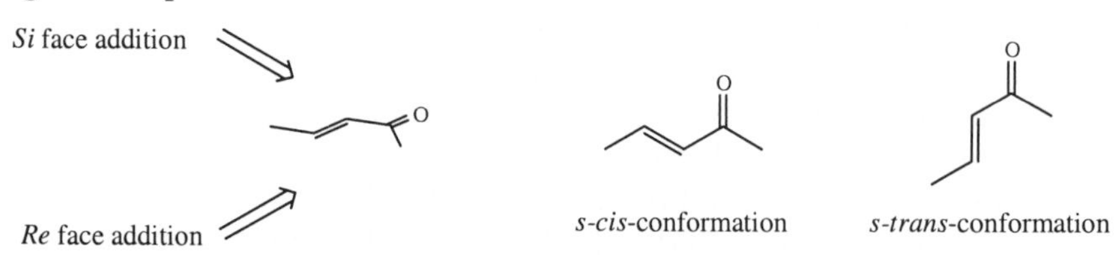

This review is divided into ten sections depending on the type of nucleophile being added to the Michael acceptor. These are: alkyl, aryl and alkenyl, hydride, carbonyl stabilised anion, aliphatic nitro, ester nitrile, silylketene acetal, thiol and nitrogen based nucleophiles. Finally, a series of natural product syntheses that contain a catalytic asymmetric conjugate addition in the reaction scheme are presented.

2. Alkyl Addition

The most active area of research in the development of methods for the catalytic asymmetric conjugate addition reaction is that of the transfer of alkyl groups. This involves the development of external chiral ligands that are used in conjunction with sub-stoichiometric amounts of copper, nickel or cobalt in the delivery of Grignard, organozinc, alkyllithium and organoaluminium reagents to the Michael acceptor. The ligands that have been developed are either mono-, di- or tridentate containing a mixture of nitrogen, oxygen, sulfur and phosphorus donor atoms. Representative examples of each of these classes of ligand and metal are outlined below.

(a) Copper in conjunction with Grignard reagents.

The first example of a conjugate addition of an alkyl group was reported in 1901 (V. Grignard, *Ann. Chim.*, 1901, **24**, 433). This involved the addition of methyl magnesium iodide to the α-ethylidene ester (1).

O CO_2Et i) MeMgI ii) H_3O^+ O CO_2Et

(1)

This was later followed by a report by Kharasch (M. S. Kharasch and P. O. Tawney, *J. Am. Chem. Soc.*, 1941, **63**, 2308; M. S. Kharasch, J. W. Hancock and P. O. Tawney, *J. Org. Chem.* 1956, **21**, 322) that showed that 1,4-addition could be promoted over 1,2-addition by the addition of catalytic amounts of copper to the reaction.

O MeMgBr 1 mol% CuCl O

(2) (3)

Thus, methyl magnesium bromide adds to isophorone (2) in the presence of 1 mol% copper chloride to give the symmetrical cyclohexan-1-one (3) in 82% yield. Since these early reports many external chiral ligands have been developed for the addition of Grignard reagents to Michael acceptors, many of which can be added in a catalytic amount.
The sulfur atom has a very high affinity for copper and the resulting complex has good stability. This has resulted in the use of numerous thiocuprates as catalysts, these can be generated from either Grignard reagents or organolithium species, and a copper(I) source in the presence of the lithium or magnesium salt of a chiral thiol ligand.
(i) Monodentate S Ligands: The monodentate sulfur containing ligand (4) derived from glucofuranose was reported by Spescha to catalyse the conjugate addition of butyl magnesium chloride to 2-cyclohexenone in the presence of copper iodide and butyllithium, albeit in only 10% e.e. An extensive study of reaction conditions and the use of additives to promote the reaction led to the use of the thiol (4) in the presence of the radical scavenger TEMPO and $[Cu(Bu_3P)I]_4$ as the copper(I) source. Using this system an e.e. of up to 60% was realised for this transformation (M. Spescha and G. Rhis, *Helv. Chim. Acta*, 1993, **76**, 1219).

(4), BuMgBr, TEMPO, BuLi, $[Cu(Bu_3P)I]_4$, Et_2O, -78°C

(4)

60% e.e., 92% yield

(ii) Bidentate N,S ligands:

MeMgI, (5) 3 mol%, Et_2O, 0°C

(5) (6)

76% e.e., 80% yield

van Koten has developed a series of bidentate S,N donor ligands that are efficient homogeneous catalysts for the asymmetric conjugate addition reaction. The arenethiolatocopper(I) complex (5) promoted the addition of methyl magnesium iodide to 4-phenylbut-3-en-2-one (6) in up to 76%

e.e. (M. van Klaveren et al., *Tetrahedron Lett.*, 1994, **35**, 6135; F. Lambert et al., *Tetrahedron: Asymmetry*, 1991, **2**, 1097).
The importance of information regarding the structure of the copper species in the solid state, as a means of defining the constitution of the transition state for a reaction in solution, was stressed by Knotter et al. (D. M. Knotter et al., *J. Am. Chem. Soc.*, 1992, **114**, 3400). Such information led to the recognition of the highly ordered species (7) and assisted in the design of subsequent ligands for the addition process.

(7)

Pfaltz applied the findings of van Koten in the design of a series of mercaptoaryl-oxazoline ligands for the addition of Grignard reagents to cyclic enones. Thus, the valinol derived cuprate (8) gives up to 87% e.e. in the addition of isopropyl magnesium chloride to 2-cycloheptenone (Q. L. Zhou and A. Pfaltz, *Tetrahedron,* 1994, **50**, 4467).

(8)

^{i}PrMgCl, (8) 5 mol%, THF, -78°C

n = 0 37% e.e., 43% yield
n = 1 72% e.e., 71% yield
n = 2 87% e.e., 55% yield

Seebach has shown that the bidentate S,N donor ligand (9), derived from

(9)

BuLi/CuI then BuMgCl, (9) 5 mol%, THF, -78°C

n = 0 40% e.e., 50% yield
n = 1 80% e.e., 77% yield
n = 2 84% e.e., 72% yield

the hugely successful TADDOL ligand is highly effective in the addition of Grignard reagents to cyclic enones; the reaction with 2-cycloheptenone proceeding in 72% yield and 84% e.e. (D. Seebach, G. Jaeschke, A. Pichota and L. Audergon, *Helv. Chim. Acta*, 1997, **80**, 2515). Interestingly, Seebach has recently reported that the actual nature of the catalyst may not be the discrete bidentate N,S copper chelate as originally thought. X-ray analysis, molecular diffusion experiments and NOSEY NMR studies all suggest the actual active copper species in the reaction is the tetra-nuclear species (10), which uses an unprecedented monodentate complexation mode of the ligand (A. Pichota et al., *Angew. Chem., Int. Ed. Engl.*, 2000, **39**, 153).

(10)

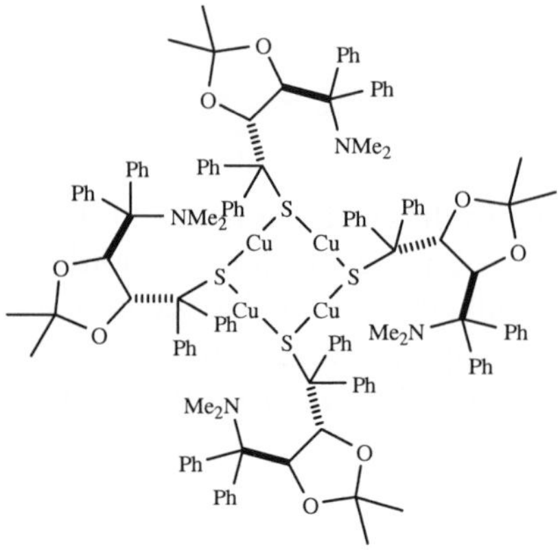

(iii) Bidentate N,N ligands: One of the first excellent examples of a catalytic asymmetric conjugate addition reaction was reported by Lippard in 1988 (G. M. Villacorta, C. P. Rao and S. J. Lippard, *J. Am. Chem. Soc.*, 1988, **110**, 3175; K. H. Ahn, B. Klassen and S. J. Lippard, *Organometallics*, 1990, **9**, 3178). Using the N,N bidentate chiral cuprate

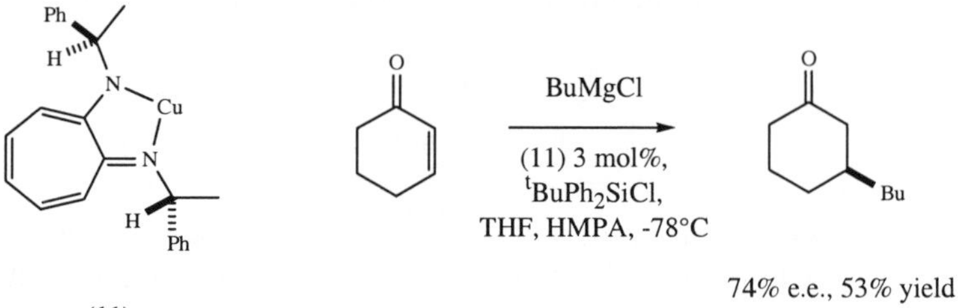

BuMgCl

(11) 3 mol%, ${}^tBuPh_2SiCl$, THF, HMPA, -78°C

74% e.e., 53% yield

(11)

complex (11), derived from the corresponding amide, copper bromide and lithium bistrimethylsilylamide, the conjugate addition of butyl magnesium chloride to 2-cyclohexenone proceeded in 96% yield and in 20% e.e. Appreciable levels of enantioselectivity (up to 74% e.e.) were obtained in the presence of HMPA and, especially, bulky silyl reagent additives. Unfortunately, this ligand system did not afford high e.e. values for a broad range of other Grignard reagents and substrates.

(iv) Bidentate P,N ligands: The P,N bidentate chiral ferrocenyl phosphine-oxazoline ligand (12) has been used to promote the conjugate addition of butyl magnesium chloride to a range of both cyclic and acyclic enones (E. L. Strangeland and T. Sammakia, *Tetrahedron*, 1997, **53**, 16503). It appears that the additional planar chirality present in (12), imparted by the ferrocenyl template, was essential for high levels of asymmetric induction. The phenyl derived phosphine-oxazoline ligand (13) gave no asymmetric induction.

(12) Ph, N, O, PPh_2, Fe

O — BuMgCl, Ligand 12 mol%, CuI 6 mol%, ether, -78°C — O, Bu

(12) 83% e.e., 97% yield
(13) 0% e.e., 6% yield

(13) O, N, Ph, PPh_2

Ph, O — BuMgCl, (12) 12 mol%, CuI 6 mol%, ether, -78°C — Ph, Bu, O

(stereochemistry undetermined)

81% e.e., 61% yield

(v) Bidentate S,O ligands: There are several S,O bidentate ligands for the copper catalysed addition of Grignard reagents to enones. The binol derived hydroxy thiol (14) was reported to catalyse the addition of Grignard reagents in only very low e.e. (<20%) to a variety of cyclic and acyclic enones (S. M. W. Bennett et al., *J. Chem. Soc., Perkin Trans. 1*, 1999, 3127). Seebach has reported the readily accessible TADDOL derived hydroxy thiol (15). With it the additions of butyl magnesium chloride to 2-cyclohexenone and 2-cycloheptenone showed stereoselectivities of 64% and 74% e.e. respectively (D. Seebach, G.

Jaeschke, A. Pichota and L. Audergon, *Helv. Chim. Acta*, 1997, **80**, 2515). Both the hydroxy thiol (15) and the amino thiol ligand (16), which have the same absolute configuration, are equally effective as catalysts of conjugate additions of Grignard reagents to enones. Surprisingly, however, these ligands induce opposite absolute configurations in the adducts. An explanation of this phenomenon will undoubtedly provide insight into the mechanistic aspects of this reaction and aid in the design of future ligands.

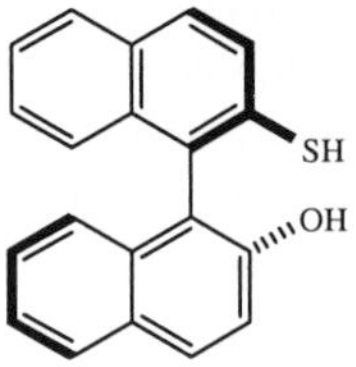

(14) 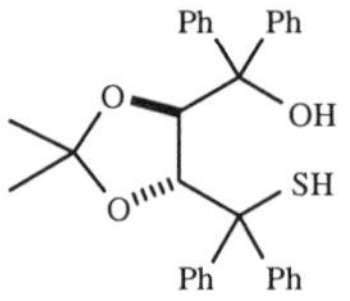 (15) 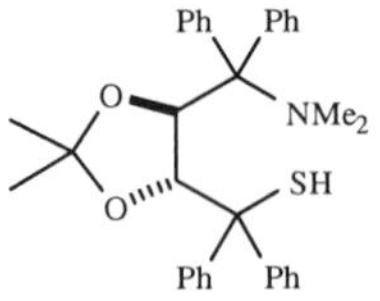 (16)

(vi) Bidentate P,O Ligands: The proline derived bidentate amido-phosphine (17) was developed by Tomioka around the concept of metal-differentiating co-ordination (M. Kanai, Y. Nakagawa and K. Tomioka, *Tetrahedron*, 1999, **55**, 3843; M. Kanai and K. Tomioka, *Tetrahedron Lett.*, 1995, **36**, 4275).

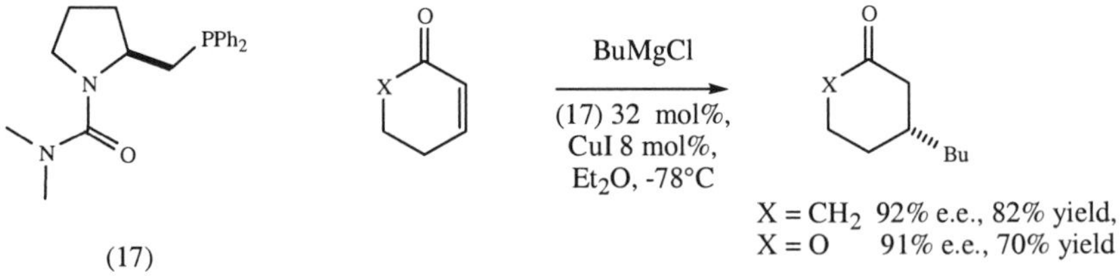

(17)

BuMgCl; (17) 32 mol%, CuI 8 mol%, Et_2O, -78°C

$X = CH_2$ 92% e.e., 82% yield,
$X = O$ 91% e.e., 70% yield

It was proposed that in the conjugate addition of butyl magnesium chloride to 2-cyclohexenone in the presence of copper(I) iodide, the carbonyl oxygen and the phosphorus atom of the ligand selectively co-ordinate to the magnesium and copper, respectively, of the reactive species. Despite the exceedingly high e.e. values reported for this system, the high loading of the external chiral ligand of 32 mol% is a major drawback to this process.

(2) Copper in conjunction with organozinc reagents.

(i) Monodentate P Ligands: One of the most successful classes of ligand for the catalytic asymmetric conjugate addition reaction for the introduction of alkyl groups comes from the use of organozinc reagents in the presence of sub-stoichiometric amounts of a copper(I) species and a monodentate phosphorus ligand.
The first major contribution in this area came from the group of Alexakis, who showed that an organozinc species effectively *trans*-metallated with copper to give the active organo copper catalyst core. The advantage of this process over the use of Grignard reagents is that the background reaction and the competing 1,2-addition are negligible. With 20 mol% of the monodentate ligand (18) in the presence of 10 mol% copper(I) iodide at room temperature, the addition of diethylzinc to 2-cyclohexenone proceeded in 70% yield and in 32% e.e. (A. Alexakis, J. Frutos and P. Mangeney, *Tetrahedron: Asymmetry*, 1993, **4**, 2427).

Ph O P N N O O Et

Et_2Zn, (18) 20 mol%, CuI 10 mol%, PhMe, r.t.

32% e.e., 70% yield

(18)

Alexakis has continued the development of these monodentate phosphorus ligands and generated a number of cyclic phosphates, based around the tartrate backbone and other commercially available chiral alcohols (A. Alexakis, J. Vastra, J. Burton and P. Mangeney, *Tetrahedron: Asymmetry*, 1997, **8**, 3193), as well as around the TADDOL backbone (A. Alexakis et al., *Tetrahedron Lett.*, 1998, **39**, 7869; A. Alexakis et al., *Eur. J. Org. Chem.*, 2000, 4011). Both the tartrate (19) and the TADDOL (20) based ligands catalysed the asymmetric conjugate addition of diethylzinc to enones, and were complimentary as regards the substrates tolerated. The tartrate ligand (19), present in 0.5 mol%, catalysed the addition of diethylzinc to acyclic enones in up to 65% e.e., but it was ineffective for cyclic substrates. The TADDOL based ligands (20) catalysed the addition of diethylzinc to cyclic enones in up to 96% e.e. but were less efficient with acyclic substrates.

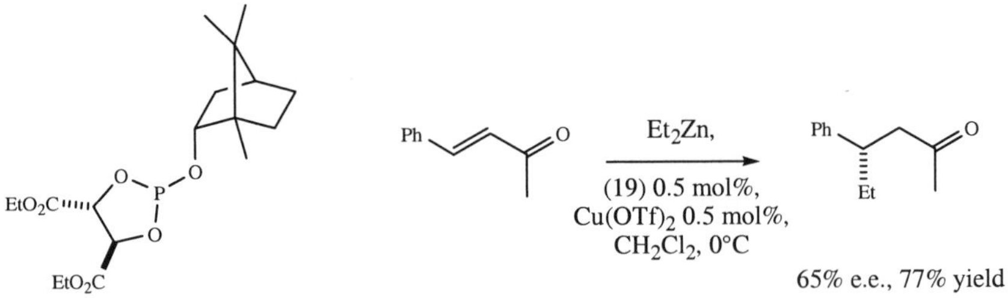

(19)

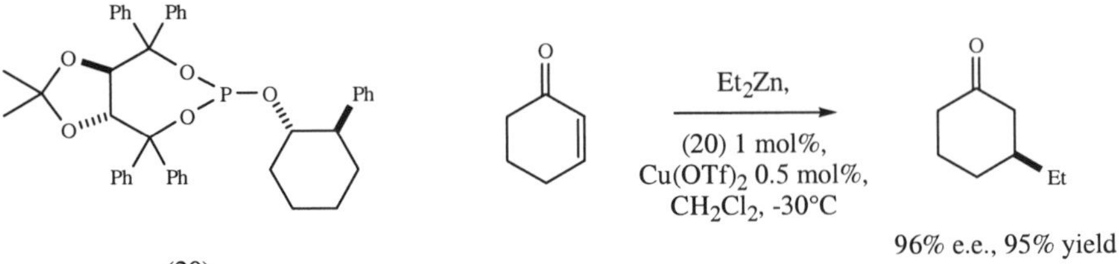

Et$_2$Zn, (19) 0.5 mol%, Cu(OTf)$_2$ 0.5 mol%, CH_2Cl_2, 0°C

65% e.e., 77% yield

(20)

Et$_2$Zn, (20) 1 mol%, Cu(OTf)$_2$ 0.5 mol%, CH_2Cl_2, -30°C

96% e.e., 95% yield

Muller has developed monodentate phosphorus amidite ligands derived from the axially chiral biaryl (-)-(a*R*)-[1,1'-binaphthalene]-8,8'-diol and showed that (21) was an effective asymmetric catalyst in a number of transformations, including the borane reduction of acetophenone which proceeded in 96% e.e. In the presence of 6 mol% (21) and copper(II) triflate the addition reaction of diethylzinc to 2-cyclohexenone proceeded in up to 50% e.e., and addition to the acyclic enone, chalcone, proceeded in up to 31% e.e. (P. Muller, P. Nury and G. Bernardinelli, *Helv. Chim. Acta*, 2000, **83**, 843).

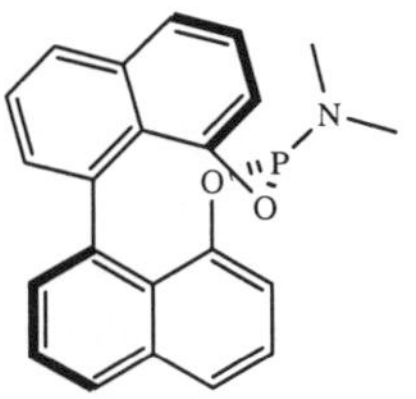

(21)

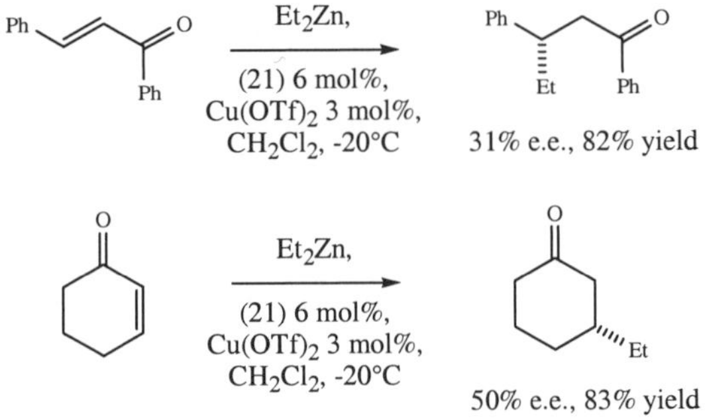

Et$_2$Zn, (21) 6 mol%, Cu(OTf)$_2$ 3 mol%, CH_2Cl_2, -20°C

31% e.e., 82% yield

Et$_2$Zn, (21) 6 mol%, Cu(OTf)$_2$ 3 mol%, CH_2Cl_2, -20°C

50% e.e., 83% yield

Recently, Faraone has reported a new monodentate phosphoramidite ligand (22) based upon 8-substituted quinolines and (*S*)-binaphthol, which catalysed the addition of diethylzinc to 2-cyclohexenone in up to 70% e.e. (C. G. Arena, G. Calabro, G. Francio and F. Faraone, *Tetrahedron: Asymmetry*, 2000, **11**, 2387).

N P Cl O O

(22)

Et_2Zn, (22) 6 mol%, $Cu(OTf)_2$ 3 mol%, PhMe, -15°C

70% e.e., 96% conversion

Feringa has realised some remarkable e.e. values using binaphthol derived phosphorus amidites. When 2-cyclohexenone was used as a substrate with the ligand (23), stereoselectivities of 60% e.e. were obtained with copper(II) triflate as the copper source. The efficiency rises to 83% e.e. when the Michael acceptor is 4,4-dimethyl-2-cyclohexenone (A. H. M. de Vries, A. Meetsma and B. L. Feringa, *Angew. Chem., Int. Ed. Engl.*, 1996, **35**, 2374). This catalyst system was also one of the first to give relatively high e.e. values for acyclic systems as well as cyclic Michael acceptors. For example, 83% e.e. was observed in the addition of diethylzinc to chalcone.

O P—N O

(23)

Et_2Zn, (23) 6.5 mol%, $Cu(OTf)_2$ 3 mol%, CH_2Cl_2, -15°C

83% e.e., 88% yield

Et_2Zn, (23) 6.5 mol%, $Cu(OTf)_2$ 3 mol%, CH_2Cl_2, -15°C

60% e.e., 78% yield

Chan has reported that the partially hydrogenated binaphthol derivative (24) catalysed the addition of diethylzinc to 2-cyclohexenone in up to

85% e.e. and 100% conversion in the presence of 5 mol% copper(II) triflate and 11 mol% ligand (24) (F. Y. Zhang and A. S. C. Chan, *Tetrahedron: Asymmetry*, 1998, **9**, 1179). This ligand was also highly effective over a range of other cyclic substrates.

Et_2Zn,
(24) 11 mol%,
$Cu(OTf)_2$ 5 mol%,
PhMe, -20°C

85% e.e., 100% conversion

(24)

Feringa further developed the ligand (23) by combing the C_2-symmetric bis(1-phenylethyl)amine, with an axially chiral binaphthol unit to form the phosphoramidite (25). This is currently the most versatile and generally applicable chiral ligand for the asymmetric conjugate addition of organozinc reagents to enones (B. L. Feringa, *Acc. Chem. Res.*, 2000, **33**, 346; L. A. Arnold et al., *Tetrahedron*, 2000, **56**, 2865; B. L. Feringa et al., *Angew. Chem., Int. Ed. Engl.*, 1997, **36**, 2620). It is essential that both the chirality of the binaphthol moiety and of the amine moiety are 'matched' otherwise there is a detrimental effect on the e.e. of the reaction.

Et_2Zn,
(25) 4 mol%,
$Cu(OTf)_2$ 2 mol%,
PhMe, -10°C

98% e.e., 94% yield

(25)

Et_2Zn,
(25) 6 mol%,
$Cu(OTf)_2$ 3 mol%,
PhMe, -10°C

97% e.e., 84% yield

Another advantage offered by dialkylzinc compounds over Grignard reagents is that they allow the introduction of reactive functional groups into the nucleophilic species. This was exemplified by Feringa, who showed that the highly functionalised bicyclic enones (26) could be synthesised in two steps from commercially available achiral enones in excellent e.e. (R. Naasz et al., *J. Am. Chem. Soc.*, 1999, **121**, 1104).

$[(EtO)_2CH(CH_2)_3]_2Zn$, (25) 4 mol%, $Cu(OTf)_2$ 2 mol%, PhMe, -20°C

HCl, THF

n=1 97% e.e., 62% yield
n=2 96% e.e., 50% yield
n=3 98% e.e., 42% yield

(26)

Feringa has also shown that the zinc enolate intermediate, derived from the conjugate addition of diethylzinc, can be trapped out with either aldehydes (B. L. Feringa et al., *Angew. Chem., Int. Ed. Engl.*, 1997, **36**, 2620), or allylic acetates, under palladium catalysed conditions (R. Naasz et al., *J. Am. Chem. Soc.*, 1999, **121**, 1104) to give highly functionalised products.

1. $Cu(OTf)_2$, (25), Et_2Zn
2. CH_2CHCH_2OAc, $Pd(PPh_3)_4$

1. CuCl, $PdCl_2$, DMF/H_2O, O_2
2. KO^tBu, THF

96% e.e., 56% yield

It has been shown by Feringa (J. P. G. Versleijen, A. M. van Leusen and B. L. Feringa, *Tetrahedron Letters*, 1999, **40**, 5803), Alexakis (A. Alexakis and C. Benhaim, *Org. Lett.*, 2000, **2**, 2579), and Sewald (N. Sewald and V. Wendisch, *Tetrahedron: Asymmetry*, 1998, **9**, 1341) that use of the phosphorus amidite (25) is not limited to the addition of

92% e.e., 95% yield — (27)

94% e.e., 100% yield — (28)

86% e.e., 100% conversion — (29)

dialkylzinc compounds to α,β-unsaturated carbonyls, the unsaturated nitro derivatives (27)-(29) also undergoing diethylzinc addition with excellent e.e. values.

Despite the huge success of the phosphorus amidite (25) in the asymmetric conjugate addition reaction with both cyclic and acyclic substrates, the ligand is rather ineffective in the addition of diethylzinc to 2-cyclopentenone (10% e.e., 75% yield). Greater success with this challenging substrate was observed by Feringa who has shown the utility of the monodentate phosphoramidite TADDOL derived ligand (30) (E. Keller et al., *Tetrahedron: Asymmetry*, 1998, **9**, 2409). Addition of diethylzinc to 2-cyclopentenone in the presence of 2.4 mol% (30) and 1.2 mol% copper(II) triflate, followed by trapping the resultant enolate, gave the adduct (31) in 37% e.e. This value was greatly improved by the addition of 4Å molecular sieves to the reaction mixture, which gave the product in up to 62% e.e. It was proposed that this effect could be due to the sieves removing trace amounts of water present in the reaction mixture. The water could form achiral zinc hydroxides that are detrimental to the selectivity of the reaction. An alternative explanation is that the reaction is taking place at the surface of the molecular sieve. Similar results were also observed using (30) in the additions to a series of other cyclic and acyclic Michael acceptors.

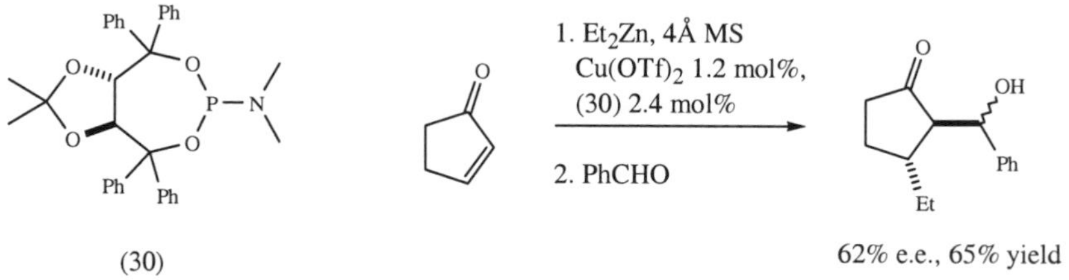
Ph
Ph
O
O
P—N
O
O
Ph
Ph
O
1. Et2Zn, 4Å MS
Cu(OTf)2 1.2 mol%,
(30) 2.4 mol%
2. PhCHO
O
OH
Ph
Et
62% e.e., 65% yield

(30)

(31)

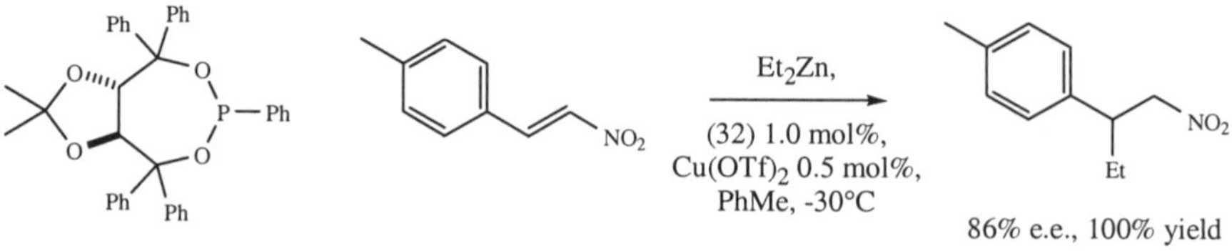
Ph
Ph
O
O
P—Ph
O
O
Ph
Ph
NO2
Et2Zn,
(32) 1.0 mol%,
Cu(OTf)2 0.5 mol%,
PhMe, -30°C
NO2
Et
86% e.e., 100% yield

(32)

Other monodentate TADDOL based systems that are highly effective in the asymmetric conjugate addition reaction include the phosphonate (32) developed by Alexakis (A. Alexakis and C. Benhaim, *Org. Lett.*, 2000, **2**, 2579). Amongst other substrates, this ligand was effective in the asymmetric conjugate addition reaction of dialkylzinc to nitroalkenes. It is noteworthy that phosphates derived directly from the unmodified diethyl tartrate ligand, in contrast to the TADDOL derivatives shown above, are poor ligands for the copper-catalysed asymmetric conjugate addition reaction of dialkylzinc species to enones (A. Alexakis, J. Vastra, J. Burton and P. Mangeney, *Tetrahedron: Asymmetry*, 1997, **8**, 8193). Once again it has been proposed that a possible catalytic cycle / reactive intermediate involves π-complexation of the copper to the double bond on the enone and sigma complexation of the zinc atom to a carbonyl lone pair electrons (L. A. Arnold et al., *Tetrahedron*, 2000, **56**, 2865).

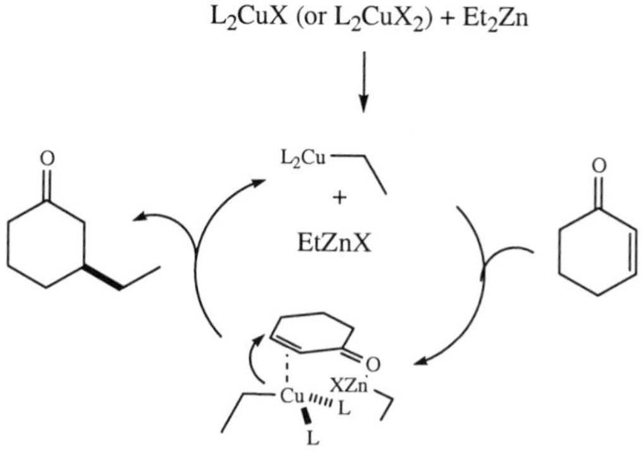

L_2CuX (or L_2CuX_2) + Et_2Zn

L_2Cu– + EtZnX

XZn, Cu, L, L

The chiral monodentate phosphorus ligand (33), based around the highly successful C_2 symmetric diphenylethylenediamine framework, was reported to act as an external ligand in the addition of diethylzinc to 2-cyclohexenone giving the (*S*)-adduct in 70% yield and 55% e.e. Changing the methyl group on the phosphorus to a phenyl group lowered

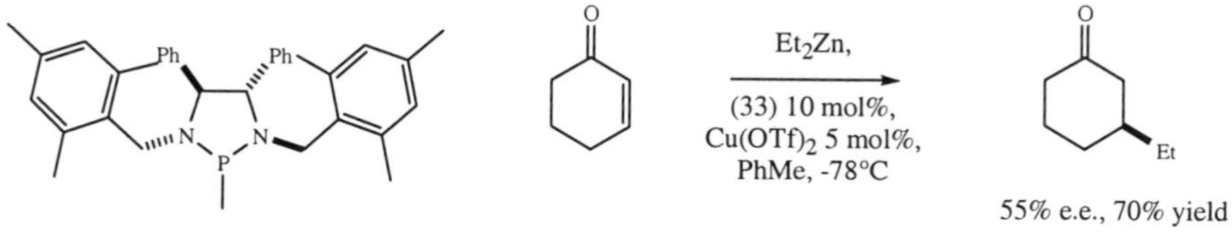

Et_2Zn, (33) 10 mol%, $Cu(OTf)_2$ 5 mol%, PhMe, -78°C

55% e.e., 70% yield

(33)

the observed e.e. value to 44% (T. Mori et al., *Tetrahedron: Asymmetry*, 1998, **9**, 3175).

(ii) Monodentate N Ligands: There are few representatives of this class of ligand for the copper catalysed addition of diethylzinc to Michael acceptors. An isolated report by Sewald (V. Wendisch and N. Sewald, *Tetrahedron: Asymmetry*, 1997, **8**, 1253) was based on Noyori's conclusion that it was necessary for both sulfonamides and a copper(I) salt to be present in order to bring about the conjugate additions of diorganozincs to 2-cyclohexenone. Only a limited number of chiral sulfonamide ligands were reported in this study, the most efficient proving to be the very simple α-methylbenzylamine derivative (34). Remarkably, for the same absolute configuration of the sulfonamide (34) altering the nature of the copper(I) salt can reverse the topicity of the conjugate addition. For example, on addition of diethylzinc to 2-cyclohexenone in the presence of copper(I) cyanide the (*R*)-adduct was obtained in 30% e.e., and in the presence of copper(I) triflate as the catalytic copper source the (*S*)-adduct was obtained in 16% e.e. Although these inductions are only small in comparison to other available ligands, the results serve to highlight the highly complex nature of the catalyst, the influence of a change in either the reaction mechanism or the structure of the reactive species.

NHTs

(34)

Et_2Zn + O

(34) CuCN → (*R*)-adduct 30% e.e.

(34) Cu(OTf) → (*S*)-adduct 16% e.e.

(iii) Bidentate N,S Ligands: The pyridyl substituted thiazolidin-4-one (35) represents a bidentate N,S ligands that has been used as an external ligand in the copper catalysed addition of diethylzinc to enones (A. H. M. de Vries et al., *Tetrahedron: Asymmetry*, 1997, **8**, 1539). The high affinity of sulfides and pyridine for copper(I) species was thought to provide a catalyst of predictable structure. This series of compounds show much promise as candidates for the asymmetric conjugate addition reaction, with e.e. values of up to 62% being observed for the addition of diethylzinc to 2-cyclohexenone.

Ph N O S N Et$_2$Zn, (35) 11 mol%, CuOTf 5 mol% PhMe, - 10°C O O Et

62% e.e., 95% yield

(35)

An elegant series of catalysts (36) have been designed by Anderson. Here the chirality situated on the backbone of an amino thiol is used to render the nitrogen donor atom chiral upon ligation to a metal. The actual catalysts (37) and (38) were used for the addition of diethylzinc to benzaldehyde giving excellent results; however, the e.e.'s obtained were not matched in the catalytic asymmetric addition of diethylzinc to enones (J. C. Anderson, R. Cubbon, M. Harding and D. S. James, *Tetrahedron: Asymmetry*, 1998, **9**, 3461). It is possible that these somewhat disappointing results could be explained by the recent work of Seebach et al. These authors also made a series of N,S bidentate ligands derived from TADDOL, and observed that only the sulfur atom bound to the copper, and the reactive species was the tetra-nuclear monodentate ligand (10), not the expected chelate (A. Pichota et al., *Angew. Chem., Int. Ed. Engl.*, 2000, **39**, 153).

R S N M S L PhMeN SH PhHN SH

(36) (37) (38)

(iv) Bidentate S,O Ligands: The thioether-based bidentate S,O ligand (39) derived from xylose participates in the copper(II) triflate catalysed addition of diethylzinc to 2-cyclohexenone.

AcO OAc AcO O S O O AcO HO O O Et$_2$Zn, (39) 1.4 mol%, Cu(OTf)$_2$ 0.07 mol%, PhMe, 0°C O Et

62% e.e., 87% yield

(39)

Enantioselectivities of up to 62% were observed, whereas other xylose and furanose derivatives were less efficient as catalysts and led to distinctly lower selectivities (O. Pamies et al., *Tetrahedron: Asymmetry*, 2000, **11**, 871).

Woodward has obtained some of his best results to date for cyclic enones using the biaryl ligand (40). The addition of diethylzinc to 2-cyclohex-enone proceeding in up to 77% e.e. and in 78% yield in the presence of 10 mol% $[Cu(MeCN)_4]BF_4$ and 20 mol% (40) (S. M. W. Bennett et al., *Tetrahedron*, 2000, **56**, 2847).

SMe OH OH SMe

Et_2Zn, (40) 20 mol%, $[Cu(MeCN)_4]BF_4$ 10 mol% THF, -20°C

77% e.e., 78% yield

(40)

(v) Bidentate P,N Ligands: The quinoline-containing bidentate P,N ligand QUIPHOS (41) was shown by Buono and co-workers to aid in the copper(I) iodide catalysed conjugate addition of diethylzinc to enones. Good selectivities were limited to cyclic substrates. Surprisingly, the addition of either water or zinc hydroxide to the system was essential in order to observe a good reaction rate and e.e. value. In the absence of water the e.e. observed was a disappointing 7%, whereas, on the addition of 50 mol% of water to the reaction mixture the e.e. value for the addition of diethylzinc to 2-cyclohexenone jumped to the more acceptable value of 61% (G. Delapierre, T. Constantieux, J. M. Brunel and G. Buono, *Eur. J. Org. Chem.*, 2000, 2507).

Et_2Zn, (41) 1 mol%, CuI 0.5mol%, PhMe, -20°C

No addative: 7% e.e., 55% yield
+50 mol% H_2O: 61% e.e., 76% yield

(41)

Pfaltz has reported that the bidentate ligands (42)-(45) are excellent catalysts for asymmetric conjugate addition reactions. This class of ligand is extremely versatile, and has proven to be complementary in efficiency to those reported by Feringa (G. Helmchem and A. Pfaltz, *Acc. Chem. Res.*, 2000, **33**, 336; I. Escher and A. Pfaltz, *Tetrahedron*, 2000, **56**, 2879; A. K. H. Knobel, I. Escher and A. Pfaltz, *Synlett*, 1997, 1429). The choice of ligand depends upon the class of substrate chosen as the Michael acceptor.
The substitution pattern of the binaphthyl moiety is also essential; thus, ligand (42) (R=H) catalyses the reaction between 2-cyclohexenone in 54% e.e., whereas with the more sterically encumbered ligand (43) (R=Me) the e.e. values rise to 90%. For the addition to acyclic enones the ligand of choice appears to be (44), and the product from the addition of diethylzinc to (*E*)-4-phenylbut-3-en-2-one being isolated in 99% yield and in 87% e.e.

R = H (42)
R = Me (43)
(44)
(45)

Et_2Zn, Ligand 2.1 mol%, $Cu(OTf)_2$ 2 mol%, PhMe, -20°C

(42) 54% e.e., 91% yield
(43) 90% e.e., 96% yield

Et_2Zn, (44) 2.1 mol%, $Cu(OTf)_2$ 2 mol%, PhMe, -20°C

87% e.e., 99% yield

Et_2Zn, (45) 2.1 mol%, $Cu(OTf)_2$ 2 mol%, PhMe, -20°C

94% e.e., 41% yield

The most significant contribution from this class of ligand, however, is in the reaction of 2-cyclopentenone as the Michael acceptor. This substrate is a notoriously difficult participant in additions of this type. Addition of diethylzinc in the presence of 2 mol% copper(II) triflate and 2.1 mol% of the ligand (45) gave the corresponding alkylated product in up to 94% e.e.; one of the best results for this ligand to date. Although the isolated yield was only a modest 41%, the high e.e. value and easy accessibility of the ligand render this a very attractive synthetic route. This type of ligand was also effective for additions to 2-cycloheptenone (up to 94% e.e.) and also for the additions of functionalised organozinc reagents. Although no one catalyst appears to afford consistently high e.e. values over a broad range of substrates it was suggested that, because of the modular construction of these ligands, it should be possible to develop tailored ligands for other substrates.

Another class of P,N ligand has recently been reported by Zhang which shows the broadest spectrum of applicability to a range of both cyclic and acyclic Michael acceptors (X. Hu, H. Chen and X. Zhang, *Angew. Chem., Int. Ed. Engl.*, 1999, **38**, 3518). Starting from 2-amino-2'-hydroxy-1,1'-binaphthyl (NOBIN) developed by Kocovsky, Zhang prepared the pyridyl containing derivative (46), with which the copper catalysed addition of diethylzinc to 2-cyclohexenone proceeded in up to 98% yield and in 92% e.e. Similarly, the addition to chalcone proceeded in up to 85% yield and in 96% e.e. More impressively, 5-methyl-2-hexenone (47) was tolerated well by this ligand, and it underwent addition of diethylzinc in 53% yield and in 86% e.e. Standard reaction conditions for this ligand use toluene as a solvent in the presence of 5 mol% copper(II) triflate, however, 10 mol% ligand is required.

O N H PPh_2 N O Et_2Zn, (46) 10 mol%, $Cu(OTf)_2$ 5 mol%, PhMe, -20°C Et O

(47)

86% e.e., 53% yield

(46)

The related ligand (48) led to much less impressive results, the copper(II) triflate catalysed addition of diethylzinc to 2-cyclohexenone proceeding in up to 95% yield and in 51% e.e. (C. G. Arena, G. Calabro, G. Francio and F. Faraone, *Tetrahedron: Asymmetry*, 2000, **11**, 2387).

Morimoto has synthesised a novel class of P,N ligands composed of (*S*)-2-alkyl-2-aminoethylphosphines and α-substituted pyridines. The ligand (49) proved to be the best in the Michael addition of diethylzinc to 2-cyclohexenone, the reaction proceeding in 91% e.e. and in 100%

Et_2Zn,
(48) 6 mol%,
$Cu(OTf)_2$ 3 mol%,
CH_2Cl_2, -15°C

51% e.e., 95% conversion

(48)

conversion. A structure-activity relationship showed that all parts of this catalyst were essential for the enantioselectivities observed. Changing to an alinol derived backbone gave the adduct in 36% e.e., and replacement of the pyridyl unit with a thiophene gave 0% e.e. Similarly, a reduction of the imine to the secondary amine also lowered the e.e. values to 30% (T. Morimoto, Y. Yamaguchi, M. Suzuki and A. Saitoh, *Tetrahedron Lett.*, 2000, **41**, 10025).

Et_2Zn,
(49) 2.5 mol%,
$Cu(OTf)_2$ 1 mol%,
CH_2Cl_2, 0°C

91% e.e., 100% conversion

(49)

The chiral lactam-bearing phosphine (50) catalysed the addition of diethylzinc to 2-cyclohexenone in a moderate 35% e.e. (Y. Nakagawa, K. Matsumoto and K. Tomioka, *Tetrahedron*, 2000, **56**, 2857).

Et_2Zn,
(50) 10 mol%,
$Cu(OTf)_2$ 5 mol%,
PhMe, 0°C

35% e.e., 84% yield

(50)

(vi) Bidentate P,P Ligands: The highly successful, low molecular weight bidentate bis(phosphine), MiniPHOS (51), developed by Imamoto, has been shown to be highly effective in the enantioselective hydrogenation of dehydroamino acids, and the asymmetric hydrosilylation of ketones. Similarly, the copper catalysed asymmetric conjugate addition reactions of diethylzinc to cyclic and acyclic α,β-unsaturated ketones are also very efficient (Y. Yamanoi and T. Imamoto, *J. Org. Chem.*, 1999, **64**, 2988). Thus, treatment of 2-cyclohexenone with 1 mol% copper(II) triflate, 1 mol% of the bidentate phosphine (51) and diethylzinc in toluene at –78°C gave the adduct in 79% yield and in 83% e.e. Addition to 2-cycloheptenone under identical conditions gave the product in 88% yield and in 90% e.e.

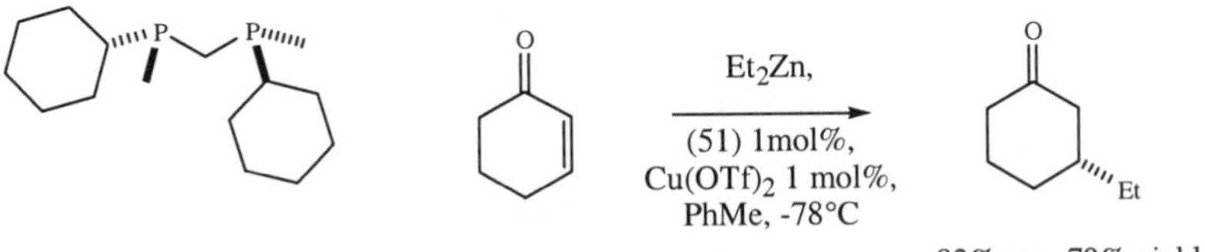

Et_2Zn, (51) 1mol%, $Cu(OTf)_2$ 1 mol%, PhMe, -78°C

(51)

83% e.e., 79% yield

The chiral bis(phosphite) ligand (52), derived from xylose, was shown to be a possible bidentate P,P ligand in the addition of diethylzinc to 2-cyclohexenone (O. Pamies, G. Net, A. Ruiz and C. Claver, *Tetrahedron: Asymmetry*, 1999, **10**, 2007). Addition of diethylzinc to 2-cyclohexenone proceeded in up to 53% e.e. and in 60% yield. To date no other substrates have been reported to undergo catalytic asymmetric conjugate addition reactions with this ligand.

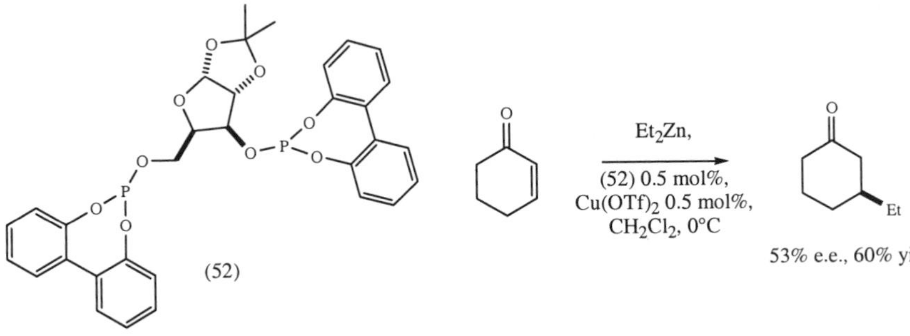

Et_2Zn, (52) 0.5 mol%, $Cu(OTf)_2$ 0.5 mol%, CH_2Cl_2, 0°C

(52)

53% e.e., 60% yield

Alexakis has systematically investigated a large library of known bidentate P,P compounds as candidates for external ligands in the asymmetric conjugate addition reaction (A. Alexakis, J. Burton and P. Mangeney, *Tetrahedron: Asymmetry*, 1997, **8**, 3987). Thirteen catalysts were investigated at just 0.5 mol% loading, and all were efficient for the asymmetric conjugate addition of diethylzinc to 2-cyclohexenone, chalcone and benzalacetone; however, the asymmetric induction observed was disappointing. The best ligand was bis(phosphine) (*S,S*)-Chiraphos (53), and when used with 2-cyclohexenone the reaction proceeded in 98% yield and in 44% e.e.

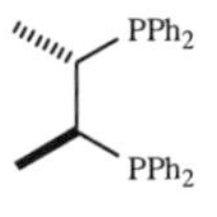

PPh_2 PPh_2

(53)

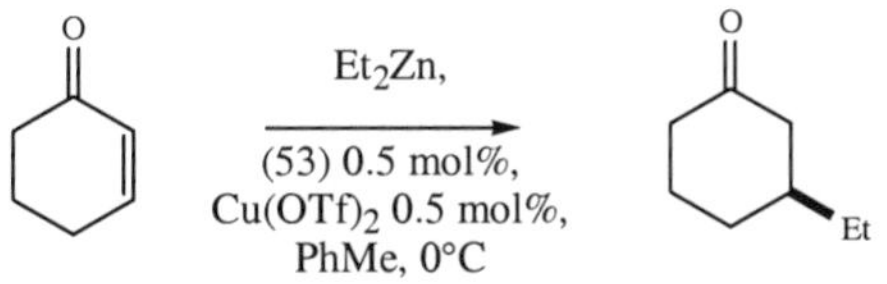

O O

Et_2Zn,

(53) 0.5 mol%,
$Cu(OTf)_2$ 0.5 mol%,
PhMe, 0°C

Et

44% e.e., 98% yield

The C_2-symmetric aryl diphosphite (54) shows good asymmetric induction in the conjugate addition of diethylzinc to cyclic enones, with up to 90% e.e. being observed in the case of 2-cyclohexenone.

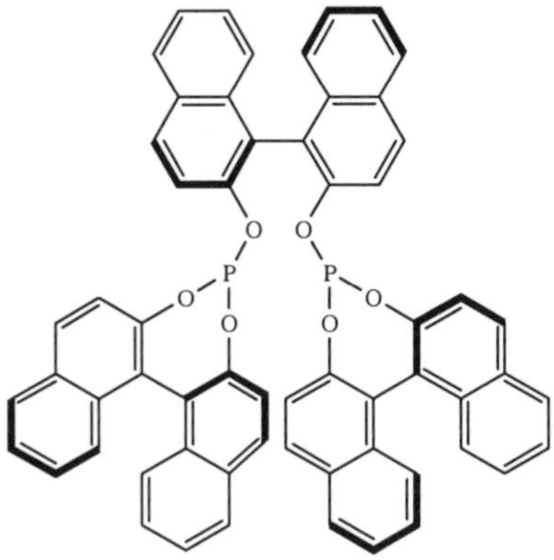

O O P O O P O O O

(54)

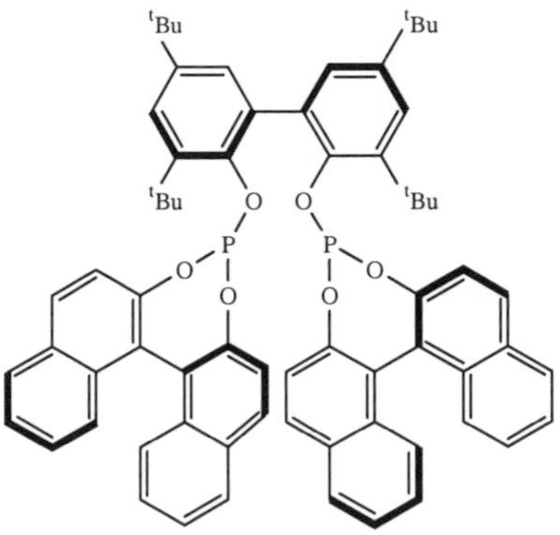

^tBu ^tBu ^tBu ^tBu

O O P O O P O O O

(55)

O O X X

Et_2Zn,

(55) 2 mol%,
$Cu(OTf)_2$ 1 mol%,
PhMe, 0°C

Et

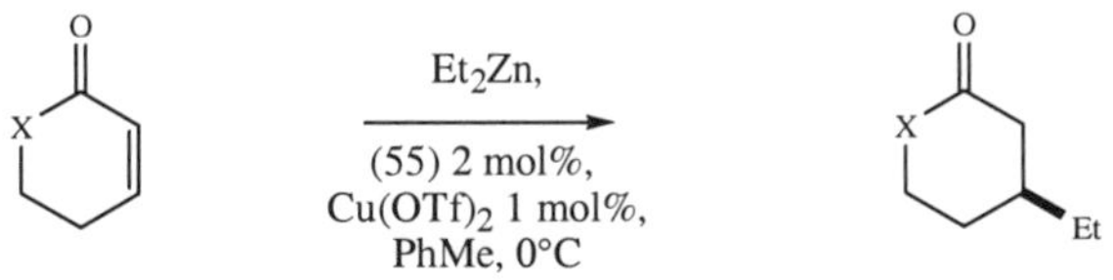

X = CH_2 (56)
X = O (57)

(56) 90% e.e., 100% yield
(57) 92% e.e., 100% yield

However, with an acyclic system, such as chalcone, only low e.e. values are observed (<10%) (M. Yan, L. W. Yang, K. Y. Wong, A. S. C. Chan, *J. Chem. Soc., Chem. Commun.*, 1999, 11). When the link between the two phosphorus atoms is changed to a hindered biaryl unit better results arise. Thus for (55), up to 90% e.e. is observed for an addition to 2-

cyclohexenone (56). Even more significantly, the e.e. for the addition to lactone (57) rises to 92% with a 100% conversion (M. Yan, Z. Y. Zhou and A. S. C. Chan, *J. Chem. Soc., Chem. Commun.*, 2000, 115; M. Yan and A. S. C. Chan, *Tetrahedron Lett.*, 1999, **40**, 6645).

(vii) Tridentate N,N,O Ligands: A library of 300 ligands was synthesised by Gennari et al. and used to catalyse the conjugate addition of diethylzinc to three different cyclic enones (I. Chataigner, C. Gennari, U. Piarulli and S. Ceccarelli, *Angew. Chem., Int. Ed. Engl.*, 2000, **39**, 916). All the compounds synthesised contained three possible metal binding sites, a phenol, an imine and a secondary sulfonamide. The library, which was made from five sulfonyl chlorides, ten primary amines and six aldehydes had the generic structure (58). Although no one catalyst was found to be the best candidate for all three enones, consistently high values were obtained. For example, with 2-cyclohexenone and 2-cycloheptenone and ligand (59), the products were isolated in 90% and in 85% e.e.'s, respectively and in >93% yield in both cases. For 2-cyclopentenone, for the best ligand (60), the yield was a disappointing (25%). However, the e.e. value of 80% was good for this challenging substrate.

(58)

R_1 = Cl R_2 = ^{i}Pr (59)
R_1 = ^{t}Bu R_2 = ^{t}Bu (60)

Et_2Zn, Ligand 2.75 mol%, $Cu(OTf)_2$ 2 mol%, PhMe, r.t.

n = 0 (60) 80% e.e., 25 % yield
n = 1 (59) 90% e.e., 93% yield
n = 2 (59) 85% e.e., 93% yield

(c) Copper in conjunction with organoaluminium reagents.

Although Grignard or dialkylzinc reagents are normally used in stoichiometric amounts in asymmetric conjugate addition reactions, there have been isolated reports where organoaluminium species containing chiral bidentate external ligands are employed

(i) Bidentate N,O Ligands: Iwata has shown that in the presence of bidentate N,O ligands, trialkylaluminiums add to symmetrical 4,4-

dimethylcyclohexa-2,5-dienones in the presence of copper(I) triflate and trialkylsilyl triflates (Y. Takemoto, S. Kuraoka, N. Hamaue and C. Iwata, *Tetrahedron: Asymmetry*, 1996, **7**, 993; Y. Takemoto et al., *Tetrahedron*, 1996, **52**, 141777; Y. Takemoto et al., *J. Chem. Soc., Chem. Commun.*, 1996, 1655). In the best case, up to 68% e.e. was observed for the addition of trimethylaluminium in the presence of the chiral oxazoline (61). It was found that in this case the addition of TBDMSOTf was essential for the stereochemical induction, for without it the e.e. fell to 5%.

OMe OMe N O

(61)

Me_3Al, (61) 20 mol%, CuOTf 5 mol%, TBDMSOTf, THF, 0°C

68% e.e., 88% yield

(ii) Bidentate S,O Ligands: For (*E*)-non-3-en-2-one, the bidentate S,O ligand (62) catalyses the addition of trimethylaluminium in the presence of a copper(I) salt (O. Pamies et al., *Tetrahedron: Asymmetry*, 2000, **11**, 871). E.e. values of up to 34% (80% yield) were reported; a good result for this acyclic enone.

Ph S HO O O O

(62)

^{n}pent O

Me_3Al, (62) 20 mol%, $[Cu(MeCN)_4]BF_4$ 10 mol%, THF, -40°C

^{n}pent O

34% e.e., 80% yield

Other S,O bidentate ligands that work in conjunction with copper to catalyse the addition of trimethylaluminium to enones are (63) and (64), both developed by Woodwood (S. M. W. Bennett et al., *Tetrahedron*, 2000, **56**, 2847; S. M. W. Bennett et al., *Tetrahedron Lett.*, 1999, **40**, 1767). With ligand (63), (*E*)-non-3-en-2-one was alkylated in up to 51% e.e., and with ligand (64) up to 71% e.e. and 79% yield was observed.

S
O NMe$_2$
OH
S
OH

(63) (64)

Me$_3$Al,

O

Ligand 20 mol%,
$[Cu(MeCN)_4]BF_4$ 10 mol%,
THF, -20°C

O

(63) 51% e.e., 47% yield
(64) 71% e.e., 79% yield

(d) Nickel in conjunction with organozinc reagents.

Although copper(II) salts are the most commonly used for the addition of organometallic reagents to Michael acceptors, there is an ever-increasing amount of interest in the use of catalytic amounts of nickel(II). Excellent e.e. values of up to 95% are now observed with this ion, and although the number of substrates used is small, and the stoichiometric reagents are of restricted functionality, further improvements in this area are expected.

(i) Bidentate N,N Ligands: The nickel(II) catalysed addition of diethylzinc to chalcone in the presence of 30 mol% of the proline derived diamine (65) gave the corresponding 1,4-adduct in 75% yield and in 82% e.e. It was found that use of nickel chloride gave higher e.e. values than the use of $Ni(acac)_2$ (M. Asami, K. Usui, S. Higuchi and S. Inoue, *Chem. Lett.*, 1994, 297).

N NHPh
R

(65) R = C_5H_{11}

Ph Ph
O

Et_2Zn,

(65) 30 mol%,
$NiCl_2$ 1.5 mol%,
CH_3CN, -10 °C

Ph Ph
Et O

82% e.e., 75% yield

Another proline derived N,N bidentate ligand that has been used to catalyse the addition of diethylzinc to chalcone is the amide (66) (A. Corma et al., *Tetrahedron: Asymmetry*, 1992, **3**, 845). Immobilisation of the ligand on a zeolite by attachment through the amide nitrogen gave a ligand (67) that could catalyse the reaction in up to 74% yield and, more significantly, in 95% e.e.

(66) R = Me
(67) R = zeolite

Et_2Zn, (67) 5 mol%, $Ni(acac)_2$ 5 mol%, THF, -10 °C

95% e.e., 74% yield

(ii) Bidentate N,O Ligands: Representative bidentate N,O ligands that have been used in conjunction with nickel(II) catalysts include the cobaltacene amino alcohol (68), which has been used to promote the addition of diethylzinc to chalcone (M. Uemura et al., *J. Org. Chem.*, 1993, **58**, 1238). The chromium moiety was essential for high enantio-selectivities in the addition of diethylzinc to benzaldehyde, and for the conjugate addition reaction with the ligand (68) enantioselectivities of up to 62% (90% yield) were obtained; the high ligand loading of 50%, however, is a major drawback to this process.

Et_2Zn, (68) 50 mol%, $Ni(acac)_2$ 5 mol%, CH_3CN, -30 °C

62% e.e., 90% yield

(68)

Soai has shown that in the presence of the ephedrine derived ligand (69), the use of either nickel bromide or $Ni(acac)_2$ lead to similar enantio-selectivities (48% and 45% e.e. respectively); however, in the presence of the bipyridyl additive (70) (1 equivalent), the e.e. for the addition of diethylzinc to chalcone rises to 90% (K. Soai, T. Hayasaka, S. Ugajin and S. Yokoyama, *Chem. Lett.*, 1988, 1571; K. Soai,

Et_2Zn, (69) 17 mol%, $Ni(acac)_2$ 7 mol%, (70), CH_3CN, -30 °C

90% e.e., 47% yield

(69) (70)

S. Yokoyama, T. Hayasaka and K. Ebihara, *J. Org. Chem.*, 1988, **53**, 4149; K. Soai, T. Hayasaka and S. Ugajin, *J. Chem. Soc., Chem. Commun.*, 1989, 516).

Interestingly, the bidentate N,O ligand (71) promotes the addition of diethylzinc to the hindered acyclic enone (72) in up to 80% e.e. in the absence of either a nickel or copper source (K. Soai, M. Okudo and M. Okamoto, *Tetrahedron Lett.*, 1991, **32**, 59).

Et_2Zn, (71) 25 mol%, hexane, r.t.

(71) (72) 80% e.e., 81% yield

The simple bidentate pyridyl ligand (73) was shown by Bolm to promote the nickel(II) catalysed addition of diethylzinc to chalcone in up to 90% e.e. (C. Bolm, M. Ewald and M. Felder, *Chem. Ber.*, 1992, **125**, 1205). The reaction was limited to diorganozinc compounds, the use of both Grignard and organoaluminium reagents leading to racemic products. As with most nickel(II) catalysed additions no asymmetric induction was observed in the addition to cyclic enones. It is thought that the mechanism of the nickel(II) addition differs from the copper-catalysed addition as the nickel co-ordinates to the carbonyl oxygen, rather than to the enone double bond. Thus, in cyclic substrates, where the enone must sit in the s-*trans* conformation, the chiral nickel complex is situated too far away from the newly forming chiral centre to exert any stereochemical influence on the course of the reaction.

Et_2Zn, (73) 20 mol%, $Ni(acac)_2$ 1 mol%, CH_3CN, -30 °C

(73) 90% e.e., 78% yield

Another class of bidentate N,O ligand is represented by the β-hydroxy-sulfoximine (74), which was shown to promote the addition of diethylzinc to chalcone in the presence of 1 mol% $Ni(acac)_2$ in up to 70% e.e. (C. Bolm, M. Felder and J. Muller, *Synlett*, 1992, 439).

(74)

Et_2Zn, (74) 20 mol%, $Ni(acac)_2$ 1 mol%, CH_3CH_2CN, -30 °C

70% e.e., 71% yield

(ii) Bidentate N,S Ligands: A series of bidentate N,S ligands has been introduced by Gibson, based on the proline framework (C. L. Gibson, *Tetrahedron: Asymmetry*, 1996, **7**, 3357). The disulfide (77) catalysed the addition of diethylzinc to chalcone in up to 50% e.e. Interestingly, the anionic thiolate (75) and the prolinol derivative (76) gave adducts with the opposite absolute configuration, although no explanation of this observation was provided.

R = SLi (75)
R = OH (76)

(77)

Et_2Zn, (77) 17 mol%, $Ni(acac)_2$ 7 mol%, CH_3CN, -30 °C

50% e.e., 85% yield

(iv) Bidentate O,S Ligands: Vast improvements on the e.e. values and yield for the nickel(II) catalysed asymmetric conjugate addition of diethylzinc to chalcone have recently been reported by Yang (Y. Yin, X. Li, D. S. Lee and T. K. Yang, *Tetrahedron: Asymmetry*, 2000, **11**, 3329), who showed that in the presence of the hydroxy sulfide (78) the adduct was isolated in up to 86% e.e. and 98% yield, although the high ligand loading of 50 mol% is a major drawback. Decreasing the ligand loading to 9 mol% caused the e.e. for the reaction to drop to 60%.

(78)

Et_2Zn, (78) 50 mol%, $Ni(acac)_2$ 2 mol%, CH_3CN, -30 °C

86% e.e., 98% yield

(v) Tridentate N,N,O Ligands: The hydroxydiamine (79) has been shown to act as a tridentate ligand in the addition of diethylzinc to chalcone (J. Spieler, O. Huttenloch and H. Waldmann, *Eur. J. Org. Chem.*, 2000,

391). In the presence of 1 mol% of 2,2'-bipyridyl (70) and 20 mol% of the ligand (79), the addition of diethylzinc to chalcone proceeds in 85% e.e. and in 86% isolated yield. It appears that for these ligands that nickel is a much better catalyst than either copper or cobalt for this transformation.

HN HO N ^{i}Pr

(79)

Ph Ph O → Et_2Zn, (79) 16 mol%, $Ni(acac)_2$ 7 mol%, (70) 1 mol% CH_3CN, -30 °C → Ph Ph Et O

85% e.e., 86% yield

The tridentate ligand (80), derived from camphor, has been used as an external chiral ligand in the nickel-catalysed conjugate addition reaction of diethylzinc to chalcone and 2-cyclohexenone (A. H. M. de Vries, R. Imbos and B. L. Feringa, *Tetrahedron: Asymmetry*, 1997, **8**, 1467). As observed with other nickel-based systems, low enantioselectivities were obtained for cyclic substrates, however, with chalcone up to 83% e.e. was realised.

N N OH

(80)

Ph Ph O → Et_2Zn, (80) 16 mol%, $Ni(acac)_2$ 7 mol%, CH_3CN, -25 °C → Ph Ph Et O

83% e.e., 83% yield

(e) Cobalt in conjunction with organozinc reagents.

There have been few reports of the use of cobalt as the catalytic metal in the asymmetric conjugate addition reaction, and those that do so only state briefly that the rate of the reaction is considerably slower than that observed for the corresponding nickel- or copper-accelerated processes. Feringa has shown that $Co(acac)_2$ and the external chiral bidentate N,O ligand (81), derived from camphor, has been employed in the enantio-selective conjugate addition of diethylzinc to chalcone. Enantio-selectivities of up to 85% were reported at 73% yield. Although the scope of the substrates was limited to chalcone, and the rate of reaction was slow compared to other metals, initial results look promising and warrant further attention (A. H. M. de Vries and B. L. Feringa, *Tetrahedron: Asymmetry*, 1997, **8**, 1377).

(81)

Et_2Zn, (81) 16 mol%, $Co(acac)_2$ 7 mol%, CH_3CN, -30 °C

85% e.e., 73% yield

3. Alkenyl and Aryl Addition.

A second class of nucleophile that are often employed in the catalytic asymmetric conjugate addition reaction are alkenyl and aryl organometallics derived from either boron or zinc. Although these systems have not received the huge amount of interest that follows alkyl organometallics, these substrates have proved to be highly successful, and e.e. values in excess of 90% are frequently obtained.

(a) Rhodium in conjunction with organoboron reagents.

A significant number of reports from Hayashi and co-workers has shown that aryl and alkenyl boron reagents, in the presence of either a (*R*)- or (*S*)-BINAP rhodium(I) catalyst, can be used in the catalytic asymmetric conjugate addition reaction. This is a general result for a number of Michael acceptors including cyclic and acyclic enones (Y. Takaya, M. Ogasawara and T. Hayashi, *J. Am. Chem. Soc*, 1998, **120**, 5579; Y. Takaya, M. Ogasawara and T. Hayashi, *Tetrahedron Lett.*, 1998, **39**, 8479; Y. Takaya, M. Ogasawara and T. Hayashi, *Tetrahedron Lett.*, 1999, **40**, 6957), 1-alkenylnitro compounds (T. Hayashi, T. Senda and M. Ogasawara, *J. Am. Chem. Soc.*, 2000, **122**, 10716), α,β-unsaturated esters and lactones (Y. Takaya et al., *Tetrahedron: Asymmetry*, 1999, **10**, 4057), and 1-alkenylphosphonates (T. Hayashi, T. Senda, Y. Takaya and M. Ogasawara, *J. Am. Chem. Soc.*, 1999, **121**, 11591). The nature of the boron reagent is quite general: it may be an aryl- or alkenylboronic acid, a 2-alkenyl-1,3,2-benzodioxaborole, generated *in-situ* from the addition of catecholborane to terminal alkynes, or arylborates, which are readily accessible by lithiation of aryl bromides, followed by treatment with trimethoxyborane. Substitution of the nucleophile also does not appear to effect enantioselectivity. For example, in the presence of 3 mol% of the rhodium catalyst and (*S*)-BINAP (82), phenylboronic acid adds to 2-cyclohexenone (97% e.e., 99% yield), the notoriously difficult 2-cyclopentenone (97% e.e., 93% yield), (*E*)-non-3-en-2-one (92% e.e., 88% yield), 1-nitrocyclohexene (99% e.e., 73% yield) and (*E*)-1-propenyl diethyl phosphonate (96% e.e., 94% yield).

PPh_2
PPh_2
(82)

$PhB(OH)_2$, (82) 3 mol%, $Rh(acac)(C_2H_4)_2$ 3 mol%, dioxane/H_2O, 100 °C
97% e.e., 99% yield

$PhB(OH)_2$, (82) 3 mol%, $Rh(acac)(C_2H_4)_2$ 3 mol%, dioxane/H_2O, 100 °C
97% e.e., 99% yield

NO_2 — $PhB(OH)_2$, (82) 3 mol%, $Rh(acac)(C_2H_4)_2$ 3 mol%, dioxane/H_2O, 100 °C
99% e.e., 73% yield

$P(OEt)_2$ — $PhB(OH)_2$, (82) 3 mol%, $Rh(acac)(C_2H_4)_2$ 3 mol%, dioxane/H_2O, 100 °C
96% e.e., 94% yield

Miyaura has also used the same (*S*)-BINAP catalyst (82) to promote the addition of arylboronic acids to acyclic α,β-unsaturated esters and

$PhB(OH)_2$, (82) 3 mol%, $Rh(acac)(C_2H_4)_2$ 3 mol%, dioxane/H_2O, 100 °C
93% e.e., 97% yield
stereochemistry undetermined

amides, again observing spectacularly high yields and enantioselectivities (S. Sakuma, M. Sakai, R. Itooka and N. Miyaura, *J. Org. Chem.*, 2000, **65**, 5951). For example, addition of phenylboronic acid to *N*-benzyl crotonamide gave the adduct in 67% yield and in 93% e.e.
It is believed that the catalytic cycle for this reaction involves an intermediate of the type (83), where the rhodium forms a π-complex with the double bond of the Michael acceptor and delivers the nucleophile selectively to the complexed face.

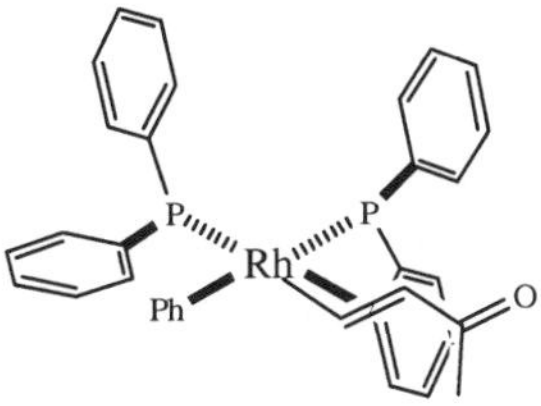

(83)

(b) Nickel in conjunction with organozinc reagents.

Ikeda has reported a highly versatile tandem reaction, which involves the nickel(II) catalysed addition of a diorganozinc reagent to a terminal or symmetrical alkyne, conjugate addition of this adduct to an α,β-unsaturated carbonyl moiety and trapping of the resultant enolate adduct with trimethylsilyl chloride (S. I. Ikeda, D. M. Cui and Y. Sato, *J. Am. Chem. Soc.*, 1999, **121**, 4712).

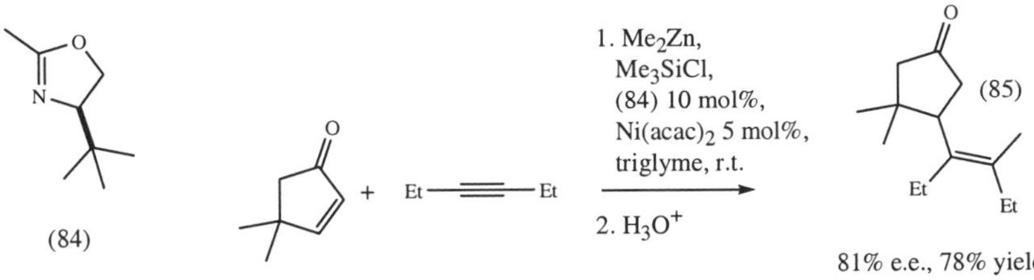

(84)

1. Me_2Zn, Me_3SiCl, (84) 10 mol%, $Ni(acac)_2$ 5 mol%, triglyme, r.t.
2. H_3O^+

(85)

81% e.e., 78% yield
stereochemistry undetermined

Despite the complexity of this cascade and the number of reagents that are involved, good yields and e.e. values for this one pot, three-step process are obtained. The chiral ligand used for this reaction is the simple monodentate oxazoline (84). Thus, addition of dimethylzinc to 3-hexyne in the presence of $Ni(acac)_2$ (5 mol%), the ligand (84) (10 mol%), 3,3-dimethylcyclopentenone and trimethylsilyl chloride in triglyme at room temperature gave the adduct (85) in 78% yield and in 81% e.e. This work was extended to a number of cyclic and acyclic enones as well as a series of terminal alkynes.

4. Hydride Addition

The simplest nucleophile that can be added in a conjugate addition reaction is hydride. Recently, Buchwald has reported a catalytic asymmetric variant of this reaction using the chiral phosphine (*S*)-*p*-tol-BINAP (86) and copper(I) chloride in the presence of the safe and inexpensive stoichiometric reductant polymethylhydrosiloxane (PHMS) (D. H. Appella et al., *J. Am. Chem. Soc.*, 1999, **121**, 9473; Y. Moritani, D. H. Appella, V. Jukauskas and S. L. Buchwald, *J. Am. Chem. Soc.*, 2000, **122**, 6797). The reaction has been applied to both acyclic α,β-unsaturated esters and cyclic ketones with e.e. values in excess of 80%. For example, the conjugate reduction of 2-cyclopentenone (87) in the presence of 5 mol% copper(I) chloride and 5 mol% ligand (86) gave the product in 98% e.e. and in 84% yield.

P*p*-tol$_2$
P*p*-tol$_2$
(86)

O
(87)

PHMS,
(86) 5 mol%,
CuCl 5 mol%,
NaO^{t}Bu 5 mol%,

O
98% e.e., 84% yield

5. Ketone and Malonate Stabilised Anion Addition

The catalytic asymmetric conjugate addition of carbonyl stabilised anions to Michael acceptors has received much attention and has resulted in a diverse number of methods being introduced in order to carry out the reaction. These include the applications of main group, transition metal and lanthanide based systems as well as, more recently, a number of metal-free alternatives.

(a) Main Group Metals: There have been a number of reports of the use of chiral crown ethers, used in conjunction with main group metals, for the addition of prochiral Michael donors to α,β-unsaturated carbonyls to give 1,5-diketones with quaternary centres present in the product. One of the first examples showed that the binaphthyl derived crown ether (88) catalysed the addition of methyl indan-1-one-2-carboxylate to methyl vinyl ketone in 48% yield and 99% e.e. (D. J. Cram and G. D. Y. Sogah, *J. Chem. Soc., Chem. Commun.*, 1981, 625).

(88)

CO_2Me

CH_2=CHCOMe,

(88) 4 mol%, KO^tBu, PhMe, -78°C

CO_2Me

99% e.e., 48% yield

The addition of prochiral acyclic Michael donors has also been thoroughly investigated, especially in the case of the addition of methyl phenylacetate to methyl acrylate. Representative examples of ligands that catalyse this reaction include the carbohydrate derived crown ether (89), developed by Toke and co-workers, that gives the adduct in up to 84% e.e. and in 82% yield (L. Toke, L. Fenichel and M. Albert, *Tetrahedron Lett.*, 1995, **36**, 5951; L. Toke et al., *Tetrahedron*, 1998, **54**,213), the camphor derived crown ether (90) that gives the adduct in up to 83% e.e. (E. Brunet et al., *Tetrahedron: Asymmetry*, 1994, **5**, 935) and the simple 1,2-dimethyl-18-crown-6 (91), which, in conjunction with potassium-*tert*-butoxide gives the adduct in 79% e.e. (S. Aoki, S. Sasaki and K. Koga, *Tetrahedron Lett.*, 1989, **30**, 7229). A number of other crown ethers have also been reported to be effective in this transformation. These reveal a complex structure-activity relationship of the chiral crown ether, and the sense of induction observed for the reaction, slight changes in the structure considerably changing the stereochemistry and e.e. observed (J. Crosby, J. F. Stoddart, X. Sun and M. R. W. Venner, *Synthesis*, 1993, 141; E. V. Dehmlow and V. Knufinke, *Liebigs Ann. Chem.*, 1992, 283; D. A. H. van Maarschalkerwaart, N. P. Willard and U. K. Pandit,

Tetrahedron, 1992, **48**, 8825; E. V. Dehmlow and C. Sauerbier, *Liebigs Ann. Chem.*, 1989, 181; M. Alonso-Lopez et al., *Tetrahedron*, 1988, **44**, 1535; M. Alonso-Lopez, M. Martin-Lomas and S. Penades, *Tetrahedron Lett.*, 1986, **27**, 3551; B. Raguse and D. D. Ridley, *Aust. J. Chem.*, 1984, **37**, 2059).

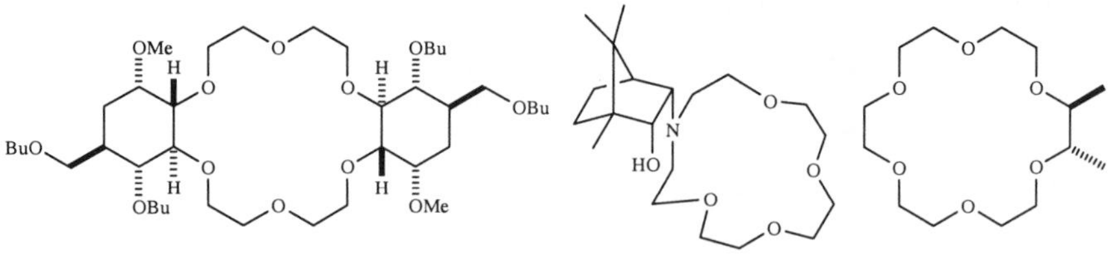
OMe
H
OBu
BuO
OBu
H
OBu
OMe
HO
N

(89) (90) (91)

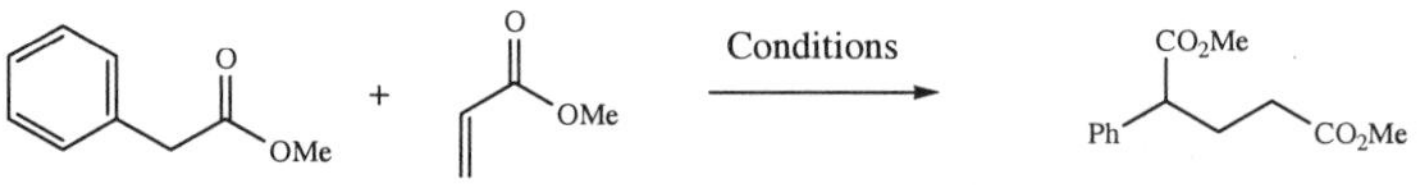
OMe
OMe
Conditions
CO_2Me
Ph
CO_2Me

(89): ^tBuOK (34 mol%), (89) 6 mol%, PhMe, -78°C, 84% e.e., 82% yield
(90): ^tBuONa, (90) 10 mol%, PhMe, -78°C, 83% e.e., 86% yield,
(91): ^tBuOK, (91) 10 mol%, PhMe, -78°C, 79% e.e., 95% yield.

Yamaguchi has shown that the rubidium salt of proline (92) is an effective catalyst for the asymmetric conjugate addition of malonates to prochiral cyclic and acyclic enones (M. Yamaguchi, T. Shiraishi and M. Hirama, *J. Org. Chem.*, 1996, **61**, 3520; M. Yamaguchi, T. Shiraishi, Y. Igarashi and M. Hirama, *Tetrahedron Lett.*, 1994, **44**, 8233; M. Yamaguchi, T. Shiraishi and M. Hirama, *Angew. Chem., Int. Ed. Engl.*,

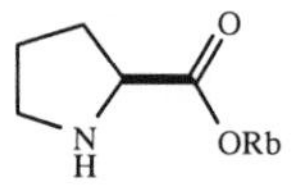
ORb
N
H

(92)

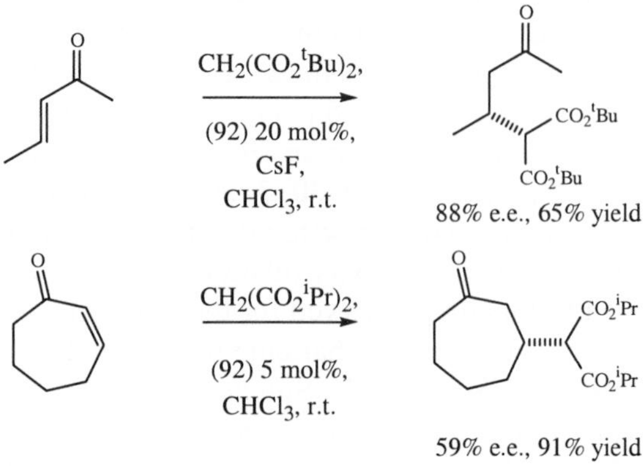
$CH_2(CO_2{}^tBu)_2$,
(92) 20 mol%,
CsF,
$CHCl_3$, r.t.
$CO_2{}^tBu$
$CO_2{}^tBu$
88% e.e., 65% yield
$CH_2(CO_2{}^iPr)_2$,
(92) 5 mol%,
$CHCl_3$, r.t.
$CO_2{}^iPr$
$CO_2{}^iPr$
59% e.e., 91% yield

1993, **32**, 1176). This catalyst possesses both an amine and a cation in the same molecule which, it was proposed, clearly defined the transition state for the transformation; the catalyst functioning as a base as well as being involved with substrate activation. It was also suggested that a reversible imine formation involving the amine group of the catalyst and the carbonyl group of the substrate provided a key intermediate in the catalytic cycle. The role of the metal is significant, changing to lithium significantly lowers the e.e. values observed for the transformation, and changing to either sodium or potassium leads to racemic product.
Corey has reported that the cinchonidine based phase-transfer catalyst (93) in the presence of caesium hydroxide catalyses the asymmetric conjugate addition of the glycinate ester (94) to acrylates as well as cyclic and acyclic ketones, with exceedingly high e.e. values. For example, the use of 10 mol% (93) and caesium hydroxide catalyses the addition of (94) to 2-cyclohexenone in 99% e.e. and in 88% yield (E. J. Corey, M. C. Noe and F. Xu, *Tetrahedron Lett.*, 1998, **39**, 5347). These catalysts are also effective for the potassium hydroxide catalysed addition of ketones to chalcone type Michael acceptors (F. Y. Zhang and E. J. Corey, *Org. Lett.*, 2000, **2**, 1097).

(93)

(94)

$CsOH.H_2O$, (93) 10 mol%, CH_2Cl_2, -78°C

99% e.e., 88% yield

PhCOMe, (93) 10 mol%, KOH, PhMe, -10°C

80% e.e., 72% yield

(b) Transition Metals: The simple diamine (95) acts as an external ligand in the nickel(II) catalysed conjugate addition of cyclic β-ketoesters to methyl vinyl ketone to give the product in reasonable yield and up to 91% e.e. The reaction proceeds at ambient temperature under neutral conditions and without the need to exclude both air and moisture (J. Christoffers, U. Robler and T. Werner, *Eur. J. Org. Chem.*, 2000, 701).

(95)

CH_2=CHCOMe, (95) 37 mol%, $Ni(OAc)_2.4H_2O$ 5 mol%, $CHCl_3$, r.t.

91% e.e., 37% yield

Addition of methyl indan-1-one-2-carboxylate to methyl vinyl ketone is catalysed by a series of 2-substituted 2-(salicylideneamino)ethanols in the presence of a copper(II) source to give the corresponding adduct in up to 75% e.e. (G. Desimoni et al., *Tetrahedron*, 1995, **51**, 4131). For example, the copper(II) acetate derived ligand (96) gave the adduct in 75% e.e. at 100% conversion.

(96)

CH_2=CHCOMe, (96) 1 mol%, CCl_4, -20°C

75% e.e., 100% conversion

A similar e.e. value was obtained when the same reaction was catalysed using a cobalt(II) catalyst in the presence of the 1,2-diphenylethylene-diamine ligand (97), giving the adduct in 50% yield and in 66% e.e. (H. Brunner and B. Hammer, *Angew. Chem., Int. Ed. Engl.*, 1984, **23**, 312; H. Brunner and J. Kraus, *J. Mol. Cat.*, 1989, **49**, 133).

(97)

CH_2=CHCOMe, (97) 5 mol%, $Co(acac)_2$ 5 mol%, PhMe, -50°C

66% e.e., 50% yield

(c) Lanthanides: One of the first reports of the use of a lanthanide based catalyst in the catalytic asymmetric Michael addition of 1,3-dicarbonyls to α,β-unsaturated ketones utilised the chiral complex $[Eu(tfc)_3]$ (F. Bonadies et al., *Tetrahedron Lett.*, 1993, **34**, 7649). Thus, addition of the ethyl acetoacetate derivative (98) to methyl vinyl ketone gave the adduct in 50% yield and in 36% e.e.

(98)

CH_2=CHCOMe, [Eu(tfc)$_3$] 30 mol%, CCl_4, -25°C

36% e.e., 50% yield, stereochemistry undetermined

Shibasaki and co-workers have reported by far the largest and most comprehensive body of work on the use of lanthanides in the catalytic asymmetric conjugate addition reaction. They have synthesised a family of heterobimetallic alkoxides based around the binaphthyl framework and used them for the addition of malonates, acetoacetates, and Horner-Wadsworth-Emmons reagents to both cyclic and acyclic enones, enals and acrylates; the products being obtained in excellent yield and e.e. (T. Arai, H. Sasai, K. Yamaguchi and M. Shibasaki, *J. Am. Chem. Soc.*, 1998, **120**, 441; H. Sasai, E. Emori, T. Arai and M. Shibasaki, *Tetrahedron Lett.*, 1996, **37**, 5561; T. Arai et al., *Angew. Chem., Int. Ed. Engl.*, 1996, **35**, 104; T. Arai et al., *Chem. Eur. J.*, 1996, **2**, 1368; H. Sasai et al., *J. Am. Chem. Soc.*, 1995, **117**, 6194; H. Sasai, T. Arai and M. Shibasaki, *J. Am. Chem. Soc.*, 1994, **116**, 1571). This work has been thoroughly reviewed previously (M. Kanai and M. Shibasaki, in *Catalytic Asymmetric Synthesis*; 2nd edn., I. Ojima Ed. Wiley-VCH: Weinheim, 2000, 569; M. Shibasaki, H. Sasai and T. Arai, *Angew. Chem., Int. Ed. Engl.*, 1997, **36**, 1236). Much attention has been made in trying to define the exact nature of the catalyst, both in solution and in the solid state, in order to try and obtain a good understanding of the catalytic cycle, and to aid the design and synthesis of new catalysts. To this end, Shibasaki has been exceedingly successful and has provided a range of catalysts for the synthetic chemists' tool-box. Representative examples of the catalysts developed by Shibasaki include (*R*)-LSB (99), (*R*)-GSB (100) and (*R*)-ALB (101) and a selection of the reactions they have been used to catalyse are shown below.

An 'ALB' catalyst has also been reported by Feringa to catalyse the asymmetric conjugate addition of α-nitroesters to α,β-unsaturated ketones (E. Keller, N. Veldman, A. L. Spek and B. L. Feringa, *Tetrahedron: Asymmetry*, 1997, **8**, 3403). Although e.e. values for the reaction were high (up to 80%), the specific structure of the catalyst differs from that reported by Shibasaki, presumably because methods for the synthesis of the catalytic species differed.

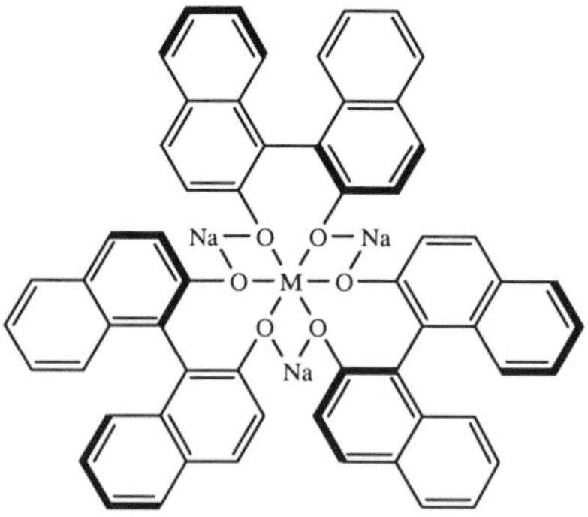

M = La (*R*)-LSB (99)
M = Ga (*R*)-GSB (100)

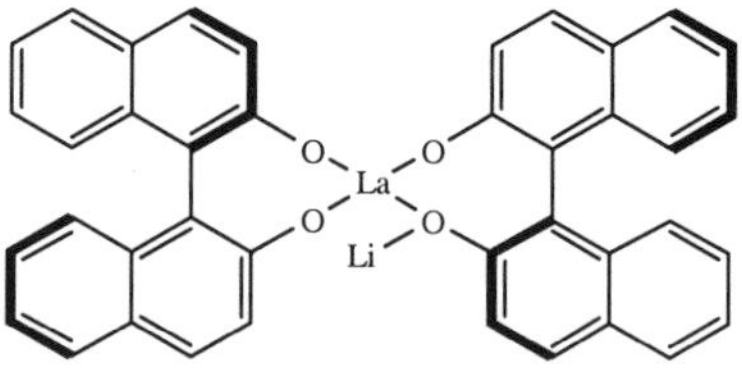

(*R*)-ALB (101)

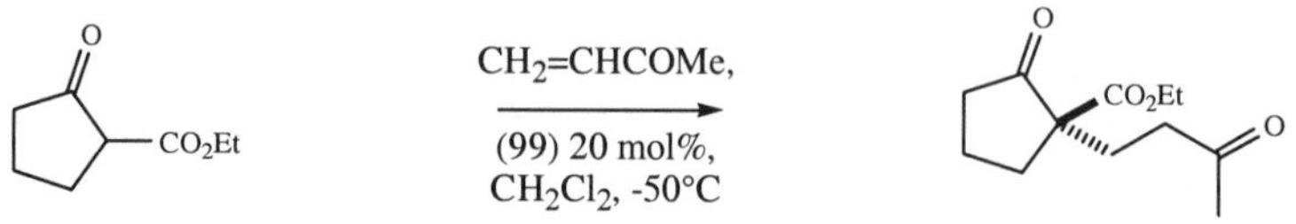

CH_2=CHCOMe, (99) 20 mol%, CH_2Cl_2, -50°C

84% e.e., 97% yield

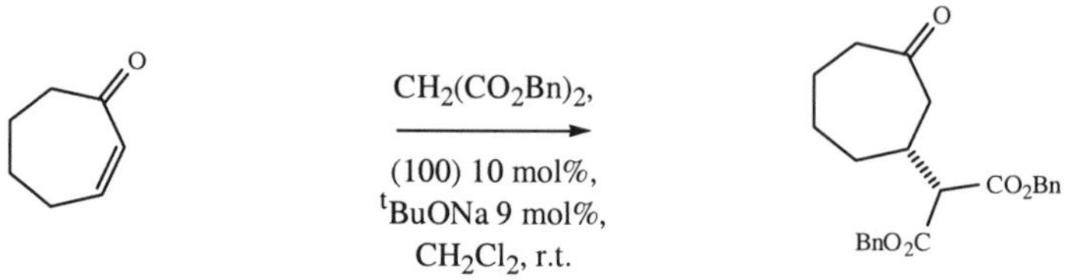

$CH_2(CO_2Bn)_2$, (100) 10 mol%, ^tBuONa 9 mol%, CH_2Cl_2, r.t.

99% e.e., 79% yield

$CH_2(CO_2Et)_2$, (101) 10 mol%, THF, r.t.

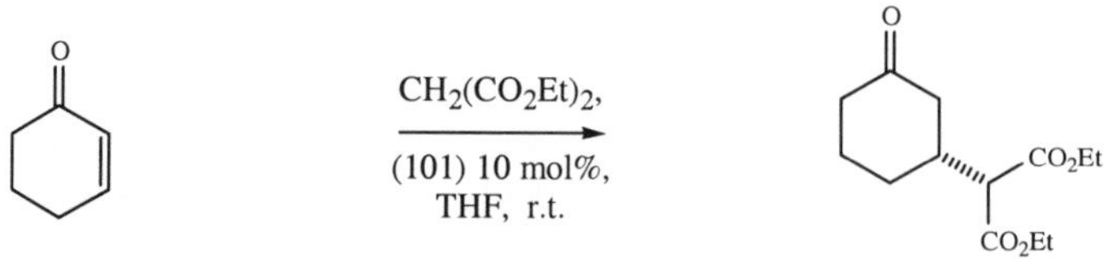

95% e.e., 87% yield

More recently Shibasaki has directed his attention towards developing reusable air-stable catalysts that provide an alternative to the ALB series. For example, the alkali metal-free catalyst (102), promotes the addition of methyl dibenzylmalonate to cyclic enones in 99% e.e. and in >85% yield (Y. S. Kim et al., *J. Am. Chem. Soc.*, 2000, **122**, 6506). The catalysts have been shown to be air-stable for up to four weeks, without loss of activity. They are readily recoverable from the reaction mixture by a simple precipitation/filtration technique, and have been recycled up to four times with only slight loss of activity. These catalysts have also been immobilised on a solid support, which lowers both yield and the e.e. value for the catalytic asymmetric conjugate addition reaction (S. Matsunaga, T. Ohshima and M. Shibasaki, *Tetrahedron Lett.*, 2000, **41**, 8473).

O La O O O H O

$MeCH(CO_2Bn)_2$, (102) 10 mol%, DME, r.t.

CO_2Bn CO_2Bn

(102)

99% e.e., 95% yield

(d) Metal-Free Catalysts: (*S*)-*N*-(2-Pyrrolidylmethyl)-*N*,*N*,*N*-trimethylammonium hydroxide (103) catalyses the asymmetric conjugate addition of malonates to both cyclic and acyclic enones through ion pair control (A. Kawara and T. Taguchi, *Tetrahedron Lett.*, 1994, **35**, 8805). For example, addition of dibenzylmalonate to 2-cyclohexenone in the presence of (103) (10 mol%) gave the adduct in 61% yield and in 71% e.e. In the absence of the additive hexafluoroisopropanol (HFIP) only low e.e. values were obtained for the reaction (< 5%); HFIP was thought to reduce the basicity of the catalyst.

NH $\oplus NMe_3$ $\ominus OH$

$CH_2(CO_2Bn)_2$, (103) 10 mol%, HFIP, PhMe, r.t.

CO_2Bn CO_2Bn

(103)

71% e.e., 61% yield

The addition of the achiral-glycine derivative (94) to ethyl acrylate in the presence of 20 mol% of the guanidine (104) gave the product in 99% yield and 30% e.e. (D. Ma and K. Cheng, *Tetrahedron: Asymmetry*, 1999, **10**, 713). Although these e.e. values were poor by modern standards, the use of guanidines to catalyse the reaction is unprecedented and opens up the possibility of developing a new family of catalysts for this reaction.

(104) (94) CH_2=$CHCO_2Et$, (104) 20 mol%, THF, -10°C

30% e.e., 99% yield

6. Aliphatic Nitro Group Addition

The aliphatic nitro moiety is a versatile synthetic group that can easily be oxidised, reduced and oxygenated making in very useful in contemporary organic synthesis. It is also an excellent candidate as a Michael donor and many groups have investigated its catalytic asymmetric conjugate addition to both cyclic and acyclic α,β-unsaturated carbonyl compounds. There are a host of methods available for this transformation, some of which are outlined below.

(a) Phase-Transfer Catalysis: Corey has reported the use of phase-transfer catalysis in the addition of nitromethane to 4-chlorobenzylidineaceto-phenone in the presence of the cinchonium salt (105) (10 mol%) and caesium fluoride, in toluene at -40°C to give the adduct in 89% yield and in 70% e.e. (E. J. Corey and F. Y. Zhang, *Org. Lett.*, 2000, **2**, 4257).

(105) CH_3NO_2, (105) 10 mol%, CsF, PhMe, -40°C

70% e.e., 89% yield

There are few reports of the reaction of aliphatic nitro compounds in the asymmetric conjugate addition reaction with Michael acceptors other than enones. When nitromethane was added to (*E*)- and (*Z*)-2-propenyl sulfone obtained in the presence of the chiral phase-transfer catalyst (106), the antipodes were obtained depending on the vinyl sulfone starting material used, however, neither the absolute configuration nor the e.e. values were reported in this communication (R. Annunzita, M. Cinguini and S. Colonna, *Chem. Ind.*, 1980, 238).

BnMe$_2$N$^+$ Br$^-$ OH

(106)

CH_3NO_2, (106) 10 mol%, KF, PhMe, r.t.

(+)-isomer (*E*)-sulfone
(-)-isomer (*Z*)-sulfone

(b) Alkali metals in conjunction with crown ethers: There have been numerous reports of chiral crown ethers that catalyse the asymmetric conjugate addition of malonates to prochiral Michael acceptors, but relatively few have exemplified the addition of alkylnitro compounds to enones. One example by Toke showed that in the presence of the mono-aza crown ether (107) (7 mol%) the addition of 2-nitropropane to chalcone proceeded in up to 78% yield and in 84% e.e. (P. Bako, K. Vizvardi, Z. Bajor and L. Toke, *J. Chem. Soc., Chem. Commun.*, 1998, 1193).

(107)

$(CH_3)_2CHNO_2$, (107) 7 mol%, KO^{t}Bu, PhMe, r.t.

84% e.e., 78% yield

(c) Amine catalysts: The rubidium proline-based catalyst (92) of Yamaguchi, discussed earlier in the addition of malonates to prochiral α,β-unsaturated carbonyl compounds, has also been used in the catalytic asymmetric conjugative addition of aliphatic nitro compounds to Michael acceptors (M. Yamaguchi, T. Shiraishi, Y. Igarashi and M. Hirama, *Tetrahedron Lett.*, 1994, **35**, 8233; M. Yamaguchi et al., *Tetrahedron*, 1997, **53**, 11223).

(92)

(92) 5 mol%, $CHCl_3$, r.t.

84% e.e., 84% yield

Once again acyclic (*E*)-enones gave (*S*)-adducts and cyclic (*Z*)-enones gave (*R*)-adducts making the mechanistic course of the reaction amenable to interpretation based on an imine intermediate discussed earlier. In the presence of 5 mol% of the catalyst (92) nitrocyclohexane underwent addition to 2-cycloheptenone in 84% yield and in 84% e.e. The nitro group can be removed with tri-butyltin hydride without loss of optical purity, to give an overall transformation that allows the asymmetric alkylation of enones at the β-carbon.

Hanessian has used proline (108) in conjunction with 2,5-dimethyl-piperazine (109) for the metal-free addition of nitroalkanes to cyclic enones (S. Hanessian and V. Pham, *Org. Lett.*, 2000, **2**, 2975). Substantial non-linear effects indicate a complex multi-component chiral catalytic system is operating here, making an interpretation of the results through a mechanistic rationale a lot more difficult than with the Yamaguchi system.

(108) 3 mol%, (109), $CHCl_3$, r.t.

(108) (109)

93% e.e., 73% yield

It appears, however, that this is a superior system for cyclic substrates; addition of 3 mol% proline (108) and 2,5-dimethylpiperazine (109) to a solution of 2-cyclohexenone and nitrocyclohexane in chloroform at room temperature gave the product in 73% yield and in 93% e.e.

Sera has reported the addition of nitromethane to chalcone using the naturally occurring amines, quinine and quinidine (A. Sera et al., *J. Org. Chem.*, 1988, **53**, 1157). Although the reaction does not occur at atmospheric pressure in aprotic solvents, raising the pressure to 900 MPa

CH_3NO_2, (110) 2 mol%, 900 MPa, PhMe, r.t.

(110)

60% e.e., 100% yield, stereochemistry undefined

gave the adduct in up to 60% e.e. (100% yield) in the presence of 2 mol% quinidine (110). Raising or lowering of the pressure from 900 MPa lowered the e.e. and the yield for the reaction.

(d) Transition Metal Complexes: The prolinamide (111), in the presence of Ni(acac)$_2$, promoted the asymmetric conjugate addition of nitromethane to chalcone in up to 61% e.e. (M. Basato et al., *J. Mol. Cat.*, 1987, **42**, 115; A. Schionato, S. Paganelli, C. Botteghi and G. Chelucci, *J. Mol. Cat.*, 1989, **50**, 11; C. Botteghi et al., *J. Mol. Cat.*, 1991, **66**, 7).

(111)

CH_3NO_2, (111) 16 mol%, Ni(acac)$_2$ 16 mol%, PhMe, r.t.

61% e.e., 33% yield

(e) Lanthanum-binol complexes: The Shibasaki series of heterobimetallic catalysts are also effective for the addition of nucleophiles such as nitromethane to Michael acceptors.

(112)

CH_3NO_2, (112)20 mol%, ^{t}BuOH 120 mol% PhMe,-20 °C

97% e.e., 59% yield

Thus, with the use of (*R*)-LPB (112) (20 mol%), in which the lanthanum works as a Lewis acid and the potassium naphthoxide as a Bronsted base, nitromethane reacts with chalcone to give the adduct in 59% yield and 97% e.e. (K. Funabashi et al., *Tetrahedron Lett.*, 1998, **39**, 7557). So far this system has only been used with nitromethane.

7. Ester Nitrile Addition.

The addition of a nucleophile containing both ester and nitrile functionality to an α,β-unsaturated ketone leads to the possibility of generating two new asymmetric centres in a product which contains three chemically distinct functional groups. This problem has been approached using two different metals, rhodium and palladium.

(a) Rhodium with a bidentate P,P ligand: Sawamura has shown that the asymmetric conjugate addition reaction of an α-cyano Weinreb amide with an acyclic enone, catalysed by a rhodium complex with *trans*-chelating chiral diphosphine PhTRAP (113), proceeds in up to 94% e.e. and in 99% yield. The ketone of the resulting adduct can be selectively protected as its acetal and this compound can be further functionalised with reducing agents and Grignard reagents. This catalyst also provides remarkably high e.e. values for a range of substituted esters as the Michael donors, and a number of acrylates and enones as Michael acceptors (M. Sawamura, H. Hamashima, H. Shinoto and Y. Ito, *Tetrahedron Lett.*, 1995, **36**, 6479; M. Sawamura, H. Hamashima and Y. Ito, *Tetrahedron* 1994, **50**, 4439; M. Sawamura, H. Hamashima and Y. Ito, *J. Am. Chem. Soc.*, 1992, **114**, 8295). It is essential that the nitrile group is present in the substrate to bind to the active catalyst, otherwise the superb e.e. values for the reaction are reduced.

Ph$_2$P Fe PPh$_2$ Fe

(113)

Ph + NC OMe N — Rh(acac)(CO)$_2$ 1 mol%, (113) 1 mol%, PhH, 3°C → Ph CN N OMe

94% e.e., 99% yield

More recently Nozaki has shown that the asymmetric Michael reaction of α-cyano esters to α,β-unsaturated ketones and acrylates is also catalysed by a chiral rhodium(I) complex with the bis(phosphine) (114), the reaction proceeding in up to 93% yield and in 72% e.e. (K. Inagaki, K.

Ph$_2$P Ph$_2$P O O

(114)

+ CN CO$_2$^{i}Pr — Rh(acac)(CO)$_2$ 1 mol%, (114) 1 mol%, PhH, 0°C → CO$_2$^{i}Pr CN

72% e.e., 93% yield

Nozaki and H. Takaya, *Synlett*, 1997, 119). It was also shown, by NMR studies, that it is the nitrile moiety of the Michael donor that binds to the rhodium metal, and the ligand-metal bond has a *trans*-geometry.

(b) Cationic Palladium(II) complexes: It is apparent that the palladium catalysed Michael addition is a lot more sluggish than its rhodium counterpart, however, it does offer yet another opportunity to explore the subtleties of the asymmetric conjugate additions, and further develop the scope of the reaction.

(115) 1 mol%,
Hunigs base 10 mol%
PhMe, r.t.

34% e.e., 86% yield

(115)

Richards has shown that the cationic bis(oxazoline) palladium(II) complex (115) catalyses the addition of ethyl α-cyanopropanoate to methyl vinyl ketone in up to 86% yield and in 34% e.e. (M. A. Stark, G. Jones and C. J. Richards, *Organometallics*, 2000, **19**, 1282; M. A. Stark and C. J. Richards, *Tetrahedron Lett.*, 1997, **38**, 5881).

8. Silylketene Acetal Addition

The addition of silylketene acetals to α,β-unsaturated carbonyls has been thoroughly investigated with a host of chiral Lewis acidic metal complexes including tin, titanium, scandium, and, most successfully, copper.
The catalytic tin(II) triflate promoted addition of ethylthiotrimethylsilane to (*E*)-4-phenyl-but-3-en-2-one in the presence of the chiral diamine (116) gave the product in up to 82% yield and in 70% e.e. (N. Iwasawa, T. Yura and T. Mukaiyama, *Tetrahedron*, 1989, **45**, 1207; T. Yura, N. Iwasawa, K. Narasaka and T. Mukaiyama, *Chem. Lett.*, 1988, 1025). It was proposed that the tin co-ordinated the sulfur and aided in breaking the rather weak sulfur-silicon bond, thus generating a chiral tin(II) ene-

thiolate, which then added selectively to one of the diastereofaces of the prochiral enone.

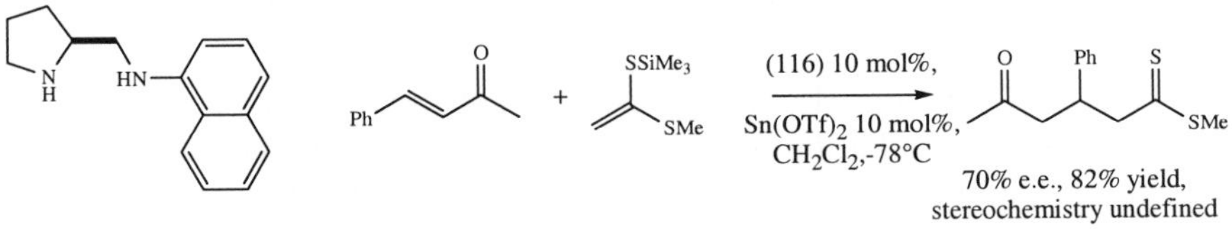

(116) 10 mol%, $Sn(OTf)_2$ 10 mol%, CH_2Cl_2,-78°C

70% e.e., 82% yield, stereochemistry undefined

(116)

The titanium oxide containing ligand (117) catalysed the asymmetric addition of the silyl enol ethers of thioesters to both cyclic and acyclic α,β-unsaturated ketones to give 1,5-dicarbonyl compounds in high yields and moderate to high e.e. (S. Kobayashi, S. Suda, M. Yamada and T. Mukaiyama, *Chem. Lett.*, 1994, 97). Addition of the *S*-benzyl thioester (118) to 2-cyclopentenone in the presence of 20 mol% of (117) in toluene at –78°C gave the product in 75% yield and in 90% e.e. E.e. values were lower for cyclic substrates (~40%).

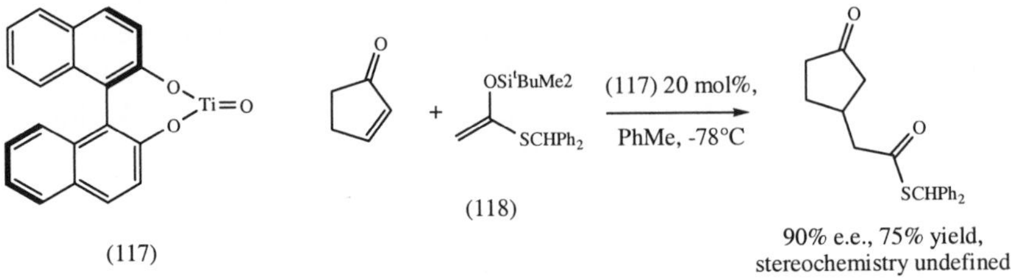

(117) 20 mol%, PhMe, -78°C

(117) (118)

90% e.e., 75% yield, stereochemistry undefined

Scolastico has also reported the use of a series of chiral titanium complexes to promote the addition of silylketene acetals to 2-cyclopentenone 2-carboxylic acids (A. Bernardi, K. Karamfilova, S. Sanguinetti and C. Scolastico, *Tetrahedron* 1997, **53**, 13009).

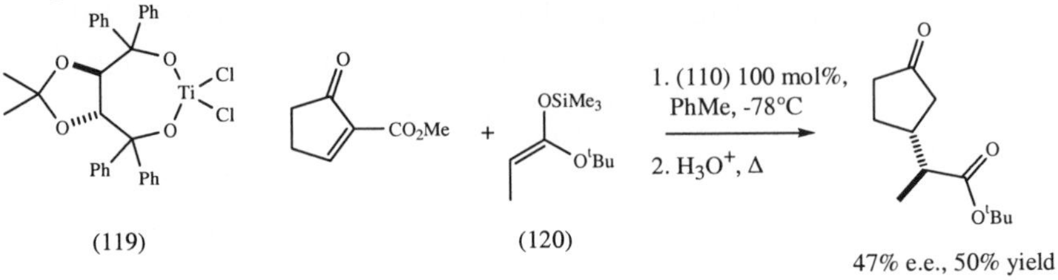

1. (110) 100 mol%, PhMe, -78°C
2. H_3O^+, Δ

(119) (120)

47% e.e., 50% yield

For example, the Lewis acidic TADDOL derived titanium complex (119) catalysed the addition of silylketene acetal (120) to 2-methoxycarbonyl-2-cyclopentenone to give, after hydrolysis/decarboxylation, the adduct in up to 47% e.e.

The most studied and successful catalysts for the addition of silylketene acetals in the conjugate addition reaction are bis(oxazolines). The scandium(II) triflate catalysed addition of 2-(trimethylsilyloxy)furans to oxazolidinone enoates in the presence of hexafluoroisopropanol proceeded stereoselectively to give 4-substituted butenolides in good yields. A 1:1 complex of scandium(II) triflate and 3,3'bis(diethylamino-methyl)-1,1'-bi-2-naphthol showed excellent *anti*-selectivity and moderate enantioselectivity (up to 68% e.e.). The copper(II) triflate bis(oxazoline) complex (121) exhibited excellent enantioselectivity (up to 95% e.e.) and good *anti*-selectivity (up to 96%) (H. Kitajima, K. Ito and T. Katsuki, *Tetrahedron*, 1997, **53**, 17051).

2^{+} 2 OTf^{-}

(121)

(121) 5 mol%, HFIP, CH_2Cl_2, 0°C

95% e.e., 89% yield

Evans has shown that the addition of silylketene acetals to alkylidene malonates can be catalysed by addition of the bis(oxazoline)-copper complex (122) in the presence of hexafluoroisopropanol to give 1,5-di-carbonyl adducts in up to 99% e.e. (D. S. Johnston and D. A. Evans, *Acc. Chem. Res.*, 2000, **33**, 325; D. A. Evans, T. Rovis, M. C. Kozlowski and J. S. Tedrow, *J. Am. Chem. Soc.*, 1999, **121**, 1994; D. A. Evans, M. C. Willis and J. N. Johnston, *Org. Lett.*, 1999, **1**, 865; D. A. Evans and D. S. Johnston, *Org. Lett.*, 1999, **1**, 595). For example, addition of the silyl-ketene acetal (124) to the cyclohexyl-substituted malonate (123) gave the conjugate addition product in 99% yield and in 95% e.e. This catalyst is also highly effective for the addition of silylketene acetals to fumaroyl oxazolidinones and azodicarboxylates.

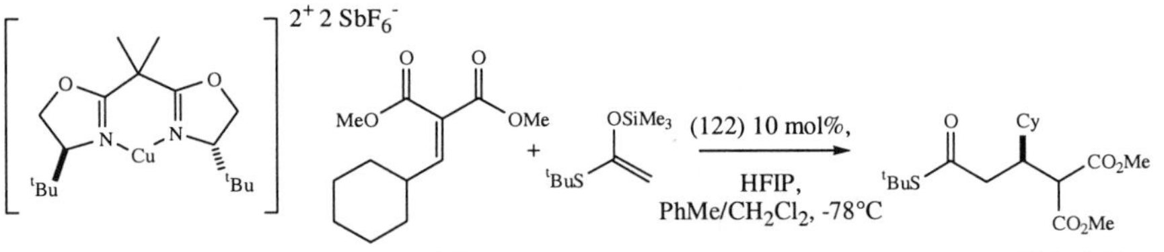

(122) (123) (124) 95% e.e., 99% yield

Evans has carried out extensive X-ray and modelling studies on these catalysts in order to provide a mechanistic rationale for the sense of asymmetric induction observed within the reaction and shown that the catalysts are amenable to design. The sense of enantioface selection by these copper(II) bis(oxazolines) (121) and (122) can be explained by mutual co-ordination of two carbonyl atoms to the copper and the resulting complex adopting the least hindered conformation (125) and (126) respectively. The silylketene acetal then approaches from the sterically less hindered bottom side leading to the observed products.

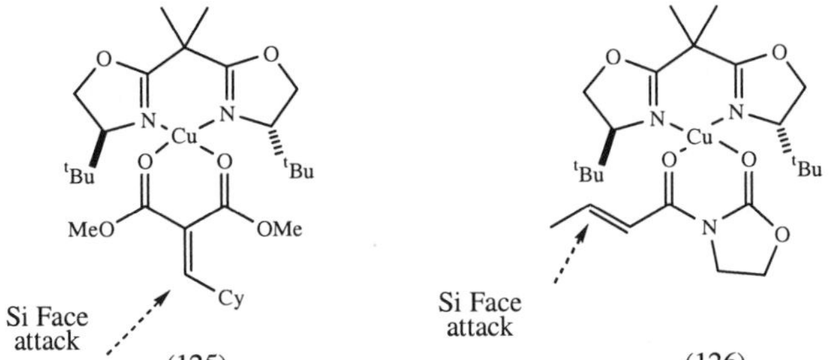

(125) (126)

It has also been shown that conjugate addition of geometrically pure propionate silylketene acetals to 2-methoxycarbonyl-2-cyclopentenone is promoted by the bis(oxazoline)-copper complex (127) with up to 63% e.e. and in 72% d.e. (A. Bernardi, G. Colombo and C. Scolastico, *Tetrahedron Lett.*, 1996, **37**, 8921).

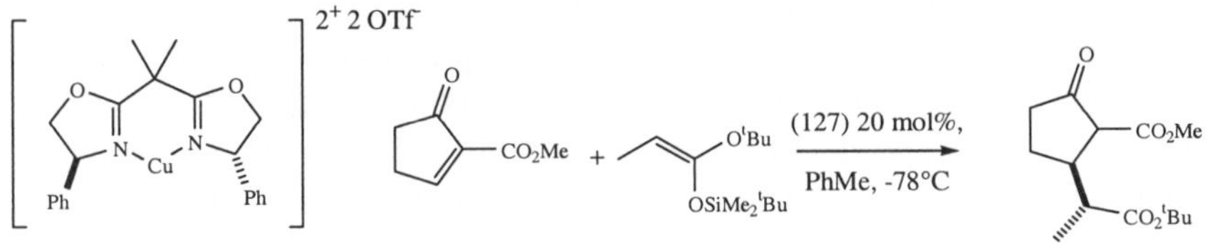

(127) 63% e.e., 65% yield

9. Thiol Addition

Addition of a thio-nucleophile to a β-substituted Michael acceptor results in the formation of a new chiral centre. There are many naturally occurring compounds with the β-thio functionality and so the catalytic asymmetric conjugate addition of thiols to α,β-unsaturated carbonyl compounds has received much attention. E.e. values in excess of 90% have been frequently reported using catalytic amounts of external chiral ligands in conjunction with metals such as lanthanides, cadmium and nickel as well as chiral amine bases.

(a) Chiral ligands in conjunction with metals: The heterobimetallic complexes developed by Shibasaki are excellent catalysts for the addition of thiols to α,β-unsaturated carbonyl compounds (E. Emori, T. Arai, H. Sasai and M. Shibasaki, *J. Am. Chem. Soc.*, 1998, **120**, 4043). The addition of benzylthiol to 2-cyclopentenone in the presence of 10 mol% of the catalyst LSB (99) proceeded in 86% yield and in 90% e.e. These catalysts also contain an acidic proton as part of their structure in the proposed catalytic cycle and since this proton is in a chiral environment it can be used as a source of asymmetric protonation of the intermediate enolate. Thus, the addition of 4-*tert*-butylbenzenethiol to the α-substituted ethyl thiomethacrylate in the presence of the samarium based catalyst SmSB (128) (10 mol%) gave the adduct in 86% yield and in 93% e.e.

M = La (99)
M = Sm (128)

$PhCH_2SH$, (99) 10 mol%, PhMe, -40°C → SCH_2Ph — 90% e.e., 86% yield

4-^{t}BuPhSH, (128) 10 mol%, CH_2Cl_2, -78°C — 93% e.e., 86% yield

The cadmium iodide complex of the chiral pyridine *N*-oxide (129) has been used for the catalytic asymmetric conjugate addition of thiols to both cyclic and acyclic α,β-unsaturated carbonyl compounds. For example, conjugate addition of methanethiol to (*E*)-2-pentenal in the presence of 1 mol% of cadmium iodide and 1 mol% of the catalyst (129) gave the adduct in 89% yield and in 69% e.e. E.e. values of up to 78% were also observed for the addition of thiols to cyclic substrates (M. Saito, M. Nakajima and S. Hashimoto, *J. Chem. Soc., Chem. Commun.*, 2000, 1851; M. Saito, M. Nakajima and S. Hashimoto, *Tetrahedron*, 2000, **56**, 9589).

(129)

MeSH, (129) 1 mol%, CdI_2 1 mol%, PhMe, r.t.

SMe

69% e.e., 89% yield

The addition of thiols to 3-(2-alkenoyl)-2-oxazolidinones is catalysed by the chiral Lewis acidic DBFOX/Ph aqua complex of nickel(II) perchlorate (130) (S. Kanemasa, Y. Oderaotoshi and E. Wada, *J. Am. Chem. Soc.*, 1999, **121**, 8675). Addition of thiophenol to 3-crotonyl-2-oxazolidinone in the presence of 10 mol% of the catalyst (130) gave the adduct in 84% yield and in 94% e.e.

M = $Ni(ClO_4)_2$ (130)

PhSH, (130) 1 mol%, CH_2Cl_2/THF, 0°C

SPh

94% e.e., 84% yield

(b) Chiral amine bases: Aida has used the chiral *N*-substituted porphyrin base (131) which has a conformationally locked asymmetric nitrogen atom to catalyse the asymmetric conjugate addition of thiophenol to 2-cyclohexenone; the reaction proceeding in 99% conversion and in 47% e.e. (A. Ito, K. Konishi and T. Aida, *Tetrahedron Lett.*, 1996, **37**, 2585). The use of substituted thiophenols led to e.e. values of up to 58% being obtained.

PhSH,
(131) 1 mol%, PhMe, -40°C

47% e.e., 99% conversion

(131)

The addition of thiophenol to 2-cyclohexenone in the presence of the naturally occurring amine bases quinine and quinidine has been investigated (A. Sera et al., *J. Org. Chem.*, 1988, **53**, 1157). In the presence of quinidine (110) the addition of thiophenol occurred in 49% e.e. and in 99% yield. It was also shown that at elevated pressures (up to 900 MPa) the e.e. value of the product decreased and that quinidine gave slightly better results than quinine for the reaction.

PhSH,
(110) 0.8 mol%, PhMe, r.t.

49% e.e., 99% yield

(110)

Tomioka has reported remarkably high level of asymmetric induction in the addition of aliphatic and aromatic thiols to both cyclic and acyclic α,β-unsaturated carbonyl compounds in the presence of the tridentate amine catalyst (132) (K. Nishimura, M. Ono, Y. Nagaoka and K. Tomioka, *J. Am. Chem. Soc.*, 1997, **119**, 12974).

(132) 10 mol%,
PhMe, -60°C

97% e.e., 99% yield

(132)

For example, addition of 2-(trimethylsilyl)thiophenol to (*E*)-methyl-crotonate gave the Michael adduct in 99% yield and in 97% e.e. This reaction was shown to be under kinetic control, since exposure of a racemic mixture of the product under the standard reaction conditions returned racemic material.

10. Addition of nitrogen based nucleophiles

Despite the huge interest in the synthesis of β-amino acids and other similar functionalities there are surprisingly few reports of the catalytic asymmetric conjugate addition of nitrogen-based nucleophiles to Michael acceptors. However, the methods that have been disclosed are exceedingly versatile and appear to be tolerant of a number of functional groups, providing the adducts in good yield and with high e.e. values. Jacobsen has had considerable success with his salen catalyst (133) for carrying out a wide variety of asymmetric transformations including the opening of epoxides with azides, water, carboxylic acids and phenols, the addition of hydrogen cyanide to imines and the catalysis of hetero-Diels-Alder reactions as well as in the epoxidation reactions of alkenes. He has also shown that this chiral template can be used for the addition of hydrazoic acid to unsaturated imides (J. K. Myers and E. N. Jacobsen, *J. Am. Chem. Soc.*, 1999, **121**, 8959). For example, addition of hydrazoic acid to the unsaturated imide (135) in the presence of 5 mol% of the catalyst (134) gave the adduct in 95% yield and in 97% e.e. The reaction is rather insensitive to the steric demands of the substrate and the reaction is tolerant of other Lewis basic functionalities in the molecule. The adducts can easily be converted to the corresponding β-amino acids by a simple two-step reduction/hydrolysis protocol.

Miller has recently shown that azides can be added to α,β-unsaturated amides using simple tri-peptides as metal-free catalysts (T. E. Horstmann,

D. J. Guerin and S. J. Miller, *Angew. Chem., Int. Ed. Engl.*, 2000, **39**, 3635). Thus, by use of the tri-peptide (136) at ambient temperature, the addition of azide to the α,β-unsaturated amide (137) proceeded in 79% yield and in 85% e.e. Although these values are not quite as high as those observed by Jacobsen, this is a very promising system that will undoubtedly be improved on.

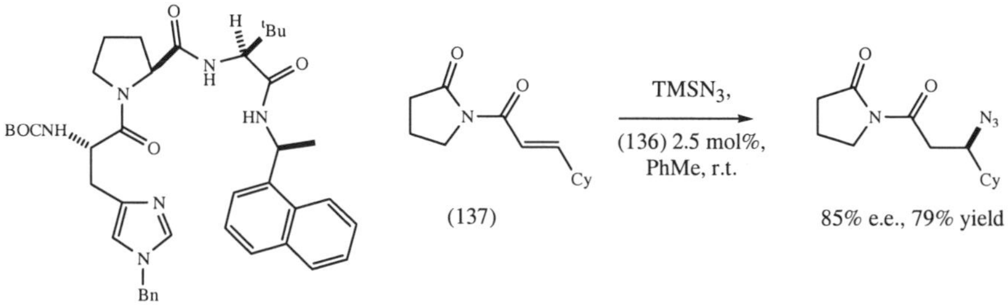
BOCNH
N
N
Bn
HN
TMSN3,
(136) 2.5 mol%,
PhMe, r.t.
Cy
N3
85% e.e., 79% yield

(136)

(137)

An asymmetric synthesis of 2-azetidinones utilising an asymmetric conjugate addition of amines to 2-phenylsulfonyl-3-phenylpropenyl chloride has been reported to be catalysed by the salen-copper(II) complex (138), the adduct being isolated in up to 54% yield and in 46% e.e. (F. Zhou, M. R. Detty and R. J. Lachicotte, *Tetrahedron Lett.*, 1999, **40**, 458).

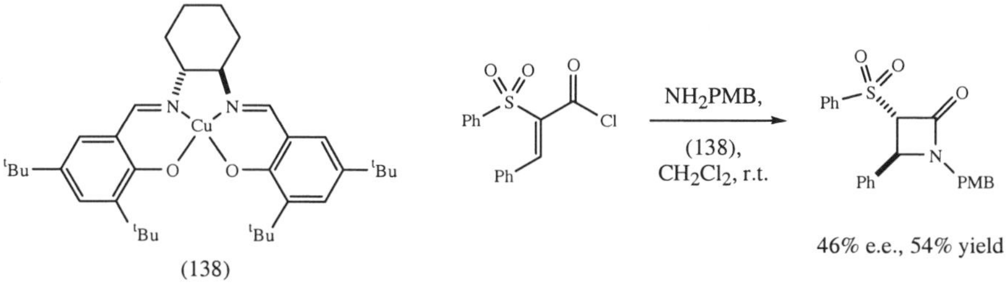
Cu
tBu
NH2PMB,
(138),
CH2Cl2, r.t.
Ph
Cl
PMB
46% e.e., 54% yield

(138)

The bis(oxazoline) (139) catalyses the addition of *O*-benzylhydroxylamine to the pyrazole-derived crotonamide (140) in the presence of magnesium bromide. Up to 95% e.e. was obtained in the presence of 10 mol% of the catalyst (M. P. Sibi, J. J. Shay, M. Liu and C. P. Jasperse, *J. Am. Chem. Soc.*, 1998, **120**, 6615).

(139)

(140)

H_2NOBn,
(139) 10 mol%,
$MgBr_2$ 10 mol%,
CH_2Cl_2, -25°C

95% e.e., 87% yield

11. Natural Product Syntheses

A true test for the applicability of any new methodology is not only the level of its uptake by the general synthetic community, but also its use in the synthesis of natural products and other compounds of biological significance. To this end there have been numerous uses of the catalytic asymmetric conjugate addition reactions discussed within this review in natural product synthesis.

One of the most accessible naturally occurring molecules from a conjugate addition reaction is (*R*)-muscone, an odiferous constituent isolated from musk, an important part of the perfume industry. Conjugate addition of a methyl anion to the macrocyclic enone (*E*)-2-cyclopentadec-enone can give directly the naturally occurring material. Alexakis has reported the synthesis of (*R*)-muscone in 53% yield and in 79% e.e. using the binaphthyl-based phosphite (141) as the source of chirality (A. Alexakis et al., *Synlett*, 1999, 1811).

(141)

Conditions

(142)

Alexakis: Me_2Zn, (141) 4 mol%, $Cu(OTf)_2$ 2 mol%, PhMe, -10°C, 79% e.e.
Tanaka: MeLi, (142) 33 mol%, CuI 33 mol%, THF, PhMe, -78°C, 99% e.e.

Tanaka has shown that use of the tridentate camphor based ligand (142) in conjunction with methyllithium and copper(I) iodide gave (*R*)-muscone in 85% yield and >99% optical purity (K. Tanaka, J. Matsui and H. Suzuki, *J. Chem. Soc., Perkin Trans I.*, 1993, 153). Yamaguchi using his proline rubidium salt (92) reported an alternative approach to the epimeric material (S)-muscone (M. Yamaguchi, T. Shiraishi and M. Hirama, *J. Org. Chem.*, 1996, **61**, 3520). Again starting with (*E*)-2-cyclopentadec-enone and treating it with di-*tert*-butylmalonate in the presence of the catalyst (92) and decarboxylation gave a β-keto acid. Reductive removal of the acid group furnished (*S*)-muscone in 36% overall yield and 82% e.e.

Iwata has used his methodology developed for the addition of dimethyl-aluminium to 2,4-cyclohexadienones in the synthesis of the phytoalexin (-)-solavetivone (145). Conjugate addition of dimethylaluminium to the dienone (144) in the presence of copper(I) triflate, *tert*-butyldimethylsilyl triflate and the external chiral ligand (143) gave (-)-solavetivone in 93% yield and in 62% e.e. (Y. Takemoto et al., *J. Chem. Soc., Chem. Commun.*, 1996, 1655).

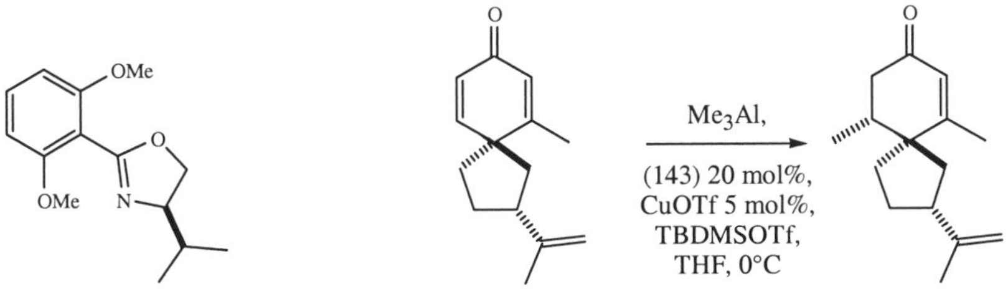

OMe OMe O N O O

(143) (144)

Me_3Al, (143) 20 mol%, CuOTf 5 mol%, TBDMSOTf, THF, 0°C

(145)

62% e.e., 93% yield

Shibasaki has used the addition of dimethylmalonate to 2-cyclohexenone in the presence of the heterobimetallic catalyst (*R*)-ALB (101) as a starting point for the total synthesis of the *Strychnos* alkaloid tubifolidine (146) (S. Shimizu et al., *J. Org. Chem.*, 1998, **63**, 7547).

The initial resultant product from any conjugate addition reaction is an enolate, which many groups have used as a nucleophile in order to further functionalise their product. Shibasaki has used this conjugate addition-enolate trapping strategy to synthesise the prostaglandin 11-deoxy-$PGF_{1\alpha}$ (150) (K. I. Yamada, T. Arai, H. Sasai and M. Shibasaki, *J. Org. Chem.*, 1998, **63**, 3666). Conjugate addition of 2-methyldibenzyl malonate to 2-cyclopentenone, catalysed by (*S*)-ALB (147), and trapping

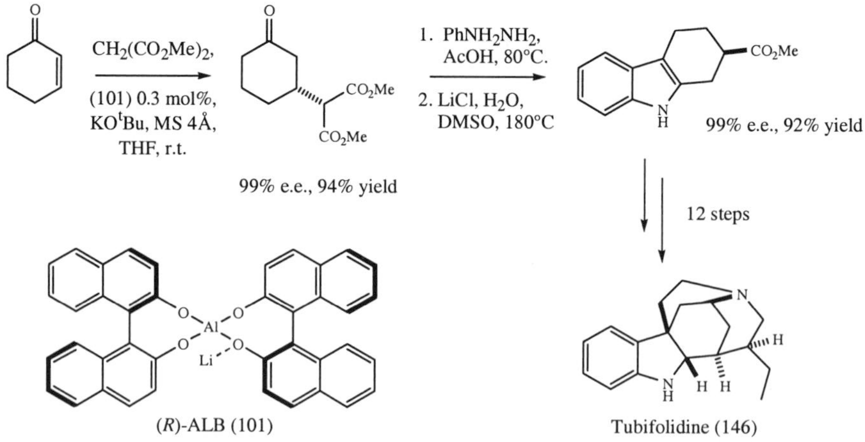

(R)-ALB (101)

Tubifolidine (146)

of the resultant enolate gave the highly functionalised adduct (148), which, after elimination gave the di-substituted cyclopentanone (149) in 73% yield and in 92% e.e. for the three steps. This was then converted into the target molecule 11-deoxy-$PGF_{1\alpha}$ (150) in 17 steps.

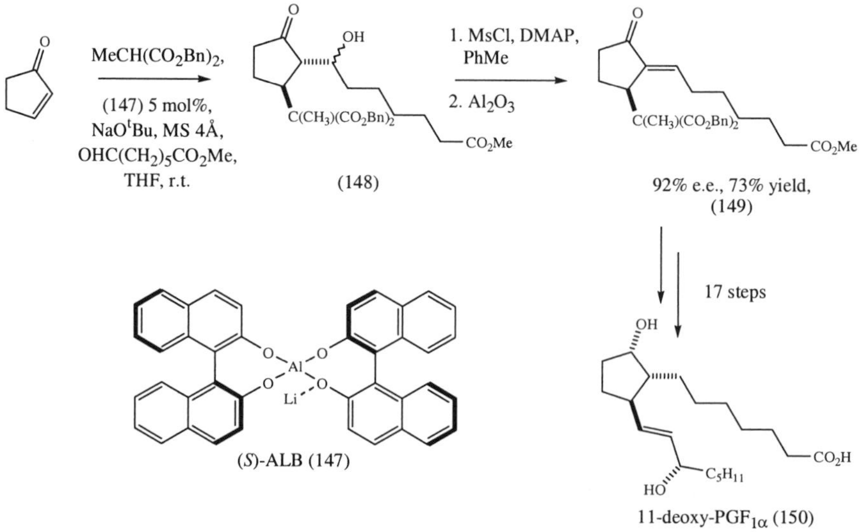

(S)-ALB (147)

11-deoxy-$PGF_{1\alpha}$ (150)

Shibasaki has also used the (*S*)-ALB catalyst (147) in the preparation of a key intermediate in the synthesis of coronafacic acid (T. Arai, H. Sasai, K. Yamaguchi and M. Shibasaki, *J. Am. Chem. Soc.*, 1998, **120**, 441).

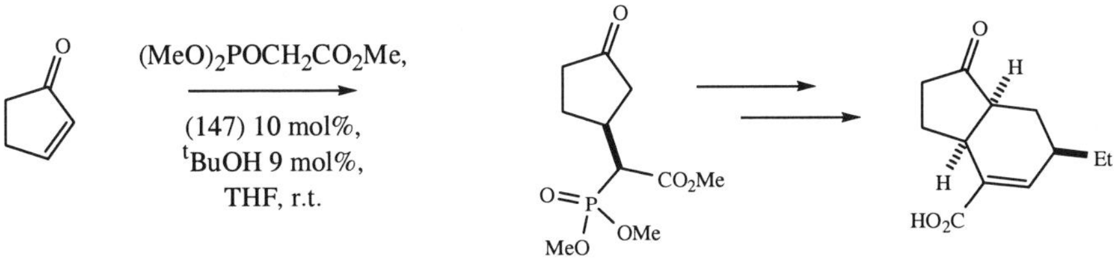

(MeO)$_2$POCH$_2$CO$_2$Me,
(147) 10 mol%,
^{t}BuOH 9 mol%,
THF, r.t.

95% e.e., 95% yield

Coronofacic acid

Corey has used his chiral phase-transfer catalysts based on the cinchona alkaloids for the synthesis of (*R*)-baclofen (154), a GABA$_β$ receptor agonist, and (*S*)-ornithine (152), a naturally occurring α-amino acid.

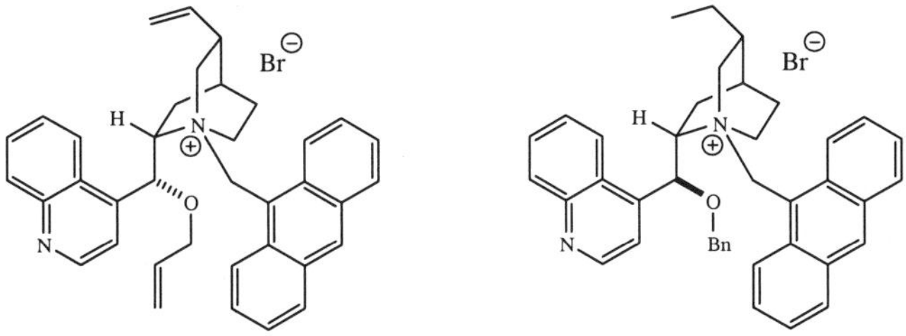

(93)

(151)

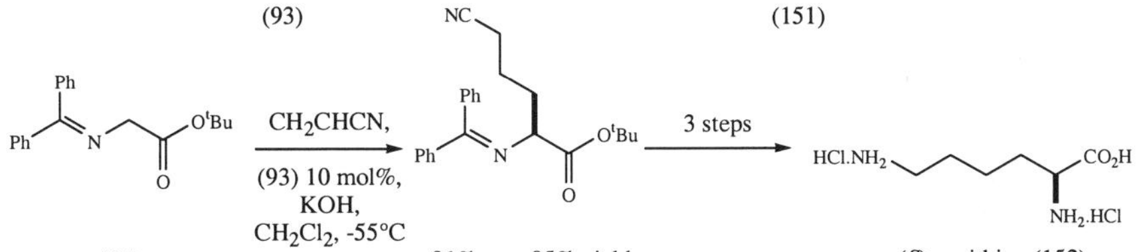

CH$_2$CHCN,
(93) 10 mol%,
KOH,
CH$_2$Cl$_2$, -55°C

(94)

91% e.e., 85% yield

3 steps

(*S*)-ornithine (152)

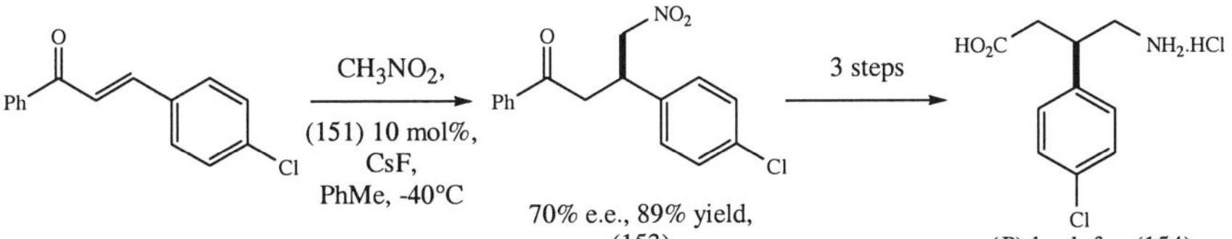

CH$_3$NO$_2$,
(151) 10 mol%,
CsF,
PhMe, -40°C

70% e.e., 89% yield,
(153)

3 steps

(*R*)-baclofen (154)

Addition of the enolate anion derived from the glycinate ester (94) to acrylonitrile in the presence of 10 mol% of the catalyst (93) gave the adduct in 85% yield and 91% e.e. Reduction of the nitrile moiety and removal of the protecting groups gave (*S*)-ornithine (152) in just four steps and an overall yield of 53% from cheap achiral precursors (F. Y.

Zhang and E. J. Corey, *Org. Lett.*, 2000, **2**, 1097). Conjugate addition of nitromethane to 4-chlorobenzylideneacetophenone in the presence of 10 mol% of the cinchonium salt (151) and caesium fluoride in toluene at -40°C gave the Michael adduct (153) in 89% yield and in 70% e.e. This adduct could easily be crystallised up to optical purity. Baeyer-Villiger oxidation, reduction of the nitro group and hydrolysis of the resultant mono-substituted pyrrolidine gave optically pure (*R*)-baclofen (154) in 52% overall yield for the four steps (E. J. Corey and F. Y. Zhang, *Org. Lett.*, 2000, **2**, 4257).

12. Conclusion

The past decade has seen a vast increase in the number of methods available for the catalytic asymmetric conjugate addition reaction. There is an array of catalysts available that can provide products with e.e. values in excess of 90% for the addition of most types of nucleophiles to a variety of Michael acceptors. It appears that although no one catalyst provides a broad scope of substrate and reagent applicability it is now possible to choose the appropriate catalyst for each particular system depending on the chosen nucleophile.

Second Edition of Rodd's Chemistry of Carbon Compounds,
Volume V, Topical Volumes
Asymmetric Catalysis, edited by M.Sainsbury

Chapter 7

COMBINATORIAL APPROACHES TO ASYMMETRIC CATALYSIS

ANDY MERRITT

1. Combinatorial Chemistry

The primary development and use of combinatorial chemistry as a research technique has been directed to drug discovery, and use is now widespread (see for example 'The combinatorial Index', B. B. Bunin, Academic Press, San Diego, 1998 or 'Combinatorial Chemistry', N. K. Terrett, Oxford University Press, Oxford, 1998). The basis of the approach is that it is more efficient to prepare and test related samples concurrently rather than through consecutive iteration. The use of combinatorial approaches gives maximum results from an input. For example in a dimer synthesis if 10 possible components are available for each half of the final target compounds then the maximum 100 compounds will be synthesised simultaneously through a combinatorial 10x10 matrix array. Synthetic technologies, chemical analysis methods and data analysis tools have all been developed to support the methodology.

Combinatorial techniques are now being applied to many other fields including development of catalytic processes. The following review highlights those research examples where a clearly defined systematic parallel or combinatorial approach has been used to identify new catalytic asymmetric processes. There are many other publications on the development of the methodology (synthetic routes, novel complexes, screening technologies for example) which do not fall under the coverage of this article. However there have been several comprehensive reviews published in the past 2 years that describe in greater detail the development of such approaches (for example see B. Jandeleit *et al., Angew. Chem. Int. Ed. Eng.*, 1999, **38**, 2494; K. W. Kuntz, M. L. Snapper and A. H. Hoveyda, *Curr. Op. Chem. Biol.*, 1999, **3**, 313; J. M. Newsam and F. Schuth, *Biotech. Bioeng.*, 1998/9, **61**, 203; P. P. Pescarmona *et al., Catalysis Lett.*, 1999, **63**, 1 and references cited within these reviews).

This review is divided into three sections relating to the identification of new metal-ligand combinations, optimising ligands around specific metals, and the identification of catalyst capability through combinatorial techniques. Within each section the examples are grouped by synthetic transformation.

2. Metal-ligand variation

(a) Epoxidation

Asymmetric epoxidation of olefins with hydrogen peroxide has been optimised (M. B. Francis and E. N. Jacobsen, *Angew. Chem. Int. Ed. Eng.*, 1999, **38**, 937). 16 Peptidic compounds on polystyrene bead were capped with one from seven aldehydes or one from five acids to give 192 peptide capped as imines or amides respectively. These were complexed with 30 metal ion sources covering 19 separate transition metals. Initial screening (on bead) of pooled peptides with discrete metal ion sources identified $VOSO_4$ and $FeCl_2$ as effective epoxidation catalysts for trans methyl styrene using H_2O_2. It was shown that $VOSO_4$ and H_2O_2 gave epoxidation in the absence of the peptide ligands, however $FeCl_2$ was not active unless coordinated. Preparation of 12 iterative libraries containing the mixed peptides capped with an individual capping group identified pyridine end caps as the most efficient promoters of epoxidation. Subsequently all the components from the $FeCl_2$ library were prepared individually, which identified 3 peptides (1-3) as significantly more active than the rest, though the best compound (2) only generated low ee (7%).

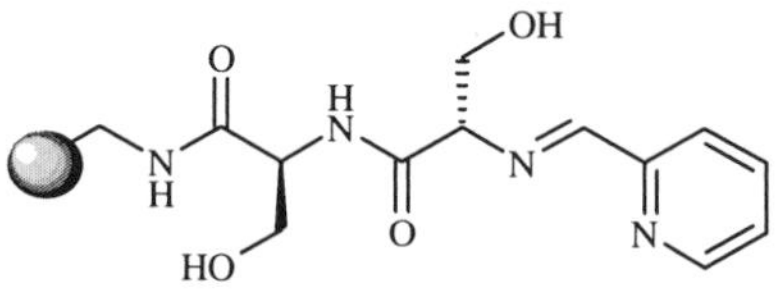

(1)

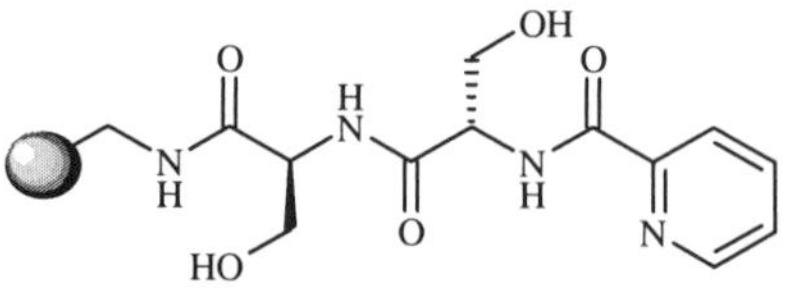

(2)

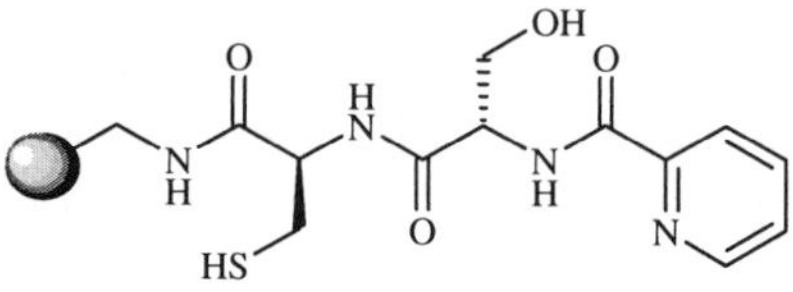

(3)

A subsequent optimisation of the peptide structure based around 12 core peptides and 8 end caps derived from the results of the initial study yielded 3 ligands (4-6) generating up to 20% ee.

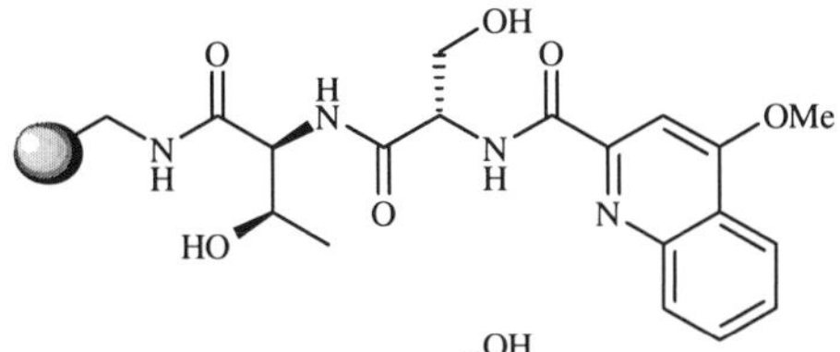

(4) 20% ee

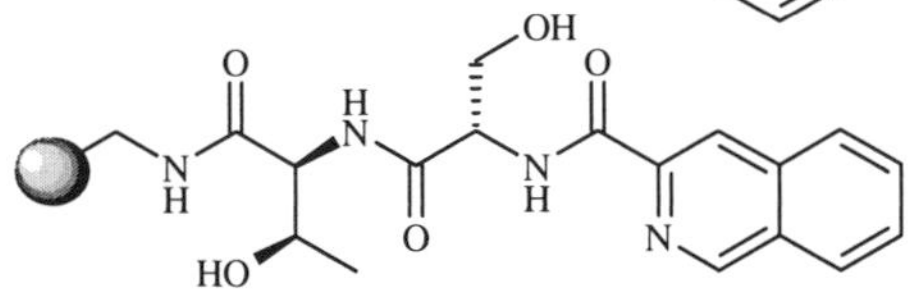

(5) 20% ee

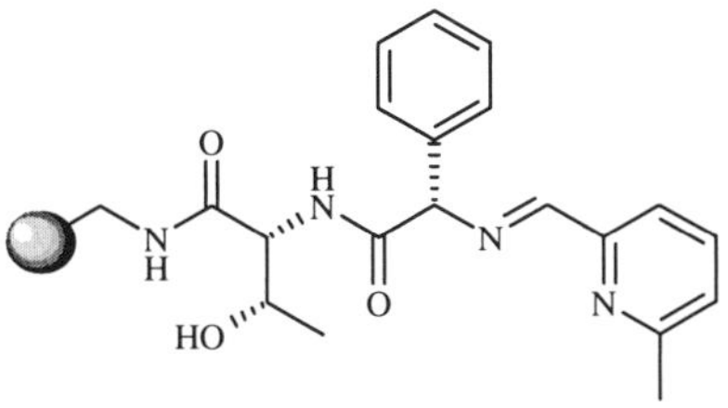

(6) 15% ee

(b) Carbenoid insertion

High throughput catalyst screening has been used to optimise metallocarbene insertion into CH bonds to generate cyclic structure (7), subsequently transformed by oxidation to the substituted indole (8) (K. Burgess *et al.*, *Angew. Chem. Int. Ed. Eng.*, 1996, **35**, 220). The menthyl substituent had previously been shown to have negligible impact on the diastereoselectivity of the reaction (H. J. Lim and G. A. Sulikowski, *J. Org. Chem.*, 1995, **60**, 2326) and therefore any selectivity could be ascribed to the catalyst-ligand combination. Five ligands, seven metal salts and four solvents were tested in a microtitre plate, though not all combinations were evaluated. The optimum ligand (9) for copper catalysts was confirmed as that previously identified, but THF was identified as a better solvent combination, and use of lower temperature gave an improved diastereomeric ration of 3.9:1. More interestingly higher yields with good stereoselectivity (2.7:1) was obtained with ligand (9) combined with $AgSbF_6$, a metal rarely associated with such reactions.

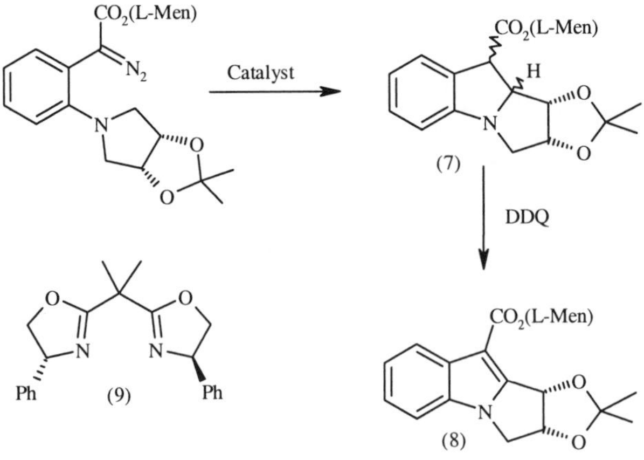

(c) Reductive Aldol reaction

192 independent catalytic systems have been screened for promotion of a reductive aldol coupling between α,β-unsaturated esters and aldehydes (S. J. Taylor and J. P. Morken, *J. Am. Chem. Soc.*, 1999, **121**, 12202). Four transition metals, 7 ligands (plus a blank) and six hydride sources were combined in glass microtitre plates, followed by benzaldehyde and methyl acrylate added as reaction substrates. Chiral GC analysis identified many trends across the combinations whilst also demonstrating where certain factors were clearly not correlated. For example activity did not correlate with diastereoselectivity as shown by the best yielding combinations (10-12) which gave syn:anti ratios between 2:1 and 23:1. Of these (12) was further evaluated against other substrates but gave no further increase in diastereoselectivity.

		Yield	Syn:Anti
(10)	$[(cod)RhCl]_2$-binap-catechol borane	100%	7:1
(11)	$Co(acac)_2$-MOP-$PhSiH_3$	94%	2:1
(12)	$[(cod)RhCl]_2$-DuPhos-Cl_2MeSiH	94%	23:1

(d) Nitrile addition to imines (Strecker reaction)

Tridentate Schiff base complexes (13) for catalysis of the asymmetric Strecker reaction have been identified and optimised through parallel methodologies (M. S. Sigman and E. N. Jacobsen, *J. Am. Chem. Soc.*, 1998, **120**, 4901). The addition of TMSCN to N-allylbenzaldimine to generate, after subsequent acetylation, the allyl amine (14) was evaluated with initially one polymer bound ligand and 11 metals. The best yields were obtained with Gd (95%) and Yb (94%) though with low enantioselectivity. However the ligand without any metal did generate moderate ee (19%) though in only 59% conversion. Subsequently libraries of ligands were screened without metal complexation, yielding ee's up to 30%, and demonstrating the significant effect of the amino acid components on the enantioselectivity. Iterative development of the best compounds through linker modification improved ee's to 55%, and subsequent library optimisation of thiourea linked ligands yielded results up to 80% ee. The best polymer bound catalyst (15) was subsequently prepared in solution as the benzylamine (16) to give reaction ee's of 91% at -78°C

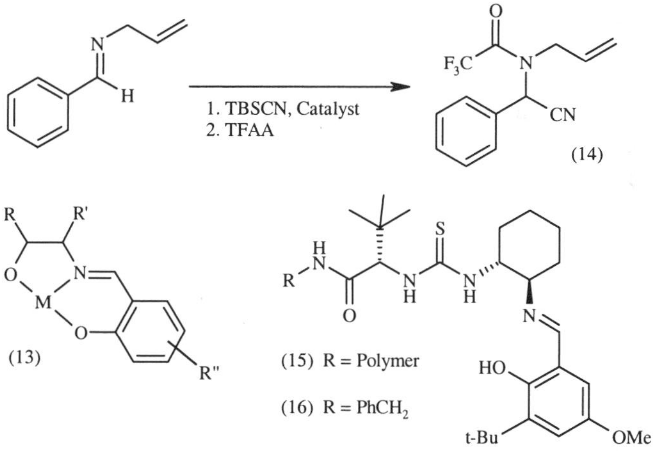

1. TBSCN, Catalyst
2. TFAA

(14)

(13)

(15) R = Polymer

(16) R = $PhCH_2$

In a similar approach peptidic imine metal complexes in solution have been demonstrated to catalyse CN addition from TMSCN reaction with benzyl imines (C. A. Krueger *et al., J. Am. Chem. Soc.*, 1999, **121**, 4284). Initial analysis of ten metals identified Titanium as the optimum species and a subsequent round of peptide optimisation identified ligands (17) and (18) as yielding high ee's (up to 97%) though with moderate conversion (up to 39%). A subsequent study adding protic sources to the reactions increased the yields significantly (conversion 93-100%), with high ee's maintained or in some cases increased.

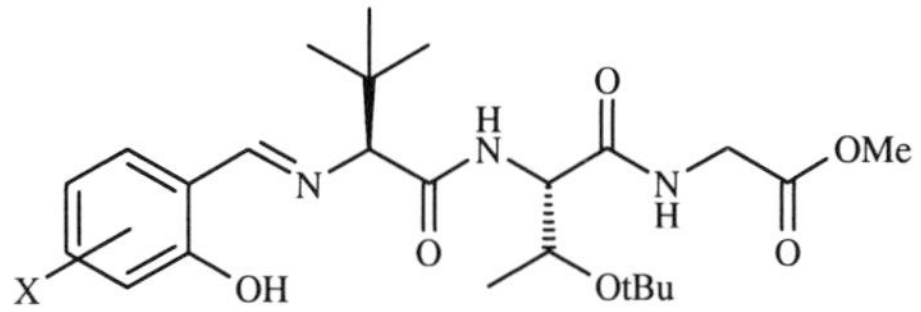

(17) X = 5-OMe

(18) X = 3,5-DiCl

(e) Epoxide opening with oxygen nucleophiles

First row transition metal salen complexes (19) have been evaluated for the opening of epoxides with carboxylic acids (E. N. Jacobsen *et al.*, *Tetrahedron Lett.*, 1997, **38**, 773). Of the 10 complexes evaluated, Cr(III) and Co(II) complexes both gave clean transformation, with 43% ee and 68% ee respectively for the reaction of benzoic acid with cyclohexene oxide. Optimum results were obtained by subsequent addition of one equivalent of tertiary amine base to the reaction. IR thermography of microtitre plate reactions has been used to confirm the cobalt (II) salen complex as the optimum catalysts for epoxide opening with water (M. T. Reetz *et al., Angew. Chem. Int. Ed. Eng.*, 1998, **37**, 2647).

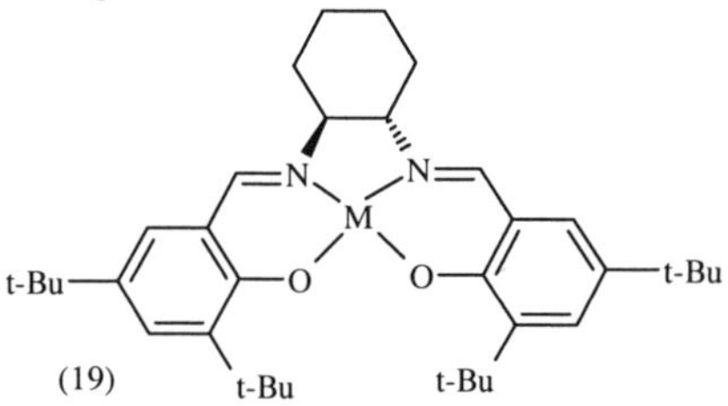

(19)

(f) Cycloaddition

Four Lewis acids have been evaluated in combination with three ligands, two additives and three solvents to identify optimum conditions for the catalysis of aza Diels Alder reactions with Danishfsky's diene (20) using microtitre plate based syntheses (S. Bromidge, P. C. Wilson and A. Whiting, *Tetrahedron Lett.*, 1998, **39**, 8905). Of the 72 combinations, 24 produced ee's greater than 40%, with all of the variable components represented within the 24 best results. The optimum yields and ee's were confirmed by subsequent larger scale preparation to be obtained from MgI_2 with ligand (21) in MeCN and lutidine (64% yield, 97% ee) and $FeCl_3$ with ligand (22) in CH_2Cl_2 with 4a molecular sieves (67% yield, 92% ee).

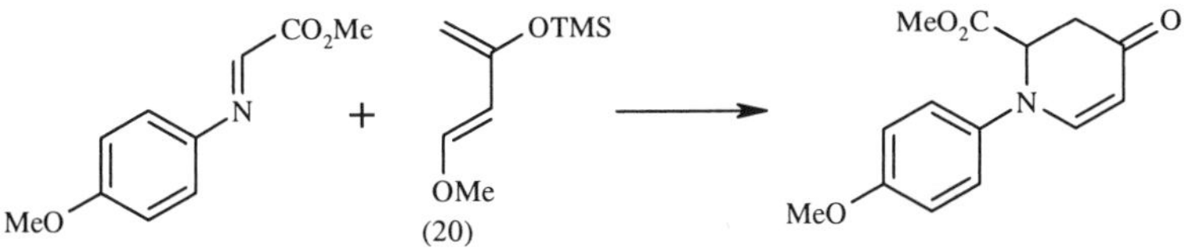

CO_2Me N MeO + OTMS OMe (20) → MeO_2C O N MeO

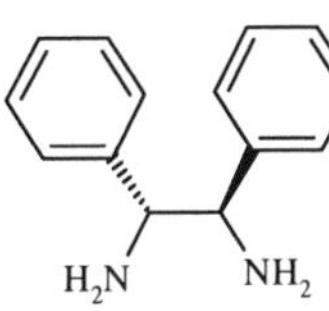

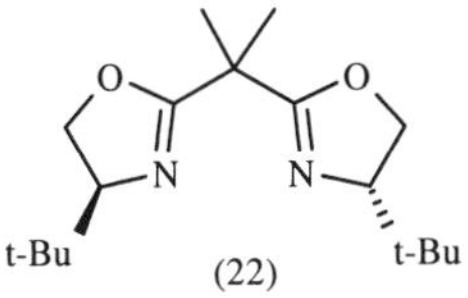

H_2N NH_2 (21) O O N N t-Bu (22) t-Bu

3. Ligand optimisation for specific metal catalysed processes

(a) Reduction

Helical peptides containing phosphine side chains spatially presented to allow co-ordination of Rhodium have been examined for the reduction of enamides. An initial study (S. R. Gilbertson and X. Wang, *Tetrahedron Lett.*, 1996, **37**, 6475) evaluated 63 peptides with the phosphine groups spaced either adjacently (36 examples) or separated by 3 amino acid residues (27 examples) (23). The phosphines were either diphenyl (24) or dicyclohexyl (25) displayed using serine derivatives, and the preparations and subsequent reactions were carried out on solid support using pin technology (A. M. Bray, N. J. Maefi and H. M. Geysen, *Tetrahedron Lett.*, 1990, **31**, 5811). Moderate enantiomeric excesses in the reduction of methyl 2-acetamido acrylate (26) were obtained. Subsequent further developments of the peptide phosphine helix increased ee's up to 38% (S. R. Gilbertson and X Wang, *Tetrahedron*, 1999, **55**, 11609) by reduction of the amino acid spacing to 2 amino acids (specifically alanines). Cleavage of the best ligand from the solid support gave equivalent enantiomeric activity in water, though in organic solvents the results were significantly different, with in some cases the opposite stereochemistry generated in the reduction (though yielding lower ee's than in the solid phase bound reactions).

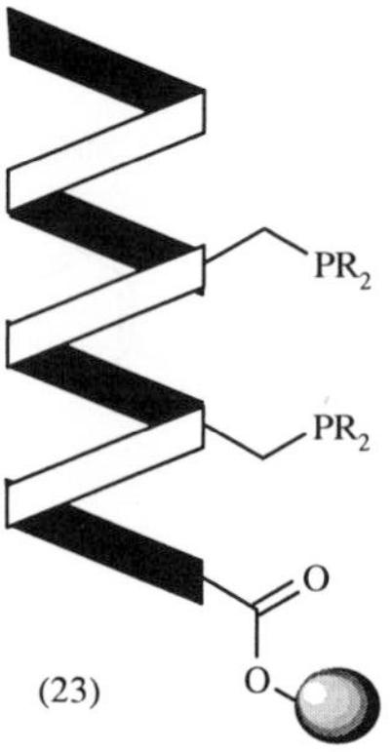

(23) 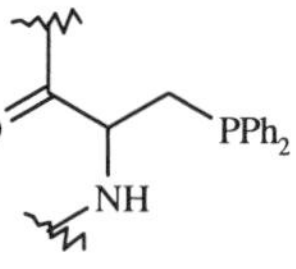 (24) 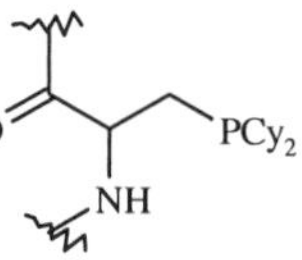 (25) 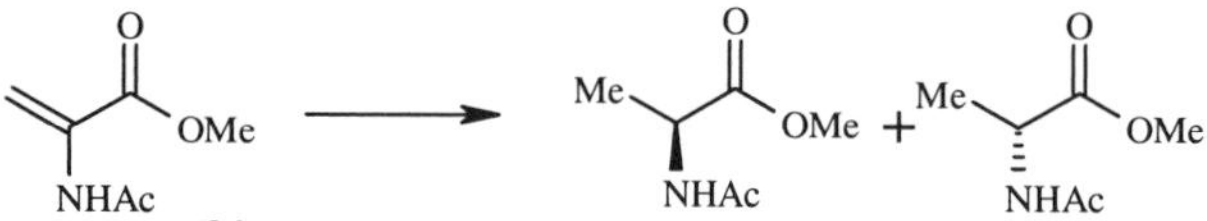 (26)

Reduction of ketones through ruthenium catalysed hydrosilylation has been developed (C. Moreau, C. G. Frost and B. Murrer, *Tetrahedron Lett.*, 1999, **40**, 5617). Mixed ligand ruthenium catalysts were prepared in parallel by combination of $[RuCl_2(C_6H_6)]_2$ with either (S) or (R) diphosphine (27) and one of 25 diamino compounds in an array format. The reduction of acetophenone with diphenylsilane was then examined by GC and chiral HPLC analysis, with the highest enantioselectivity (63% ee) achieved by complex (28). Additional information on ligand trends and on the impact of matched and mismatched ligands was also identified. For example the complex derived from the same phosphine but the enantiomeric amine to complex (28) produced racemic products.

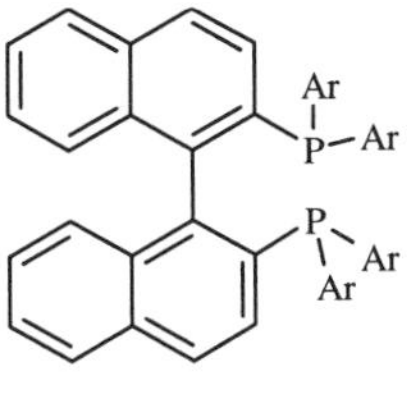

(27) Ar = p-MePh

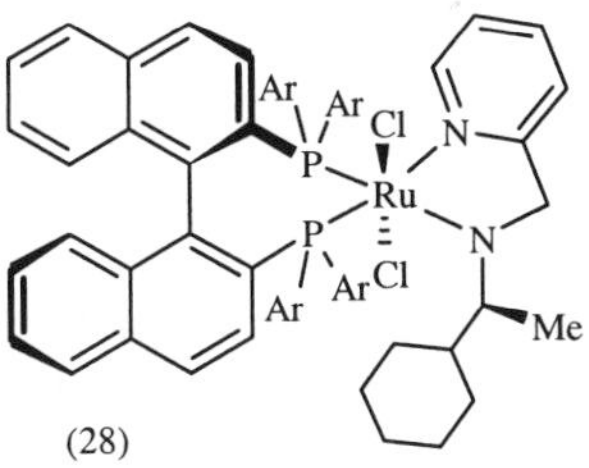

(28)

(b) Carbenoid insertion

Parallel screening of dirhodium carboxylate complexes for the asymmetric insertion of carbenoids into silicon hydride bonds has been investigated (R. T. Buck *et al.*, *Tetrahedron Lett.*, 1998, **39**, 7181). In this study 80 chiral carboxylate derived catalysts were prepared through ligand exchange with previously prepared complexes (typically in parallel batches of 20). Of these 69 successfully generated the desired catalysts and these were screened for activity in promoting the insertion of the carbenoid derived from methyl 2-(diazo)phenylethanoate (29) into 3 silanes. 47 complexes catalysed the desired reaction, with two acids (30) and (31) giving the best results (ee's of 53 and 48% respectively for insertion into tri-isopropylsilane). Subsequent screening of a second set of complexes based on these results gave some enhancement of selectivity, with ee's up to 76% obtained.

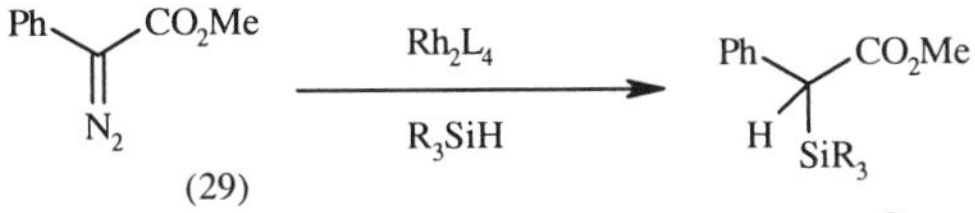

(29)

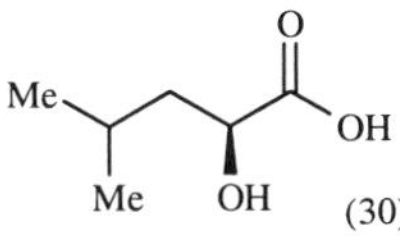

Rh_2L_4, R_3SiH

(30)

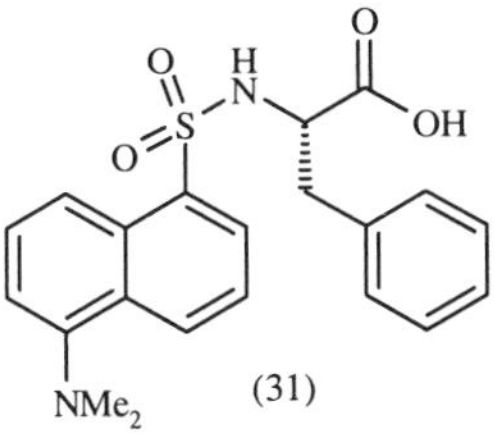

(31)

(c) Palladium mediated allylation

The asymmetric addition of dimethyl malonate to 1,3-diphenylprop-2-enyl acetate (32) utilising new phosphine ligands has been evaluated in two separate studies. Ten phosphinodihydrooxazoles (33) capable of forming a six membered ring chelate on complexation with palladium were synthesised using a modular building block approach (S. R. Gilbertson and C-W. T. Chang, *Chem. Commun.*, 1997, 975). The catalysis was evaluated in two separate experiments, the first with ligands with only one chiral centre (33, R1=H) and the other with those containing two centres. In both cases high ee's were obtained, with little effect noticed for variation of the additional chiral centre. Highest ee's were found for ligands (34) (90% ee) and (35) (97% ee), with the use of $(C_6H_{13})N_4^+$ as counter ion and bis(trimethylsilyl)acetamide as base. An alternative type of oxazaline phosphine ligands (36) prepared through solution phase divergent synthesis methods have been examined in the same reaction (A. M. Porte, J. Reibenspies and K. Burgess, *J. Am. Chem. Soc.,* 1998, **120**, 9180). In this example parallel screening of 13 ligands under various conditions evaluated solvent effects, ligand substituent electronic and steric effects, the effect of chloride ions on the reaction and the effect of ligand to metal ratios, and served to demonstrate the complexities of evaluating any multivariate reaction set up. Highest enantioselectivities (up to 94% ee) were obtained with ligand (37).

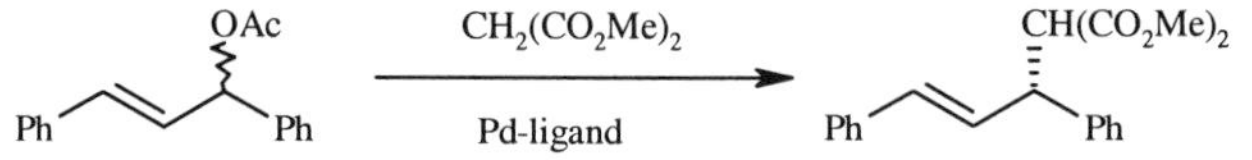

OAc
Ph Ph
(32) 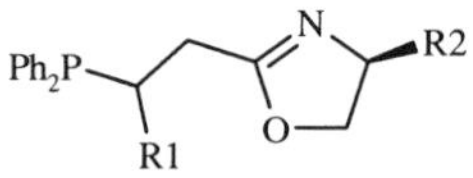
$CH_2(CO_2Me)_2$
Pd-ligand
$CH(CO_2Me)_2$
Ph Ph

Ph_2P N R2 O R1
(33)

	R1	R2
(34)	H	i-Pr
(35)	(R)-Ph	i-Pr

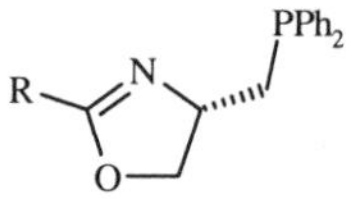

PPh_2
R N O
(36)

(37) R = 1-adamantyl

The development of new C_3 symmetric triarylphosphines (38) as asymmetric catalyst ligands for allylation reactions has also been studied using parallel techniques (M. T. Powell, A. M. Porte and K. Burgess, *Chem. Commun.*, 1998, 2161). A limited set of 6 ligands was evaluated in parallel in the reaction of phthalimide with allylic benzoate (39). An ee of 82% was obtained with ligand (40), with the level of enantioselectivity ascribed to the ordering of the aromatic rings related to the propeller shape of the ligand complex.

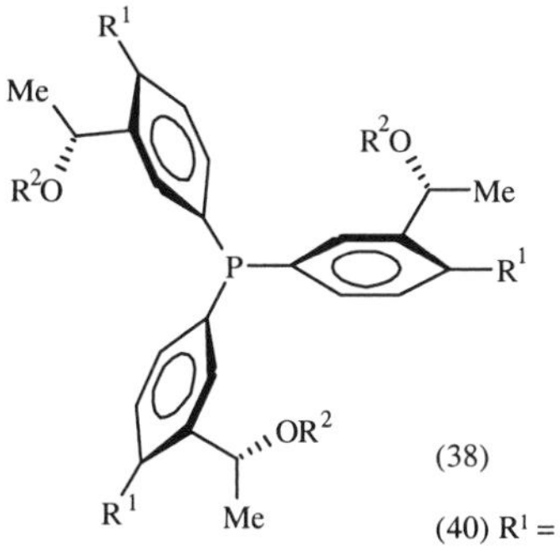

(38)

(40) R^1 = OMe, R^2 = Me

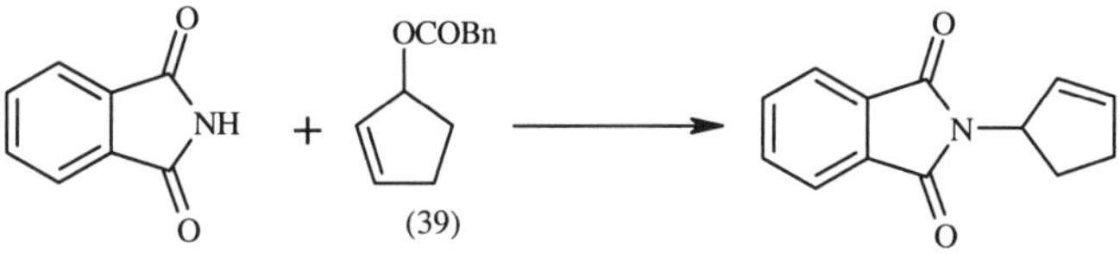

(39)

(d) Diethyl Zinc addition to aldehydes

Pyrrolidinemethanol ligands prepared on solid phase have been evaluated in the enantioselective catalysis of Et_2Zn addition to benzaldehyde (G. Liu and J. A. Ellman, *J. Org. Chem.*, 1995, **60**, 7712). The ligands were prepared in 3 or 4 steps by solid phase synthesis, and initially screened still bound to the resin. Subsequent evaluation involved cleavage from the support, with similar enantiomeric excesses obtained from the polymer free ligand, with the best ee of 94% obtained with ligand (41).

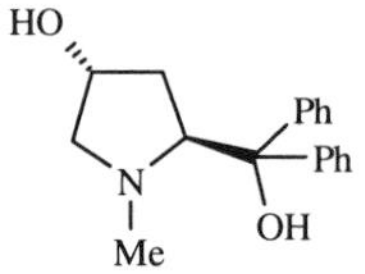

(41)

Mixed complexes derived from diol diamine combination have been evaluated for the same reaction (K. Ding, A. Ishii and K. Mikami, *Angew. Chem. Int. Ed. Eng.*, 1999, **38**, 497). In this study 5 dinaphthyl derived diols and 5 diamine or diimine ligand combinations were evaluated, giving 65% ee with the best combination (42). The optimum diol was then further evaluated with 12 diimines to give 90% ee with complex (43). Cooling of the reaction to –78°C gave enhancement to 99% ee with quantitative yield.

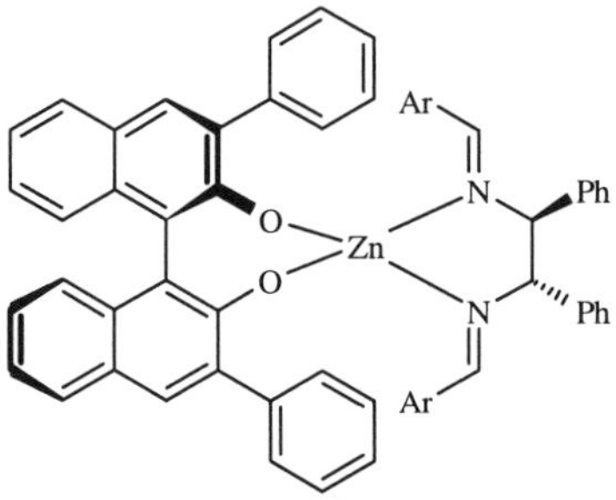

(42) Ar = Ph

(43) Ar = 2,4,6-Me_3Ph

$Ti(OiPr)_4$ complexed with bissulphonamide (44) is known to mediate the enantioselective addition of Et_2Zn to aldehydes with high enantioselectivity (H. Takahashi *et al., Tetrahedron*, 1992, **48**, 5691). Two separate studies have used combinatorial approaches to develop new complexing agents for this reaction. A combinatorial preparation of bissulphonamides containing additional complexing amines has been evaluated against a range of aldehydes (C. Gennari *et al., J. Org. Chem.*, 1998, **63**, 5312). The reactions were carried out in solution, and the best combination (45) gave ee's in the range of 86-96% with aromatic and aliphatic aldehydes. In a separate study 10 solid phase bound pyrrolidine 'tweezers' (46) (5 as shown with SS pyrrolidine diamine central core, 5 with RR) were evaluated for a similar range of Et_2Zn/ $Ti(OiPr)_4$ additions to aldehydes (A. J. Brouwer, H. J. van der Linden and R. M. J. Liskamp, *J. Org. Chem.*, 2000, **65**, 1750). It was found necessary to use poly(ethyleneglycol)-polystyrene resins as simpler polystyrene resins gave no catalytic activity. The best polymer bound reactions gave ee's in the range of 30-35% for the reaction of benzaldehyde with ligand (47), though this ratio was raised to nearly 60% on release of the ligand from the support and screening the subsequent solubilised ligand.

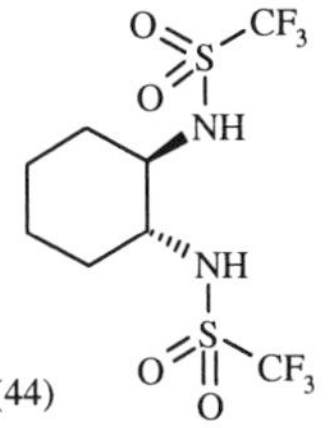

(44) 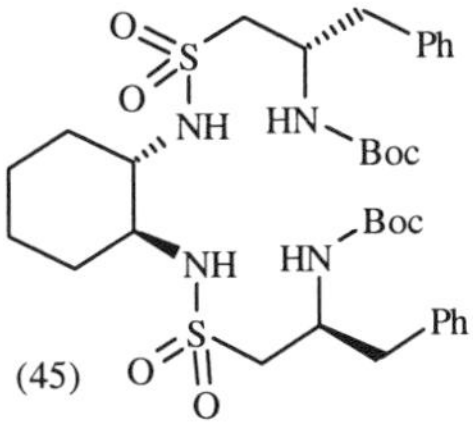 (45)

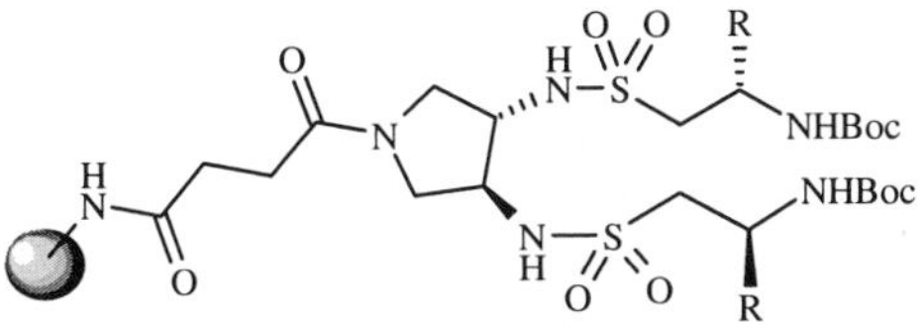

(46) (47) R = Me_2CHCH_2

(e) Nitrile addition to imines (Strecker reaction)

Further optimisation of the reactions described in section 2(d) have been described. In an extension to the Schiff base catalysis without additional metal species, 70 potential ligands derived from seven amino acids with large substituents and ten salicaldehydes were prepared and evaluated on resin support (M. S. Sigman, P. Vachal and E. N. Jacobsen, *Angew. Chem. Int. Ed. Eng.*, 2000, **39**, 1279). A soluble analogue (48) of the best resin bound catalysts was subsequently evaluated against a wide range of imine substrates for the addition of HCN with subsequent trifluoromethyl acetylation, and ee's ranging from 77 to 97 % were obtained.

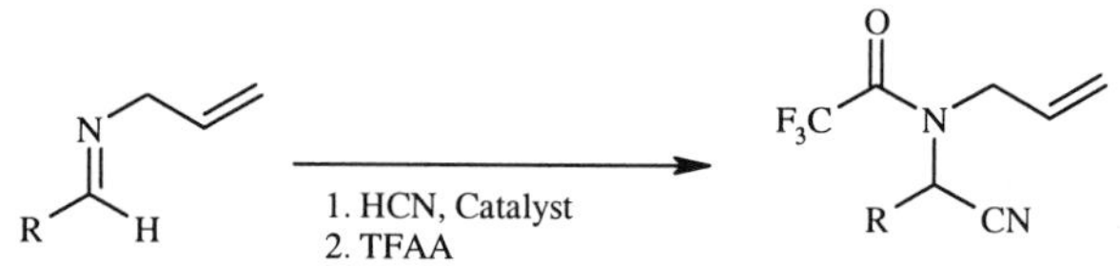

N R H → 1. HCN, Catalyst 2. TFAA → O F_3C N R CN

Ph H N O N H O N H N HO t-Bu O O t-Bu

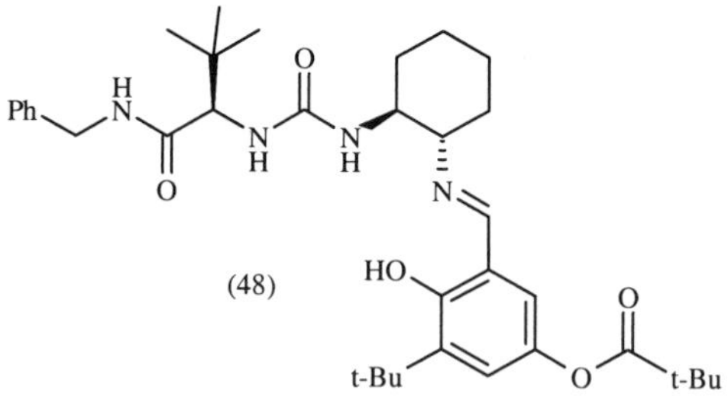

(48)

The imine catalysed reaction of TMSCN with $Ti(OiPr)_4$ has also been extended to the reaction with α,β-unsaturated imines to generate unsaturated α-amino nitriles (J. R. Porter *et al.*, *J. Am. Chem. Soc.*, 2000, **122**, 2657). Several tripeptide ligand libraries led to three optimum ligands (18), (49) and (50) yielding ee's of up to 97%, structurally similar (in one case identical) to those described for addition to simple imines as detailed in section 2(d). In all cases the regioselectivity was maintained with addition only at the imine centre.

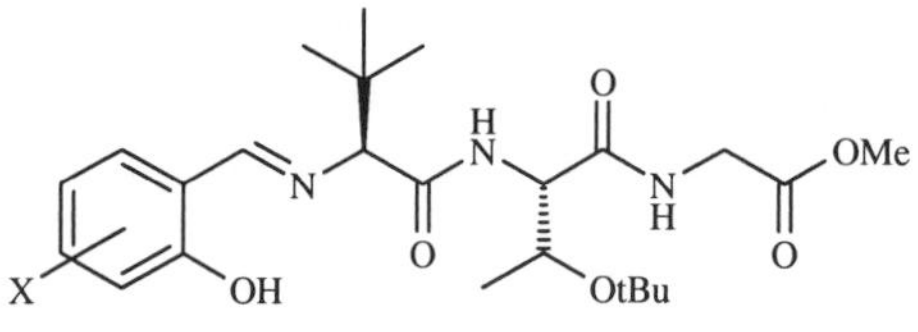

X OH N O H N O OtBu N H OMe O

(18) X = 3,5-DiCl

(49) X = 3,5-DiBr

(50) X = 1-Naphthyl

(f) Nitrile addition to epoxides

With a ligand approach similar to the Strecker reaction described in previous sections, two studies of parallel screening techniques have been used to optimise the asymmetric opening of meso epoxides with TMSCN, a reaction known to be catalysed by $Ti(OiPr)_4$ and Schiff bases (M. Hayashi, M. Tamura and N. Oguni, *Synlett*, 1992, 663). In initial work a cycle of three generations of libraries was used to identify the optimum ligand in an approach analogous to positional scanning methods in peptide recognition studies (B. M. Cole *et al., Angew. Chem. Int. Ed. Eng.*, 1996, **35**, 1668). The ligands were prepared on support but cleaved prior to analysis. The first round of 10 peptides identified t-butyl glycine as the best amino acid for the first position, followed by a second library which identified O-t-butyl threonine as the second position optimum acid. A third cycle identified the optimum aldehydic compound, leading to the best ligand (51), which gave an ee of 86% on the opening of cyclohexene oxide with TMSCN. In a subsequent study (K. D. Shimizu *et al., Angew. Chem. Int. Ed. Eng.*, 1997, **36**, 1704) the same approach was taken but with screening whilst the ligand was still polymer bound. No significant discrepancies were discovered between polymer bound asymmetric induction and the ligand in solution, and the same structures as in the previous work were shown to be the most efficient enantioselective ligands.

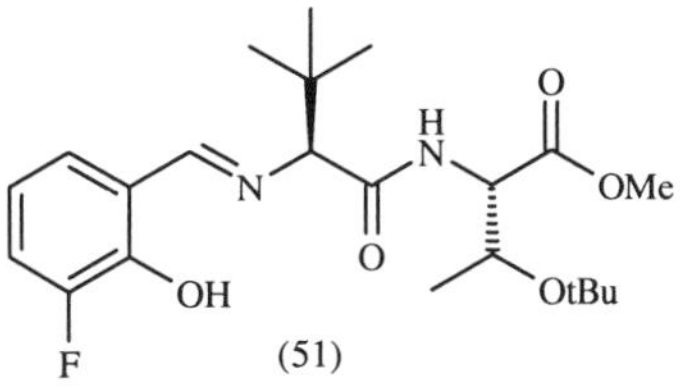

H N O OMe N O OH OtBu F

(51)

4. Combinatorial substrate evaluation of catalysts

(a) Borane reduction

Single pot multi substrate screening has been used to evaluate the scope of oxazaborolidine catalysed borane reductions (X Gao and H. B. Kagan, *Chirality*, 1998, **10**, 120). Various mixtures of ketones were evaluated in the reduction with the oxazaborolidine (52), with monitoring of the resultant conversion and enantioselectivity through chiral HPLC. The experiments demonstrated the capability of such single reaction vessel assessment, though limitations in terms of the number of components were identified.

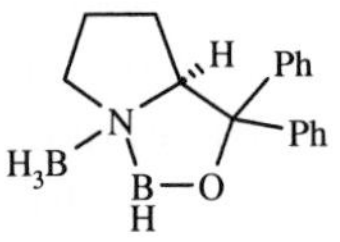

(52)

(b) Phenol additions to epoxides.

Using the same Co salen complex (19, M = Co) described in section 2(e) above, the scope of asymmetric catalysis for this complex has been evaluated with parallel arrays derived from ten phenols and five epoxides (S. Peukert and E. N. Jacobsen, *Organic Lett.*, 1999, **1**, 1245). All reactions gave enantiomeric excesses greater than 80%. Subsequent elaboration of the product structures led to potentially valuable enantiomerically enriched libraries of important pharmacologically active compounds.

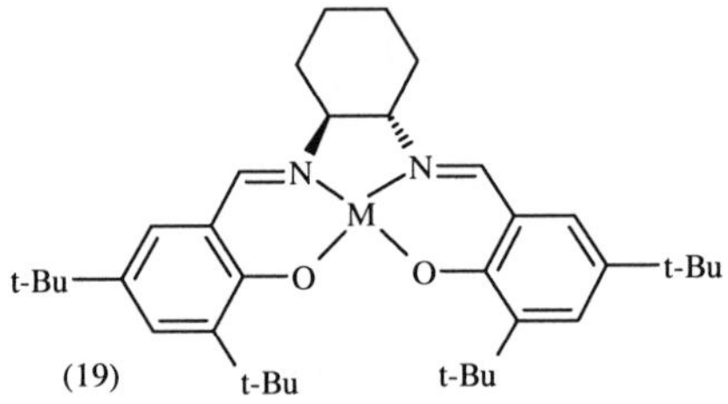

(19)

Second Edition of Rodd's Chemistry of Carbon Compounds,
Volume V, Topical Volumes
Asymmetric Catalysis, edited by M.Sainsbury

Chapter 8

ENANTIOSELECTIVE CYCLOADDITION REACTIONS

MICHAEL C. WILLIS

1. Introduction

Cycloaddition reactions are some of the most popular and widely used reactions employed in organic synthesis and constitute an enormous body of chemistry; this chapter will consider Diels-Alder reactions (intermolecular, intramolecular and hetero-), 1,3-dipolar cycloaddition reactions and [2+2] cycloaddition reactions. Due to space considerations a comprehensive review of the area has not been attempted, rather, the focus is on landmark achievements as well as state-of-the-art examples of relevant chemistry. Particular attention has also been made to examples in which the synthetic utility of the system has been explored.

2. Diels-Alder Reactions

The numerous examples of Diels-Alder reactions in successful syntheses of natural and non-natural products are testament to its utility and practicability (for general reviews, see; W. Oppolzer in "Comprehensive Organic Synthesis", Vol. 5, pp315, Eds. B. M. Trost and I. Fleming, Pergamon, Oxford, 1991; W. R. Roush in "Comprehensive Organic Synthesis", Vol. 5, pp513, Eds. B. M. Trost and I. Fleming, Pergamon, Oxford, 1991). It is therefore not surprising that the development of enantioselective variants of the reaction has been at the forefront of the development of enantioselective catalysis. Several excellent reviews of enantioselective Diels-Alder reactions have recently appeared (D. A. Evans and J. S. Johnson in "Comprehensive Asymmetric Catalysis", Vol. 3, Eds. E. N. Jacobsen, A. Pfaltz and H. Yamamoto, Springer, Berlin, 1999; L. C. Dias, *J. Braz. Chem. Soc.*, 1997, **8**, 289; H. B. Kagan and O. Riant, *Chem. Rev.*, 1992, **92**, 1007). Historically, one of the first examples of a catalytic enantioselective reaction that delivered products with moderate to high levels of selectivity was a Diels-Alder reaction. Koga used the chiral aluminium complex (1) to catalyse the reaction between methacrolein and cyclopentadiene and obtained the bicyclic product with 72% ee (S.-I. Hashimoto, N. Komeshima and Koga, *J. Chem. Soc., Chem. Commun.*, 1979, 437). An example of a hetero-Diels-Alder reaction between silyloxy diene (2) and benzaldehyde quickly followed (M. D. Bednarski, C. J. Maring and S. J. Danishefsky, *Tetrahedron Lett.*, 1983, **24**, 3451). In this example the commercially available chiral shift reagent $Eu(hfc)_3$ was employed as the chiral catalyst. The dihydropyran product (3) was obtained with 55% ee.

Me OAlCl$_2$ CHMe$_2$ (1)

+ Me O H — 15 mol% (1), -78 °C, toluene → CHO Me — *exo*:*endo*, 98:2; 72% *exo* ee

Me OtBu TMSO Me (2) + O H Ph — 1. Eu(hfc)$_3$ (1 mol%), -10°C, neat; 2. TFA → Me O O Ph Me (3) — 55% ee

In common with the majority of enantioselective cycloadditions both of the above examples use Lewis acidic metal complexes as catalysts. More recently considerable effort has been expended on the rational design of chiral Lewis acid complexes capable of acting as catalysts. The majority of these complexes achieve substrate activation by coordinating to a carbonyl-unit and thus lowering the energy of the LUMO. The complexity of designing a chiral Lewis acid is evident when the variables operating in the complexation of a simple achiral Lewis acidic metal to a carbonyl group are considered. For example, η_1 versus η_2 binding of the metal centre to the carbonyl group is possible. The formation of regioisomeric complexes between the carbonyl and the Lewis acid can also occur. A further variable applies to the activation of α,β-unsaturated carbonyl systems in which the possibility of *s-cis* or *s-trans* isomers is possible. Schreiber has comprehensively reviewed carbonyl-Lewis acid complexation (S. Shambayati, W. E. Crowe and S. L. Schreiber, *Angew. Chem., Int. Ed. Engl.*, 1990, **29**, 256). To develop an enantioselective system the further and not inconsiderable complication of differentiation of enantiofaces must also be achieved. The intense research in this area has resulted in the development of many highly successful catalyst systems; in the remainder of this section we will group the catalysts according to metal type.

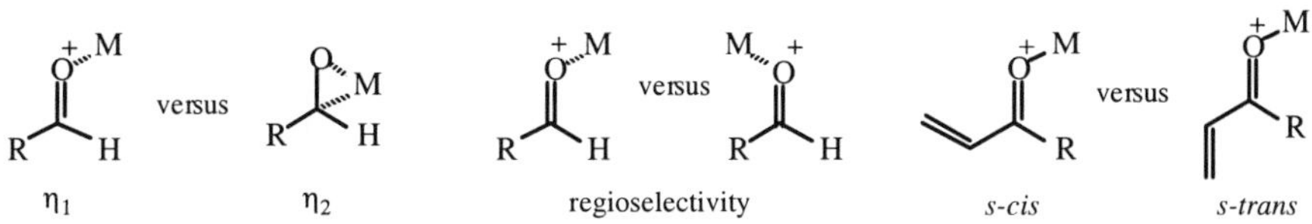

η_1 versus η_2 — regioselectivity — *s-cis* versus *s-trans*

(a) Boron Lewis acids

Yamamoto and co-workers have developed a family of Lewis acidic boron complexes derived from modified tartrate esters. The chiral acyloxy borane (CAB) catalysts, such as (4), are particularly effective for enal dienophiles and, as shown below, can be used with cyclic and acyclic dienophiles to deliver products with extremely high levels of selectivity (K. Ishihara, Q. Gao and H. Yamamoto *et al.*, *J. Org. Chem.*, 1993, **58**, 6917; K. Ishihara, Q. Gao and H. Yamamoto *et al.*, *J. Am. Chem. Soc.*, 1993, **115**, 10412). The reaction between cyclopentadiene and α-bromocrotonaldehyde delivers the *exo* product (5) as the major isomer. This diastereomer preference is common in Lewis acid catalysed reactions between α-methyl- and α-bromo enal substrates when cyclopentadiene is employed as the diene. With the majority of alternative dienophiles the *endo* isomer predominates.

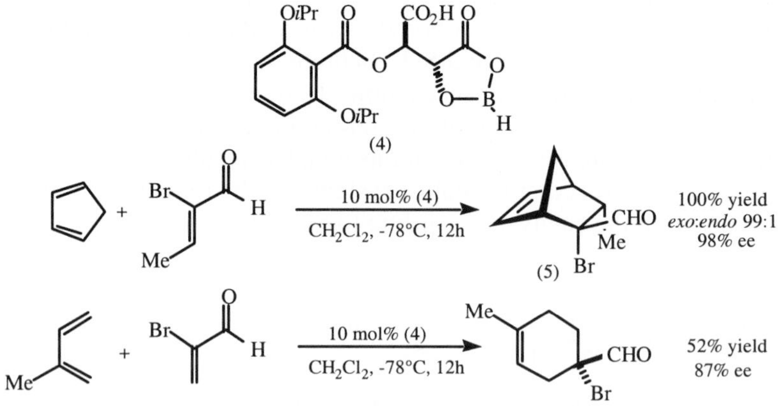

The CAB catalyst (4) has also been employed in enantioselective intramolecular Diels-Alder reactions. Treatment of triene (6) with 10 mol% (4) delivers the fused ring product in high yield and with excellent selectivity. The α-methyl group was found to be crucial to obtain a highly enantioselective reaction; the unsubstituted triene delivered the corresponding product with only 46% ee (K. Furuta, A. Kanematsu and H. Yamamoto, *Tetrahedron Lett.*, 1989, **30**, 7231).

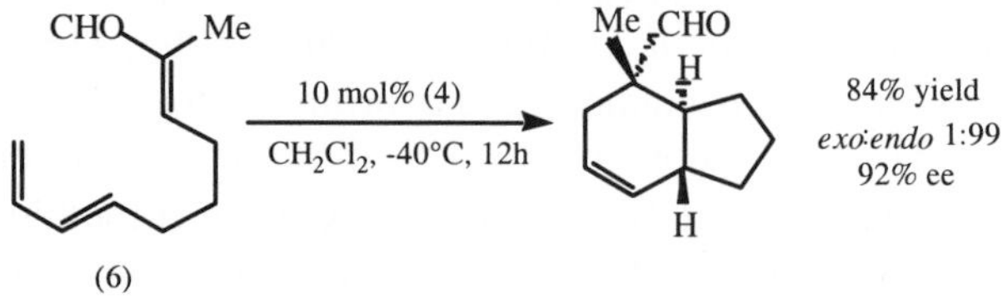

The Corey group have developed a series of chiral oxazaborolidine based catalysts for the Diels-Alder reaction (E. J. Corey *et al*, *J. Am. Chem. Soc.*, 1992, **114**, 8290). A common limitation with a number of reported enantioselective systems is that they are only demonstrated on a limited range of standard substrates and in particular cyclopentadiene is often the only diene employed. The Corey group have demonstrated the utility of their catalyst by using several more complex diene substrates. For example, reaction between substituted triene (7) and methacrolein catalysed by oxazaborolidine (8) provided the highly functionalised cyclohexene product (9) in good yield and with an excellent 97% ee. Compound (9) then served as a key intermediate in a synthesis of cassiol (10) (E. J. Corey, A. Guzman-Perez, T.-P. Loh, *J. Am. Chem. Soc.*, 1994, **116**, 3611). The same publication also details the enantioselective preparation of a suitable intermediate for conversion to gibberellic acid.

Ts Me N Bu—B O O NH

(8)

OTIPS Me OHC + Me (7) O Me O Me

25 mol% (8)

1:1 PhMe:Hexane, -78°C, 42 h

OTIPS Me H Me O (9) O Me O Me

83% yield
97% ee
single diastereomer

steps

O Me HO Me OH OH

(10), cassiol

With issues of catalyst recovery and recycling becoming an important concern the report from Itsuno and co-workers on the development of a polymer supported oxazaborolidine catalyst is significant. The valine-derived oxazaborolidine was attached *via* the sulfonyl group to a crosslinked polystyrene backbone to generate catalysts of type (11). The catalysts were evaluated on the reaction between methacrolein and cyclopentadiene and the nature of the crosslinking agent was shown to have a considerable effect on the performance of the catalyst. Polymers generated with an oxyethylene cross-linkage delivered products with enhanced enantioselectivities (K. Kamahori, K. Ito and S. Itsuno, *J. Org. Chem.*, 1996, **61**, 8321). The supported catalysts (11) could be easily recovered and reused without further activation. In a further development the polymeric catalysts could be used in a continuous flow reactor.

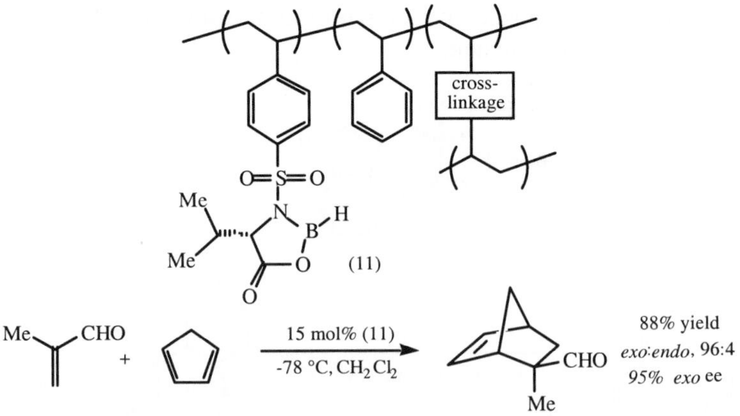

(11)

15 mol% (11)
-78 °C, CH_2Cl_2

88% yield
exo:*endo*, 96:4
95% *exo* ee

Hawkins has reported a successful dichloroborane derived catalyst (12) for the enatioselective Diels-Alder reaction between enones and activated dienes; enones are significantly less reactive than the more commonly employed enals. One of the key design features of the catalyst was the ability of dipole-induced interactions between the polar carboalkoxy group of the dienophile and the electron rich arene system of the catalyst to control the orientation of the dienophile (J. M. Hawkins, S. Loren and M. Nambu, *J. Am. Chem. Soc.*, 1994, **116**, 11657). A related dichloroarylborane catalyst (13) has been reported by the Yamamoto group (K. Ishihara *et al.*, *Synlett*, 1998, 1053).

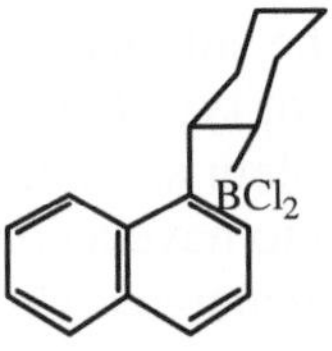

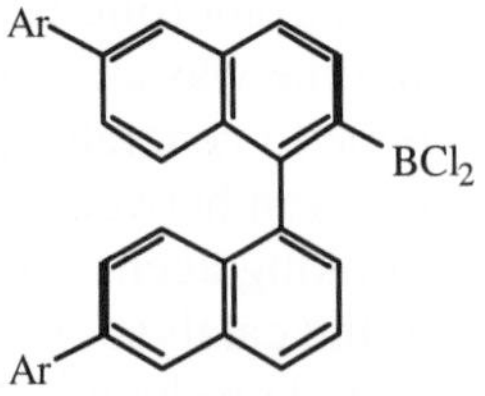

(12)

Ar = H, Ph, Mesityl
(13)

Yamamoto and co-workers have also developed a very successful Brønsted acid-assisted Lewis acid catalyst system for asymmetric Diels-Alder cycloadditions (K. Ishihara *et al.*, *J. Am. Chem. Soc.*, 1998, **120**, 6920). Catalyst (14) is generated by treating the relevant tetrol with trimethoxy borane while removing methanol. Complex (14) is an extremely selective catalyst for use with cyclic and acyclic, α-substituted as well as with α-unsubstituted enal dienophiles. The unsubstituted dienophiles are substrates notorious for providing products with poor enantioselectivities. The pendent Ph-groups on (14) where found to be crucial to obtaining high selectivities with the unsubstituted enals. The success of the system is attributed to a combination of intramolecular hydrogen bonding and attractive π-π donor-acceptor interactions in the transition state.

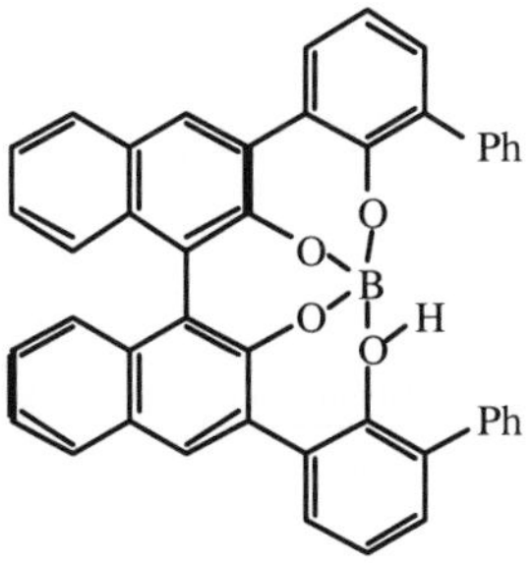
Ph
O
O
B
O
O
H
Ph

(14)

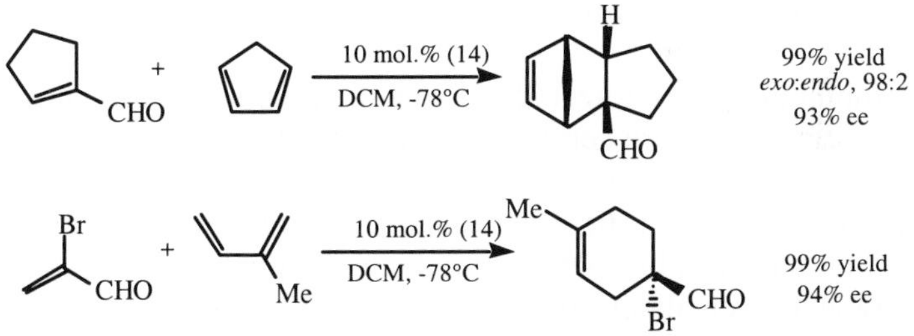
CHO
+
10 mol.% (14)
DCM, -78°C
H
CHO
99% yield
exo:endo, 98:2
93% ee
Br
CHO
+
Me
10 mol.% (14)
DCM, -78°C
Me
CHO
Br
99% yield
94% ee

The final boron based catalyst that we will consider has been reported by Corey and co-workers (Y. Hayashi, J. J. Rhode and E. J. Corey, *J. Am. Chem. Soc.*, 1996, **118**, 5502). The "super-Lewis acid catalyst" (15) is based on a chiral amino-alcohol core and features a non-coordinating tetraaryl borate counterion. The catalyst efficiently and selectively catalyses cycloadditions between a range of substituted enals and several different dienes. Significantly, less reactive dienes such as cyclohexadiene and isoprene can be successfully employed.

$\bar{B}[C_6H_3\text{-}3,5\text{-}(CF_3)_2]_4$

(15)

10 mol.% (15)
DCM, -94°C

99% yield
exo:endo, 4:96
93% ee

10 mol.% (15)
DCM, -94°C

99% yield
96% ee

(b) Aluminium based catalysts

The combination of a protected stilbene diamine ligand (Stein ligand) with a Lewis acidic aluminium centre produces an extremely effective Diels-Alder catalyst system (E. J. Corey *et al.*, *J. Am. Chem. Soc.*, 1989, **111**, 5493). The C_2-symmetric catalyst (16) is particularly effective with unsaturated imide dienophiles such as (17); a compound class popular as substrates in enantioselective catalysis due to their ability to undergo two-point binding with the Lewis acidic metal centre. In the example shown, reaction between (17) and substituted cyclopentadiene (18) selectively provides the Diels-Alder adduct (19) which was advanced to the useful building block (20) and ultimately to the prostaglandin precursor (21) (E. J. Corey, N. Imai and S. Pikul, *Tetrahedron Lett.*, 1991, 7517). The catalyst system has been extended to include maleimides such as (22) as substrates. In order to produce an asymmetric product when employing C_2-symmetric substrates such as (22) the corresponding diene must be dissymmetric. As can be seen the expected product (23) can be isolated in good yield with high levels of selectivity (E.J. Corey, S. Sarshar, D.-H., Lee, *J. Am. Chem. Soc.*, 1994, **116**, 12089).

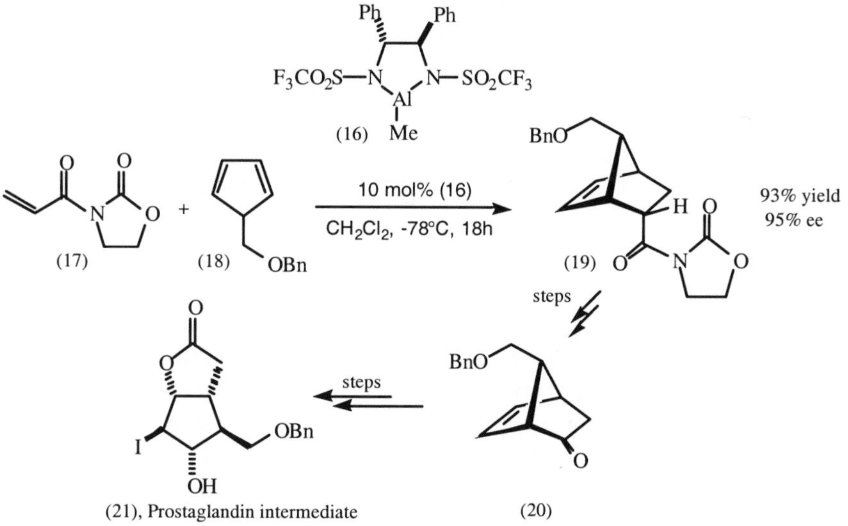

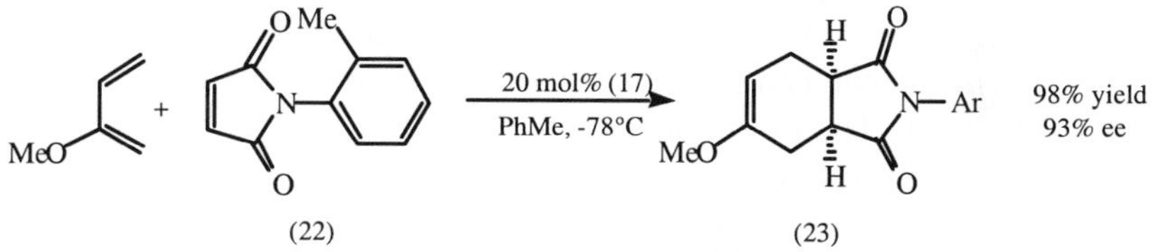

Wulff and co-workers have reported the use of the vaulted biaryl-diols such as (24) (VAPOL ligands) in combination with dialkylaluminium chlorides as effective enantioselective Diels-Alder catalysts (J. Bao and W. D. Wulff, *J. Am. Chem. Soc.*, 1993, **115**, 3814.). The remarkable feature of these complexes is the extremely low catalyst loadings that are possible. For example, the reaction between methacrolein and cyclopentadiene (illustrated below) requires only 0.5 mol.% of the catalyst to deliver material of 98% enantiomeric excess.

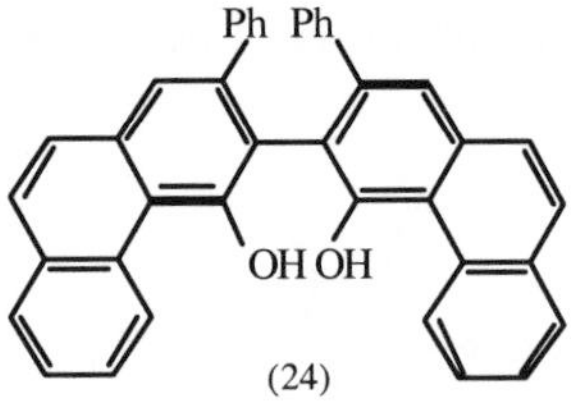

Ph Ph

OH OH

(24)

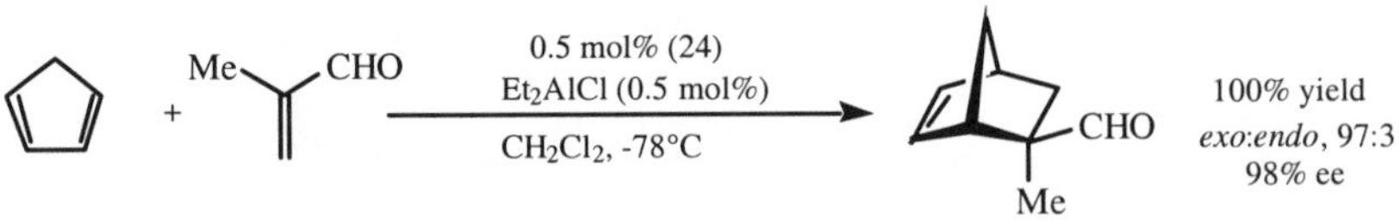

Me CHO

+

0.5 mol% (24)
Et_2AlCl (0.5 mol%)
CH_2Cl_2, -78°C

CHO
Me

100% yield
exo:endo, 97:3
98% ee

(c) Copper based catalysts

One of the most versatile and effective chiral Lewis acid systems has been developed by Evans and co-workers (D. A. Evans *et al., J. Am. Chem. Soc.*, 1999, **121**, 7559). The successful catalyst family, for example (25) and (26), is based on a Cu(II) metal centre bound to a C_2-symmetric bis(oxazoline) ligand. These complexes were initially found to be effective catalysts for Diels-Alder reactions between unsaturated imide dienophiles and range of simple dienes. As can be seen from the examples shown below the nature of the anionic counterion was found to have a significant effect upon the efficiency of the catalyst (D. A. Evans *et al., Angew*. Chem., *Int. Ed. Engl.*, 1995, **34**, 798.). The most reactive and selective catalyst (25) was formed when the SbF_6 counterion was employed. This combination of a bis(oxazoline) ligand, a Cu(II) metal centre and SbF_6 counterions provides an extremely powerful chiral Lewis acid catalyst. This is evident in the range of products that can be produced using this catalyst system; as can be seen, cyclic as well as acyclic dienes, hetero-substituted dienes and β-substituted dieneophiles are all tolerated (D. A. Evans *et al., J. Am. Chem. Soc.*, 1999, **121**, 7582).

(25) X = SbF_6
(26) X = OTf

10 mol% catalyst
CH_2Cl_2 +25 °C

counterion	time	endo ee	yield
X = TfO^-	48 h	82% ee	90%
X = SbF_6^-	5 h	93% ee	90%

94% ee
cis:trans 83:17

X = OAc, 98% ee
X = NHCbz, 90% ee
X = SPh, 98% ee

99% ee
endo:exo 85:15

95% ee
endo:exo 94:6

Evans has exploited the functionalised products available with these catalysts in a number of total syntheses. Reaction of furan with imide dienophile (17) under the action of 2 mol.% Cu(II) complex (25) produces the bicyclic product (27) in excellent yield and with a 97% enantiomeric excess. Bicycle (27) was advanced through six steps to realise *ent*-shikimic acid (D. A. Evans and D. M. Barnes, *Tetrahedron Lett.,* 1997, **38**, 57). Combining imide (17) with 1-acetoxy-3-methylbutadiene (28), again under the action of catalyst (25), provided the corresponding substituted cyclohexene (29) with exceptional levels of selectivity. Cyclohexene (29) was converted in only four synthetic operations into *ent*-tetrahydrocannabinol (D. A. Evans, E. A. Shaughnessy and D. M. Barnes, *Tetrahedron Lett.,* 1997, **38**, 3193).

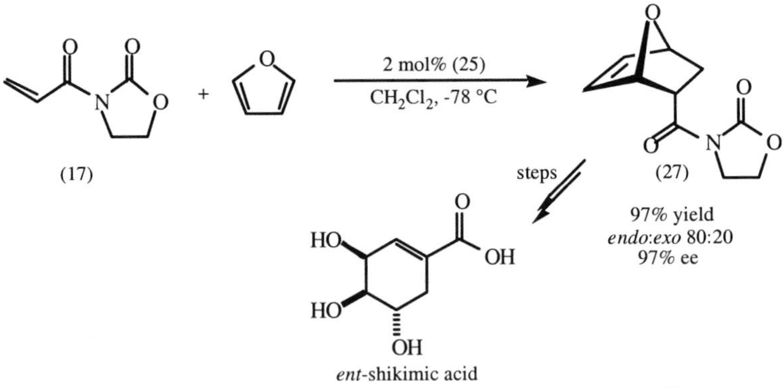

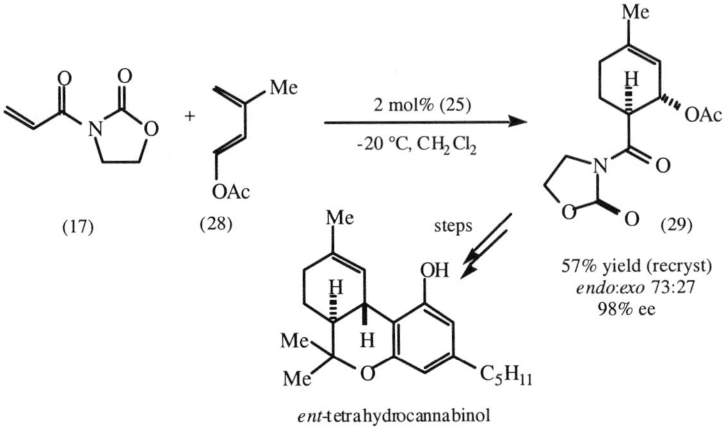

(17) + furan → 2 mol% (25), CH_2Cl_2, -78 °C → (27)

97% yield
endo:*exo* 80:20
97% ee

(27) → steps → *ent*-shikimic acid

(17) + (28) → 2 mol% (25), -20 °C, CH_2Cl_2 → (29)

57% yield (recryst)
endo:*exo* 73:27
98% ee

(29) → steps → *ent*-tetrahydrocannabinol

The high reactivity of the Cu(II)-bis(oxazoline) catalysts is demonstrated well by their use in enantioselective intramolecular Diels-Alder reactions. Intramolecular Diels-Alder cycloadditions provide a stern examination of the reactivity of chiral Lewis acids as they essentially require the complexes to catalyse reactions between substituted dienes and substituted dienophiles. As can be seen in the example presented below, the Cu(II) systems deliver the fused-ring products in high yield with good levels of selectivity. The illustrated product (30) was utilised in a total synthesis of isopulo'upone (D. A. Evans and J. S. Johnson, *J. Org. Chem.*, 1997, **62**, 786).

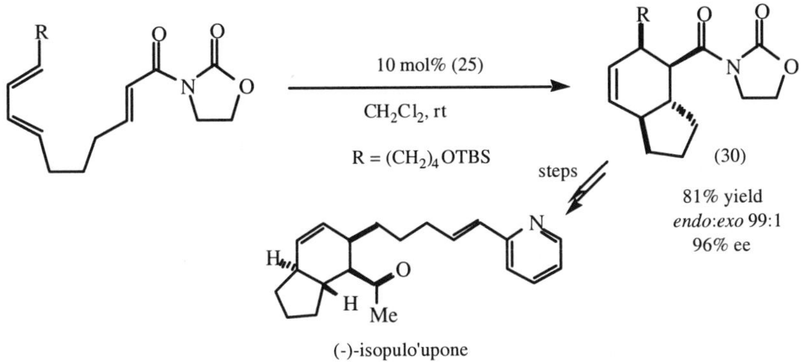

R O O N O

10 mol% (25)

CH_2Cl_2, rt

R = $(CH_2)_4$OTBS

R O O N O

steps

(30)

81% yield
endo:exo 99:1
96% ee

H N O H Me

(-)-isopulo'upone

Several alternative Cu(II) systems have proved to be effective enantioselective catalysts for Diels-Alder cycloadditions; among the more notable are the phosphine-oxazoline (31) (I. Sagasser and G. Helmchem, *Tetrahedron Lett.*, 1998, **39**, 261), the bis(imine) (32) (D. A. Evans, T. Lectka and S. J. Miller, *Tetrahedron Lett.*, 1993, **34**, 7027) and the modified bis(oxazoline) (33) (A. R. Ghosh, H. Cho and J. Cappiello, *Tetrahedron Asymmetry*, 1998, **9**, 3687).

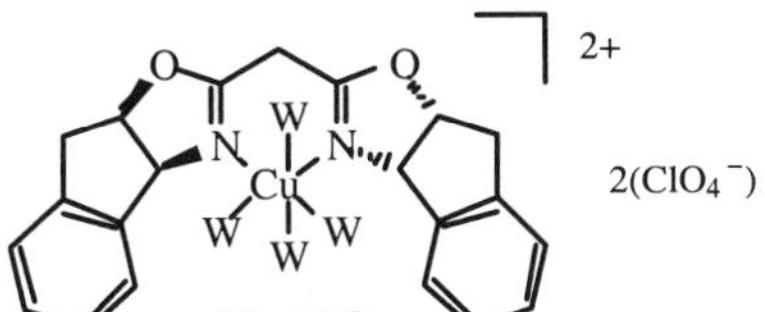

O O N W N Cu W W W

2+

$2(ClO_4^-)$

W = H_2O
(33)

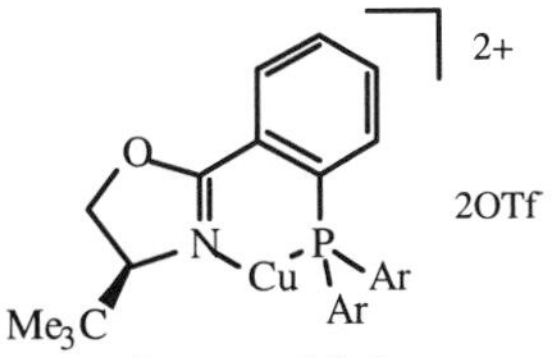

O N Cu P Ar Ar

Me_3C

2+

2OTf

Ar = α-naphthyl
(31)

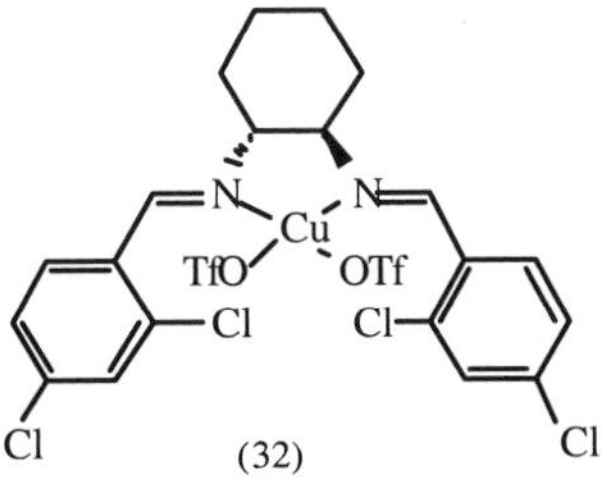

N Cu N TfO OTf Cl Cl Cl Cl

(32)

(d) Titanium based catalysts

TADDOL ligands in combination with Ti(IV) metal centres have provided a range of versatile chiral Diels-Alder catalysts. A recent and extensive review by Seebach provides a comprehensive account of the use of these complexes in a range of asymmetric transformations (D. Seebach, A. K. Beck and A. Heckel, *Angew. Chem., Int. Ed.*, 2001, **40**, 92). With regard to enantioselective Diels-Alder reactions the Ti-TADDOL systems have most commonly been used in combination with the standard imide dienophiles such as (17). A more unusual dieneophile subjected to catalysis with a Ti-TADDOL complex is the boronic ester substituted imide (34). The catalysed cycloaddition of (34) and 1-acetoxy-3-methylbutadiene using 10 mol.% of catalyst (35) provided functionalised cyclohexene (36) with high selectivity and in good yield. The boron substituent could be readily converted to the corresponding hydroxyl group (K. Narasaka and I. Yamamoto, *Tetrahedron*, 1992, **48**, 1947).

Ph Ph Ph Me O O O O $TiCl_2$ Ph Ph (35)

Me O B O Me (34) O O N O + OAc Me

10 mol% 35

4Å MS, Tol/Hex, 0°C

Me Me O B O Me O OAc N O O (36)

71% yield
95% ee

The Narasaka group have developed intramolecular Diels-Alder reactions catalysed by Ti-TADDOL systems. The dithiane-substituted imide triene (37) undergoes intramolecular cycloaddition in the presence of 20 mol.% (35) at 0°C. The corresponding decalin product (38) is obtained in 70% yield with >95% ee (K. Narasaka, M. Saitou and N. Iwasawa *Tetrahedron Asymmetry*, 1991, **2**, 1305). The dithiane unit was necessary to allow efficient production of the decalin system. Decalin (38) was subsequently advanced to provide the dihydrocompactin core structure.

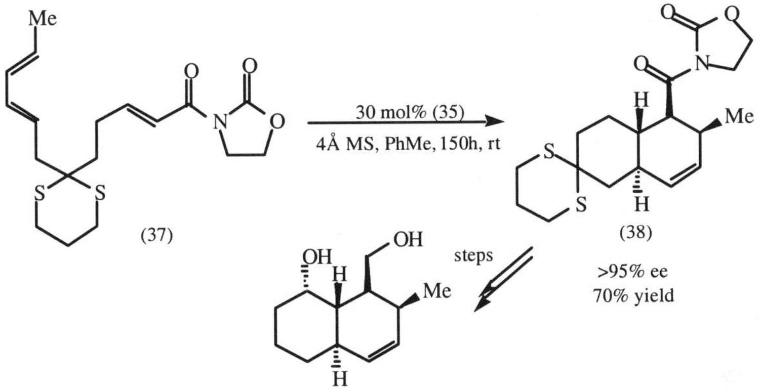

(37)

30 mol% (35)
4Å MS, PhMe, 150h, rt

(38)
>95% ee
70% yield

steps

dihydrocompactin core

The Ti-TADDOL catalysts can also promote selective cycloadditions with non-imide dienophiles. For example, in the presence of 20 mol.% (39) diketone (40) undergoes reaction with diene (41) to provide the steroid fragment (42) with good selectivity (G. Quinkert *et al.*, *Helv. Chim. Acta.*, 1995, 78, 1345).

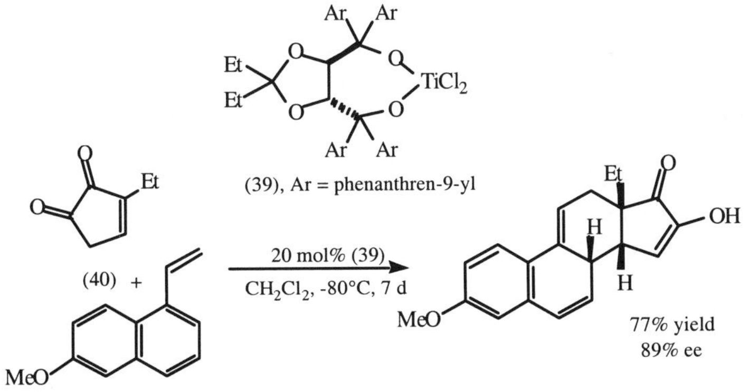

(39), Ar = phenanthren-9-yl

(40) + (41)

20 mol% (39)
CH_2Cl_2, -80°C, 7 d

(42)
77% yield
89% ee

Titanium-BINOL complexes have also been employed as chiral catalysts for enantioselective Diels-Alder reactions (K. Mikami, Y. Motoyama and M. Terada, *J. Am. Chem. Soc.*, 1994, **116**, 2812; K. Maruoka, N. Maurase and H. Yamamoto, *J. Org. Chem.*, 1993, **58**, 2938).

(d) Other metal based catalysts

Several Lewis acid catalysts featuring a magnesium metal centre have appeared in the literature; among the more notable are the bis(oxazoline) system (43) reported by Corey (E. J. Corey and K. Ishirara, *Tetrahedron Lett.*, 1992, **33**, 6807) and sulfonamido-oxazoline system (44) of Fujisawa (T. Ichiyanagri, M. Shimizu and T. Fujisawa, *J. Org. Chem.*, 1997, **62**, 7937). Both systems have been used successfully with imide dieneophiles.

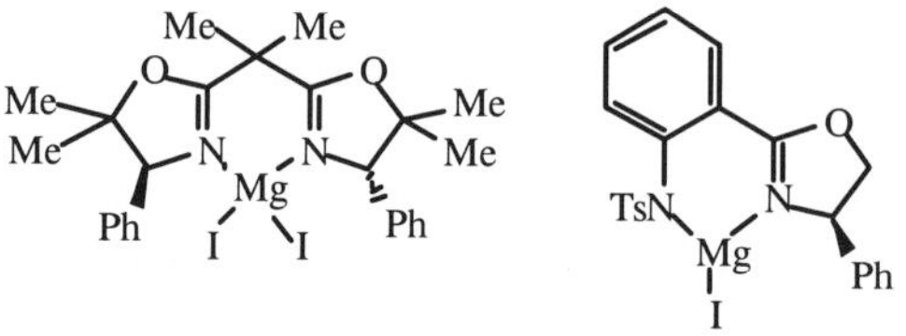

(43) (44)

Salen ligands, more usually associated with processes such as epoxidation or epoxide-opening, have been used in combination with chromium salts to provide effective Lewis acids for enantioselective Diels-Alder reactions. Rawal and co-workers have used complex (45) to catalyse the cycloadditions between amine-substituted dienes such as (46) and (47) and enals. The corresponding amino-substituted products are obtained in good yield with high selectivity (Y. Huang, T. Iwama and V. Rawal, *J. Am. Chem. Soc.*, 2000, **122**, 7843). The SbF_6 counterion was found to be crucial to obtain the optimum selectivities.

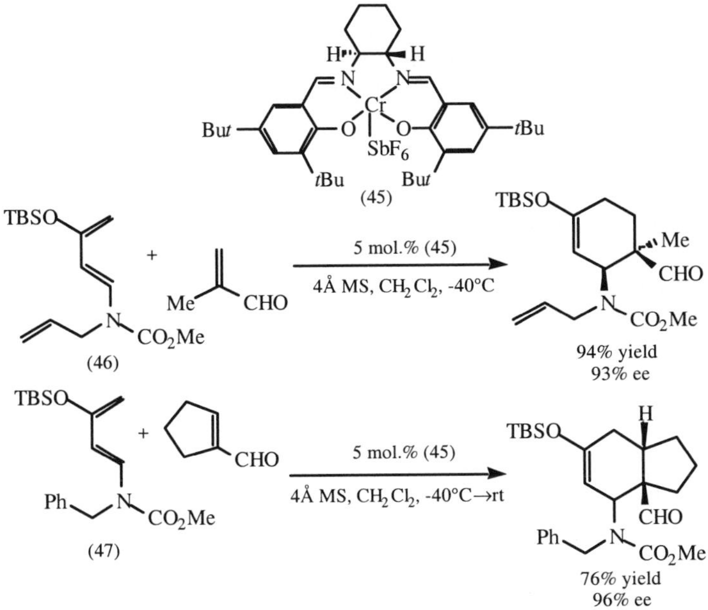

The Kundig group have developed a ruthenium based catalyst system (48) capable of promoting cycloadditions between enals and cyclic and acyclic dienes with excellent selectivity (E. P. Kundig, C. M. Saudan and G. Bernardelli, *Angew. Chem.,Int. Ed. Engl.*, 1999, **38**, 1220). The ligand system featured in this catalyst was originally developed for use in an earlier Fe(II) Lewis acid (M. E. Bruin and E. P. Kundig, *Chem. Commun.*, 1998, 2635).

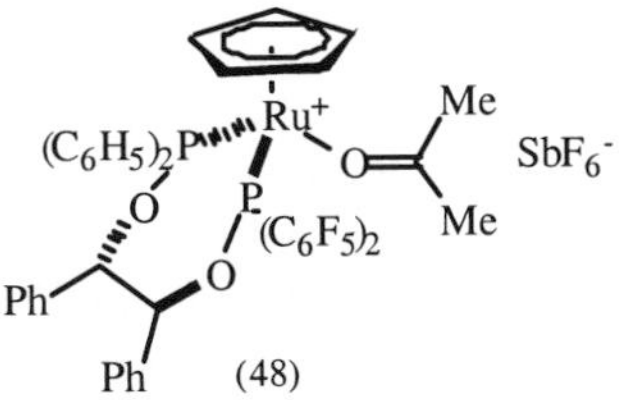

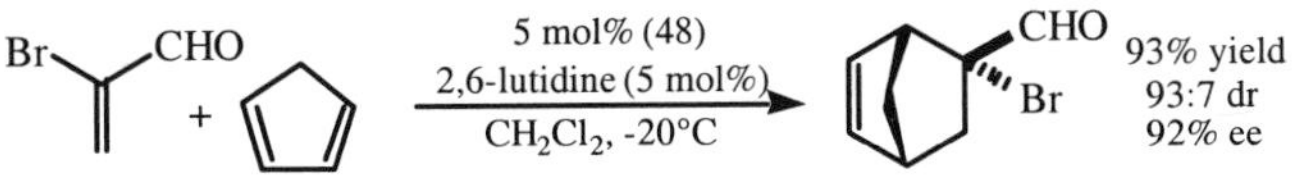

Ligands based around the binapthyl group core structure are common in most aspects of asymmetric catalysis; several chiral Lewis acid catalysts for the Diels-Alder reaction feature this "privileged ligand" motif. Nakagawa and co-workers have combined the BINOL derived chiral amide (49) with $Yb(OTf)_3$ to produce an effective catalyst (A. Nishida, M. Yamanaka and M. Nakagawa, *Tetrahedron Lett.*, 1999, **40**, 1555). The Ghosh group partner BINAP with palladium and platinum salts to generate their chiral Lewis acids (50) (A. K. Ghosh and H. Matsuda, *Organic Lett.*, 1999, **1**, 2157.). As can be seen, both catalysts effectively promote the cycloaddition between imide dienophiles and cyclopentadiene delivering the products with very high levels of enantioselectivity.

(49)

(50)

10 mol% (49), *i*PrNEt$_2$
Yb(OTf)$_3$ (10 mol.%)
CH$_2$Cl$_2$, 0°C

64% yield
endo:*exo*, 91:9
98% ee

10 mol% (50)
CH$_2$Cl$_2$, -78°C°C

75% yield
endo:*exo*, 97:3
99% ee

The majority of the Lewis acids we have considered have combined a specific ligand (or ligand type) with a single metal. The Kanemasa group have developed a ligand group (51) (known as DBFox) that can form highly efficient and selective catalysts with a variety of metal salts. As can be seen, effective catalysts are formed by the combination of (51) with Cu(II), Ni(II), Mg(II) and Fe(II) salts. Hydrated metal salts can also be employed (S.

Kanemasa *et al.*, *J. Am. Chem. Soc.*,1998, **120**, 3074; S. Kanemasa *et al.*, *J. Org. Chem.*, 1997, **62**, 6454).

(51)

10 mol% (51)
10 mol% metal salt
CH_2Cl_2

M	Yield (%)	ee (%)
$Mg(ClO_4)_2$	97	99
$Ni(ClO_4)_2$	100	96
$Ni(ClO_4)_2.6H_2O$	96	99
$Fe(ClO_4)_2$	90	98
$Fe(ClO_4)_2.3H_2O$	92	97
$Cu(ClO_4)_2$	97	92

(e) Non-metal catalysts

Chiral Lewis acid catalysts have been used to promote all of the cycloadditions that we have so far considered. In an exciting development MacMillan and co-workers have reported an "organocatalytic" system that promotes enantioselective Diels-Alder cycloadditions (K. A. Ahrendt, C. J. Borths and D. W. C. MacMillan, *J. Am. Chem. Soc.*, 2000, **122**, 4243). The MacMillan system functions by the transient formation of an iminum species (52); this activated species corresponds to the charged aldehyde-metal complexes formed from the combination of an aldehyde and a Lewis acidic metal and similar to a Lewis acid, achieves catalysis by LUMO-lowering activation. Amine hydrochloride (53) is the key catalyst and it is capable of promoting a range of Diels-Alder reactions between β-substituted enals and cyclic and acyclic dienes. The cycloadducts are formed with uniformly high yields and selectivities.

(53) (52)

5 mol.% (53)
MeOH-H_2O, rt

99% yield
exo:endo, 1.3:1
93% *exo* ee, 93% *endo* ee

5 mol.% (53)
MeOH-H_2O, rt

75% yield
exo:endo, 1:5
93% ee

3. Hetero-Diels-Alder reactions

(a) Normal electron demand

Although not as widespread as enantioselective Diels-Alder cycloadditions leading to carbocyclic ring systems, the heteroatom-version of this important process, employing either hetero-dienes or dienophiles, and leading to heterocyclic products is becoming an increasingly common transformation. Accordingly there has recently been an increase in the number of reports detailing new catalysts and substrates for this class of reaction. The area has recently been reviewed (T. Ooi and K. Marouka, in "Comprehensive Asymmetric Catalysis", Vol. 3, Ed. E. N. Jacobsen, A. Pfaltz and H. Yamamoto, Springer, Berlin, 1999).

The most common variant of the hetero-Diels-Alder reaction is that employing an aldehyde as the dienophile and an electron rich diene; Danishefsky's diene performs extremely well in this type of process. Many of the chiral Lewis acids developed for traditional Diels-Alder reactions have also been applied to the aldehyde-electron rich diene variant, including CAB (Q. Gao *et al.*, *Tetrahedron*, 1994, **50**, 979), Ti(IV)-BINOL (G. E. Keck *et* *al.*, *J. Org. Chem.*, 1995, **60**, 5598; B. Wang *et al.*, *Chem. Commun.*, 2000, 1605) and Pd(II)-BINAP (S. Oi *et al*, *J. Org. Chem.*, 1999, **64**, 8660).

The Jacobsen group have developed two Cr(III) based catalyst systems for enantioselective hetero-Diels-Alder reactions. The first of the complexes (54) features a chiral Salen ligand and is an effective catalyst for a range of aromatic aldehydes such as (55) and Danishefsky's diene (56) (S. Schaus, J. Branalt and E. N. Jacobsen, *J. Org. Chem.*, 1998, **63**, 403). The second catalyst system (57) incorporates a bulky tridentate Schiff base as the ligand. Significantly, catalysts generated with this new ligand promote reactions between aromatic aldehydes and dienes of weaker nucleophilicity than Danishefsky's diene. As can be seen in the example below, combination of diene (58) and aldehyde (59) delivers the relevant product in excellent yield and ee when using only 3 mol% of the catalyst (A. G. Dossetter, T. F. Jamison and E. N. Jacobsen, *Angew. Chem., Int. Ed. Engl.*, 1999, **38**, 2398).

H H N N Cr^+ But O O tBu BF_4^- tBu But

(54)

N O Cr O SbF_6

(57)

TMSO OMe + H O O

(55) (56)

2 mol% (54)
4Å MS, TBME, -10°C

O O O

63% yield 99% ee
after recrystallisation

Me TESO Me + H O OTBS

(58) (59)

3 mol% (57)
4Å MS, acetone, rt

O Me Me O OTBS

97% yield, 99% ee

The Salen-Cr catalyst (54) has been used to prepare building blocks for incorporation into a synthesis of the natural product muconin (S. E. Schaus, J. Branalt and E. N. Jacobsen, *J. Org. Chem.*, 1998, **63**, 4876). The second generation Cr catalyst has also been employed in a total synthesis; both the C_1-C_6 and the C_{11}-C_{16} pyran rings of FR901464 were prepared using enantioselective hetero-Diels-Alder reactions catalysed by complex (57). Each pyran was constructed with an alkyne substituent; the first (60) features an alkyne-substituted diene and the second (61) an alkyne-substituted aldehyde (C. F. Thompson, T. F. Jamison and E. N. Jacobsen, *J. Am. Chem. Soc.*, 2000, **122**, 10482).

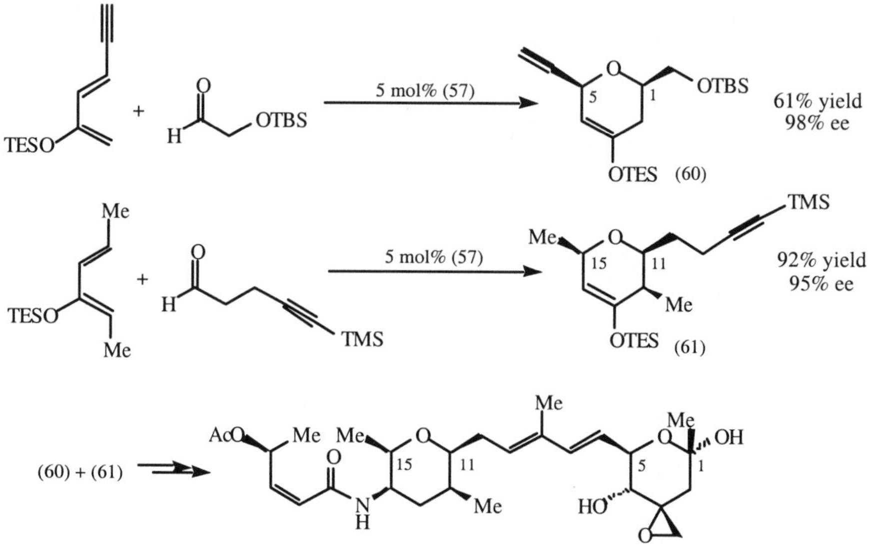

TESO + H OTBS → 5 mol% (57) → OTBS, OTES (60) 61% yield 98% ee

TESO, Me, Me + H TMS → 5 mol% (57) → Me, TMS, Me, OTES (61) 92% yield 95% ee

(60) + (61) ⇒ FR901464

The Jorgensen group have demonstrated the ability of Al(III)-BINOL type complexes to be extremely effective catalysts for the combination of benzaldehyde and Danishefsky's diene (K. B. Simonsen *et al.*, *Chem. Eur. J.*, 2000, **6**, 123). The same group also reported the first example of a hetero-Diels-Alder reaction catalysed by a chiral polymeric catalyst (M. Johannsen and K. A. Jorgensen, *J. Org. Chem., 1999*, **64**, 299). The catalyst generated from the combination of polymeric BINOL (62) and $AlMe_3$ delivered cycloadducts with comparable selectivities to those obtained using the monomeric catalysts. For example, adduct (63) was obtained in 80% yield with 88% ee. The polymeric catalyst system offers the advantage of easier recovery and reuse.

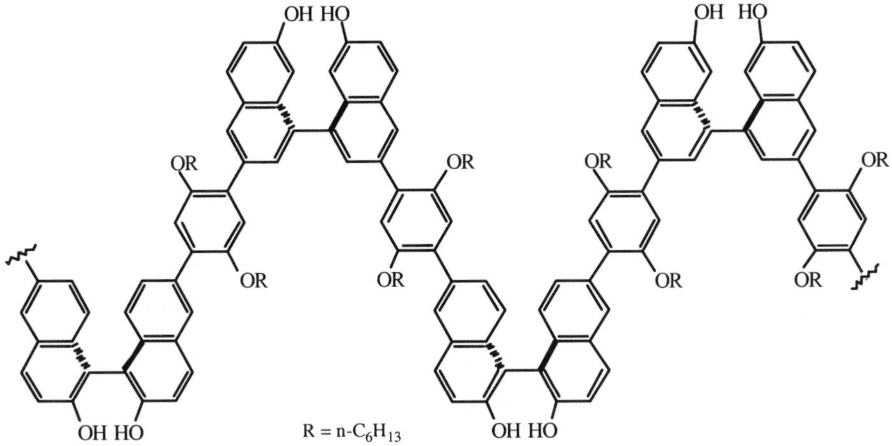

$R = n\text{-}C_6H_{13}$

(62)

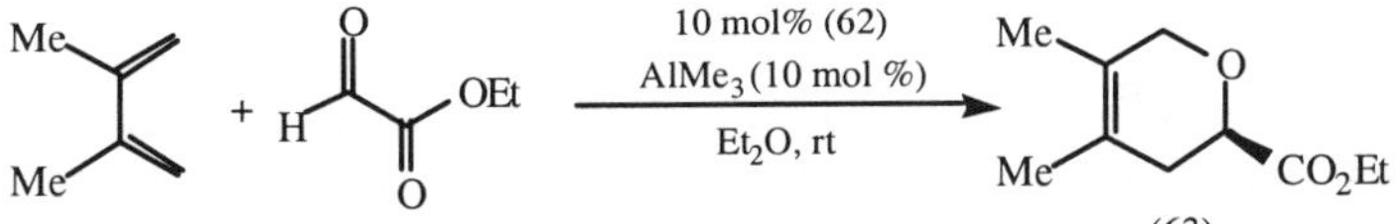

10 mol% (62)
$AlMe_3$ (10 mol %)
Et_2O, rt

(63)

80% yield
88% ee

The Jorgensen group have also investigated the use of glyoxylate and pyruvate esters and closely related derivatives as dienophiles in hetero-Diels-Alder reactions. A range of chiral Lewis acids were evaluated and for the majority of the substrates the Cu(II)-bis(oxazoline) complexes pioneered by Evans were found to offer the best efficiencies and selectivities (S. Yao *et al.*, *J. Am. Chem. Soc.*, 1998, **120**, 8599; S. Yao *et al.*, *J. Org. Chem.*, 1999, **64**, 6677). The glyoxylate Diels-Alder reaction has been applied to the total synthesis of (*R*)-dihydroactinidiolode and (*R*)-actinidiolide (S. Yao *et al.*, *J.*

Org. Chem., 1998, **63**, 118). The key step in the synthesis of actinidiolide involves the Cu(II) complex (25) catalysed reaction between ethyl glyoxylate (64) and substituted cyclohexene (65). The bicyclic adduct (66) was obtained in 90% yield with an enantioselectivity of 97%. Advancement of (66) through four synthetic steps provided the natural product.

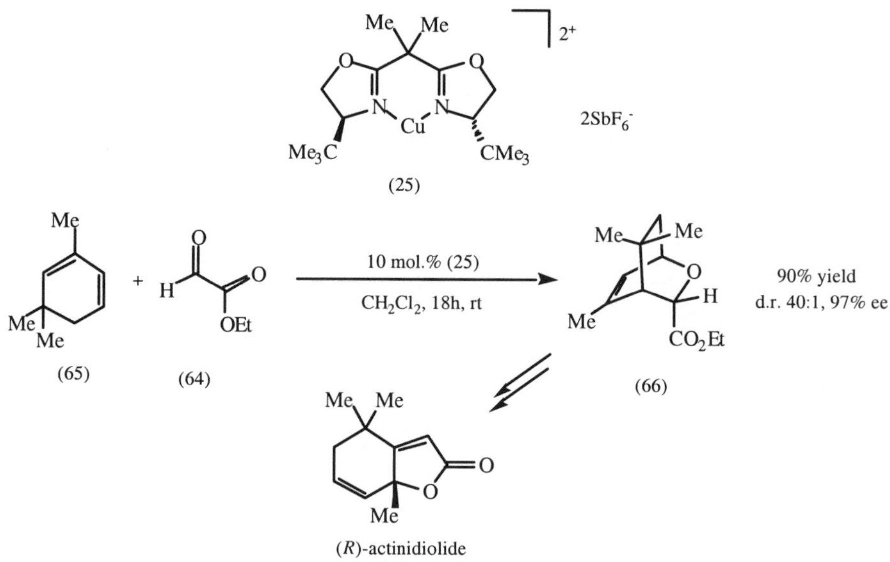

$2SbF_6^-$

(25)

10 mol.% (25)

CH_2Cl_2, 18h, rt

90% yield
d.r. 40:1, 97% ee

(65) (64) (66)

(*R*)-actinidiolide

All of the hetero-cycloadditions so far considered have concerned the formation of pyrans; Ghosez and co-workers have developed a route to the preparation of enantiomerically enriched δ-lactams (E. Jnoff and L. Ghosez, *J. Am. Chem. Soc.*, 1999, **121**, 2617). The key reaction involves the Lewis acid catalysed Diels-Alder reaction between imino-diene (67) and imide (17). Use of the chiral Cu(II) Lewis acid catalyst (26) developed by Evans delivers the cycloadduct in good yield and with an excellent 96% ee. The piperidone products can be manipulated through simple oxidation state adjustments to provide a range of useful enantiomerically enriched building blocks.

Sulfur containing dienes have also recently been employed in combination with imide dienophiles and related Cu(II) catalysts to afford enantiomerically enriched dihydrothiopyrans (T. Saito, K. Takekawa and T. Takahashi, *Chem. Commun.*, 1999, 1001).

(26)

(67) (17)

1. 8 mol.% (26)

CH_2Cl_2, 4Å MS, -45°C

2. MeOH

96% yield

d.r. 99:1, 98% ee

Nitrogen containing Diels-Alder adducts can also be produced by the combination of an imine dienophile and an electron rich diene. Jorgensen and co-workers have developed such a route to the synthesis of a family of non-proteinogenic α-amino acids (S. Yao *et al.*, *Angew. Chem., Int. Ed. Engl.*, 1998, **37**, 3120; S. Yao *et al*, *Chem. Eur. J.*, 2000, **6**, 2435). In order to obtain a suitably reactive imine the glyoxylate derived imine (68) was employed as the dienophile. This could be successfully combined with a range of cyclic and acyclic electron-rich dienophiles. Complex (69), initially developed by Leckta and co-workers as a catalyst for the imino ene and imino aldol reactions (D. Ferraris *et al.*, *J. Am. Chem. Soc.*, 1998, **120**, 4548), was employed as the chiral Lewis acid. As can been seen from the examples below the cycloadducts could be obtained with excellent selectivities. Interestingly the authors found that the *N*-substituent on the imine could be tuned to provide either enantiomer of product: *N*-tosyl and *N*-*p*-methoxyphenyl substituents were found to afford adducts of opposite absolute configuration.

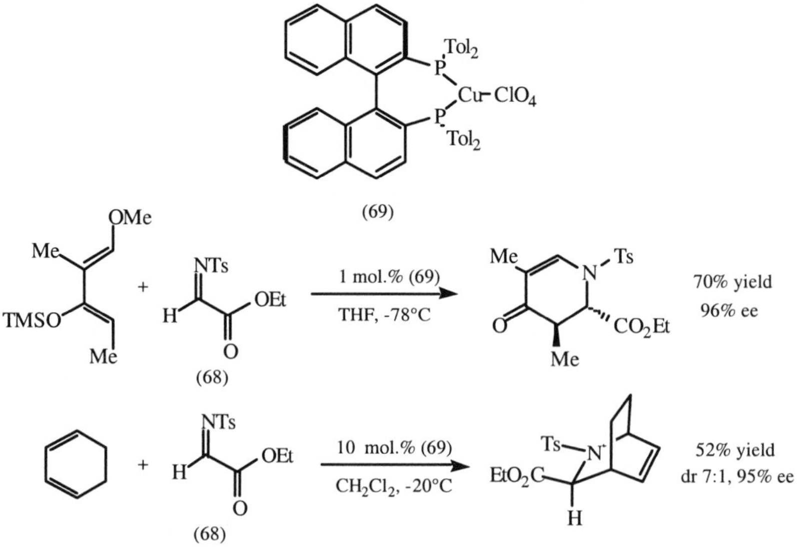
Tol2
P
Cu-ClO4
P
Tol2
(69)
OMe
Me
TMSO
Me
+
NTs
H
OEt
O
(68)
1 mol.% (69)
THF, -78°C
Me
N
Ts
O
CO2Et
Me
70% yield
96% ee
+
NTs
H
OEt
O
(68)
10 mol.% (69)
CH2Cl2, -20°C
Ts
N
EtO2C
H
52% yield
dr 7:1, 95% ee

(b) Inverse electron demand hetero-Diels-Alder reactions

Several publications have appeared in which enantioselective hetero-Diels-Alder reactions are completed using components possessing the opposite electronic properties to those used in standard Diels-Alder reactions. The groups of both Evans and Jorgensen have reported Diels-Alder reactions of β-γ-unsaturated α-keto esters and enol ethers to give dihydropyran products (D. A. Evans *et al*, *Angew. Chem.*, *Int. Ed. Engl.*, 1998, **37**, 3372; J. Thorhauge, M. Johannsen, K. A. Jørgensen, *Angew. Chem., Int. Ed. Engl.*, 1998, **37**, 2404). In both reports the optimal catalyst is the *t*-butyl substituted Cu(II) bis(oxazoline) complex (26); importantly the complex including triflate counterions is employed in contrast to the hexafluoro antimonate counterions used in many of the standard Diels-Alder reactions. Both groups report systems that tolerate a variety of substituents on both the diene and dienophile components including amino substituted dienes such as (70). The Evans group report that as little as 0.5 mol% catalyst can be employed to deliver adducts with excellent selectivity; incorporation of Florisil® as an adsorbent in the reaction solutions allows efficient catalyst recycling. The

Jorgensen group have utilised the amino-substituted products such as (71) in the synthesis of a range of amino-sugar derivatives (W. Zhuang, J. Thorhauge and K. A. Jorgensen, *Chem. Commun.*, 2000, 459).

The Evans group have also utilised α,β-unsaturated acyl phosphonates such as (72) as hetero-dienes in Diels-Alder reactions with enol ethers. The three examples presented below illustrate the high degree of functionality that is incorporated in these enantiomerically enriched Diels-Alder adducts.

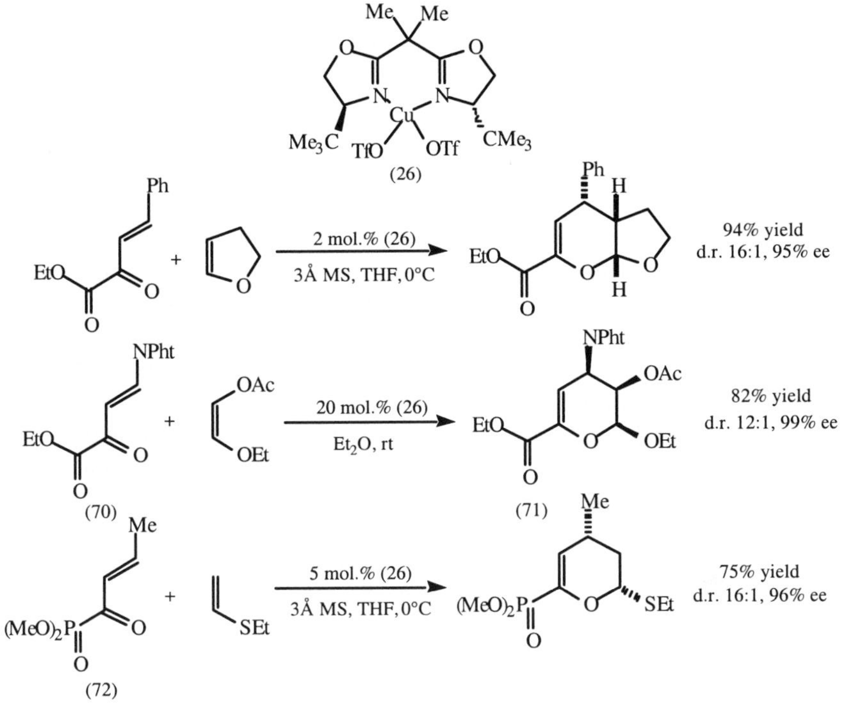

Me Me O O N N Cu Me_3C TfO OTf CMe_3 (26)

Ph EtO O O + O → 2 mol.% (26), 3Å MS, THF, 0°C → Ph H EtO O O H O — 94% yield, d.r. 16:1, 95% ee

NPht EtO O O + OAc OEt → 20 mol.% (26), Et_2O, rt → NPht OAc EtO O O OEt — 82% yield, d.r. 12:1, 99% ee

(70) (71)

Me $(MeO)_2P$ O O + SEt → 5 mol.% (26), 3Å MS, THF, 0°C → Me $(MeO)_2P$ O O SEt — 75% yield, d.r. 16:1, 96% ee

(72)

4. 1,3-Dipolar cycloaddition reactions

Similar to Diels-Alder reactions, 1,3-dipolar cycloadditions are examples of $[4\pi + 2\pi]$ cycloaddition reactions. The variety of molecular-arrangements that can be used to construct the 4π component, termed the 1,3-dipole, is considerable and a large number have been explored in cycloaddition chemistry (for general reviews, see; R. D. Little in "Comprehensive Organic Synthesis", Vol. 5, pp239, Eds. B. M. Trost and I. Fleming, Pergamon, Oxford, 1991; R. M. T. Chan in "Comprehensive Organic Synthesis", Vol. 5, pp271, Eds. B. M. Trost and I. Fleming, Pergamon, Oxford, 1991). The number of 1,3-dipoles that have featured in asymmetric synthesis is considerably smaller and an excellent review has recently appeared (K. V. Gothelf and K. A. Jorgensen, *Chem. Rev.*, 1998, **98**, 863). Examples of catalytic enantioselective 1,3-dipolar cycloaddition reactions feature an even smaller range of dipoles and among these nitrones have received the greatest attention (for a brief review, see: K. V. Gothelf and K. A. Jorgensen, *Chem. Commun.*, 2000, 1449).

(a) Nitrones as (1,3)-dipoles

An impressive recent example of the use of a nitrone as a 1,3-dipole in an enantioselective cycloaddition has been reported by the Kanemasa group (S. Kanemasa *et al., J. Am. Chem. Soc.*,1998, **120**, 12355). In common with the majority of Lewis acid catalysed normal-electron demand systems a bidentate alkene has been employed as the dipolarophile. This is necessary in order to counteract the binding of the metal catalyst to the nitrone and allow preferential binding of the Lewis acid to the alkene and thus achieve the required activation. The Kanemasa group employ a catalyst generated from the DBFox ligand (51) and $Ni(ClO_4)_2$; significantly the hydrated metal salt can be employed without any loss of reaction selectivity. A variety of aromatic nitrones such as (73) can be employed to deliver cycloadducts (74) in good yield with enantioselectivities over 90%.

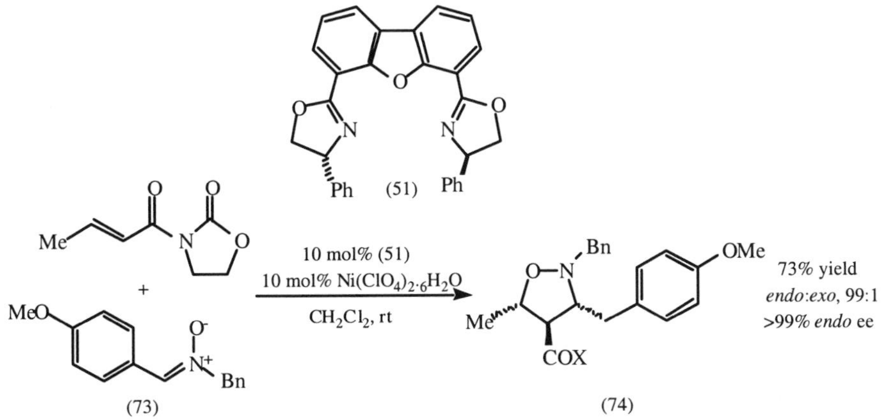

A number of other chiral Lewis acids have been applied to related processes all featuring the combination of a nitrone and an electron poor bidentate alkene. Some of the more successful examples are shown below and include the Pd(II)-(Tol-BINAP) complex (75) (K. Hori *et al.*, *J. Org. Chem.*,1999, **64**, 5017), Yb(III)-BINOL complex (76) (S. Kobayashi and M. Kawamura, *J. Am. Chem. Soc.*, 1998, **120**, 5840) and the Mg(II)-bis(oxazoline) (77) (K. V. Gothelf, R. G. Hazell and K. A. Jorgensen, *J. Org. Chem.*, 1998, **63**, 5483). To obtain high selectivities the ytterbium complex (76) must incorporate a C_2-symmetric amine as a further ligand. Interestingly, the sense of absolute asymmetric induction obtained when using magnesium-complex (77) as the catalyst was shown to be dependent on the presence of 4Å molecular sieves.

Ti(IV)-TADDOL complexes that have been employed extensively in enantioselective Diels-Alder reactions have also been employed in nitrone (1,3)-dipolar cycloadditions (K. V. Gothelf, I. Thomsen and K. A. Jorgensen, *J. Am. Chem. Soc.*, 1996, **118**, 59). More recently Seebach and Heckel have reported the use of the immobilised Ti-TADDOL complex (78) as a selective (1,3)-dipolar cycloaddition catalyst (A. Heckel and D. Seebach, *Angew. Chem., Int. Ed. Engl.*, 2000, **39**, 163).

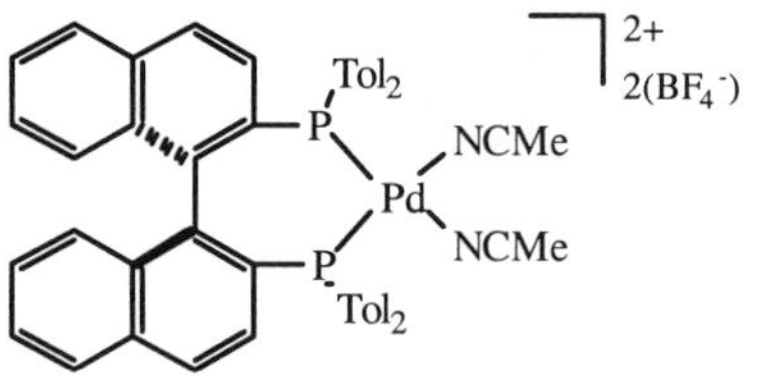

(75)

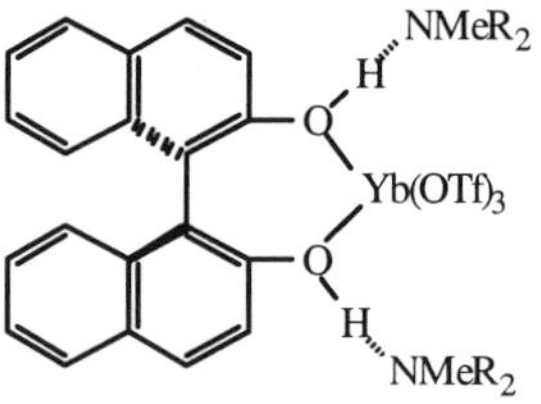

(76)

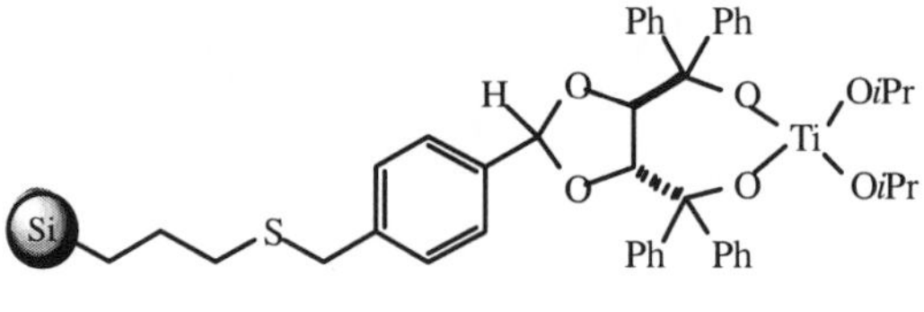

(78)

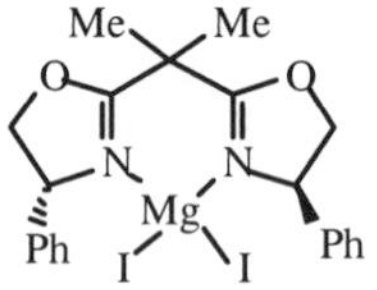

(77)

Several examples of inverse-electron demand nitrone (1,3)-dipolar cycloadditions have also been reported. In one example the use of glyoxylate derived nitrones such as (79), enol ethers and Cu(II)-bis(oxazoline) catalysts such as (26) produces a highly enantioselective reaction (K. B. Jensen, R. G. Hazel and K. A. Jorgensen, *J. Org. Chem.*, 1999, **64**, 2353). A variety of metal salts and bis(oxazoline) ligands were screened with complex (26) proving to be the most selective. The authors propose that the key transition state features coordination of both the alkene and nitrone to the Lewis acidic metal centre.

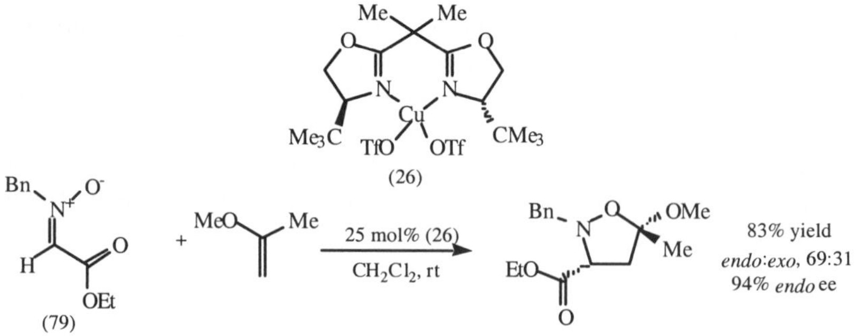

(26)

(79)

25 mol% (26)

CH_2Cl_2, rt

83% yield
endo:*exo*, 69:31
94% *endo* ee

The same group have pioneered the use of chiral Al(III) complexes as catalysts for inverse-electron demand reactions (K. B.Simonsen *et al.*, *J. Am. Chem. Soc.*, 1999, **121**, 3845). The combination of various BINOL derivatives and $AlMe_3$ produces a range of highly active catalysts. For acyclic nitones such as (80) the phenyl-substituted ligand (81) proved optimal, delivering the cycloadducts with excellent levels of selectivity. The use of the methoxyphenyl-substituted ligand (82) allowed the protocol to be extended to include cyclic nitrones such as (83). In general the use of cyclic nitrones resulted in slightly less selective reactions.

Complexes generated from chiral polybinaphthyl ligands and Al(III) salts have also been shown to be effective catalysts for the combination of acyclic nitrones and enol ethers (K. B. Simonsen *et al.*, *Chem. Commun.*, 1999, 811). In analogy to the application of these ligand systems to enantioselective Diels-Alder reactions the catalysts could be conveniently isolated and reused.

(81), R = H

(82), R = MeO

(80) + H–C(=CH₂)–O*t*Bu → 20 mol% (81), PhMe, Hexane, rt → 74% yield, *endo*:*exo*, 5:95, 93% *exo* ee

(83) + H–C(=CH₂)–OEt → 20 mol% (82), CH_2Cl_2, rt → 85% yield, *endo*:*exo*, 4:96, 85% *exo* ee

The final example of an enantioselective nitrone (1,3)-dipolar cycloaddition features LUMO-lowering activation not by a chiral Lewis acid but by the catalytic generation of an iminium ion. The MacMillan group have applied the use of amine catalyst (53) to the combination of nitrones and enals (W. S. Jen, J. J. Wiener and D. W. C. MacMillan, *J. Am. Chem. Soc.*, 2000, **122**, 9874). Both acrolein and crotonaldehyde could be combined with a range of aromatic nitrones to provide the [3 + 2] cycloaddition products with excellent levels of diastereo- and enantioselectivity. A significant advantage of this methodology over the Lewis acid catalysed systems is that monodentate dipolarophiles such as (84) can be successfully employed.

Me N Me Me N H Ph .HCl O

(53)

N+ O- Ph + Me (84) H O

20 mol.% (53)

$MeNO_2$-H_2O, -20°C

N–O Ph Me CHO

73% yield
exo:endo, 93:7
98% *endo* ee

(b) Other (1,3)-dipoles

Suga *et al* have reported a formal [3 + 2] cycloaddition between alkoxyoxazoles and aldehydes (H. Suga, K. Ikai and T. Ibata, *J. Org. Chem.*, 1999, **64**, 7040). The chiral complex generated from the combination of BINOL and $AlMe_3$ proved to be effective at promoting the reaction between aromatic aldehydes (85) and 5-alkoxy-2-aryloxazoles such as (86). Importantly, in order to obtain highly selective reactions 200 mol% of the catalyst must be employed. The synthetic utility of the chiral 2-oxazoline-4-carboxylate products was demonstrated by their conversion to β-hydroxy-α-amino acids.

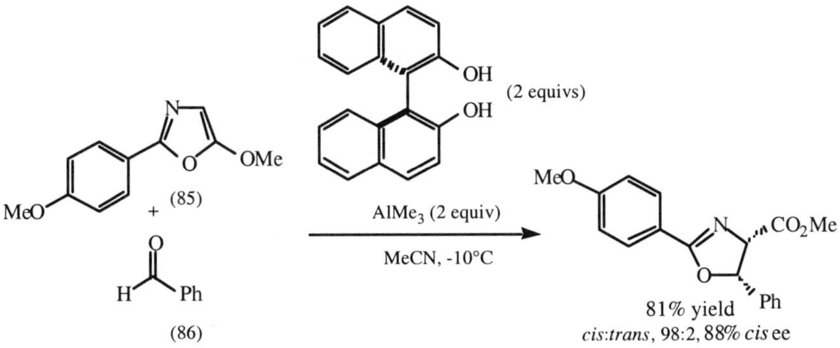

Hashimoto and co-workers have utilised chiral dirhodium(II) carboxylate catalysts to achieve an enantioselective (1,3)-dipolar cycloaddition between carbonyl ylides and alkynes (S. Kitigaki *et al.*, *Tetrahedron Lett.*, 2000, **41**, 5931). The required carbonyl ylides (87) were generated by the rhodium catalysed decomposition of α-diazo ketones such as (88); reaction with electron-poor alkynes afforded the cycloadducts with excellent enantioselectivity. A range of chiral dirhodium catalysts were screened in the process however all were inferior to (89). Extension of the chemistry to substrates other than (88) resulted in reduced selectivity.

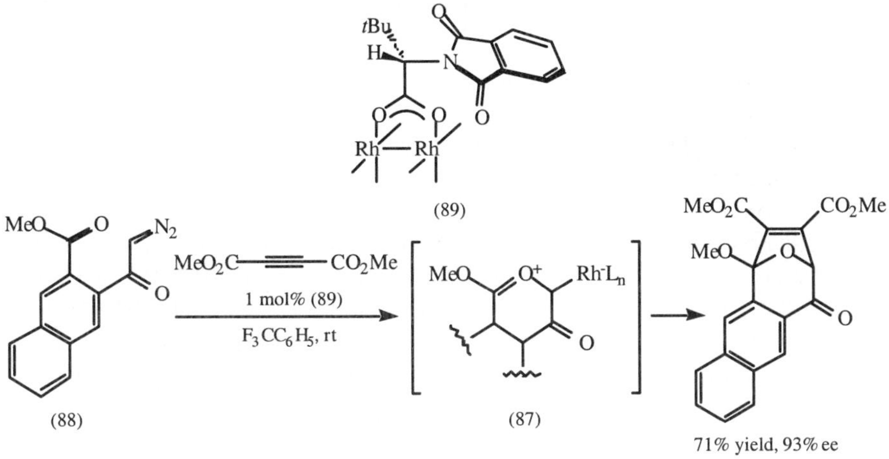

The Hodgson group have reported a catalytic intramolecular enantioselective carbonyl ylide-alkene cycloaddition (D. M. Hodgson, P. A. Supple and C. Johnstone, *Chem. Comm.*, 1999, 2185). Rh(II) binaphthol phosphate complex (90) catalyses the formation of carbonyl ylide (91) which then undergoes intramolecular reaction with the pendent alkene to generate cycloadduct (92) with 90% ee. Only 0.5 mol% of the catalyst was needed to obtain this level of selectivity.

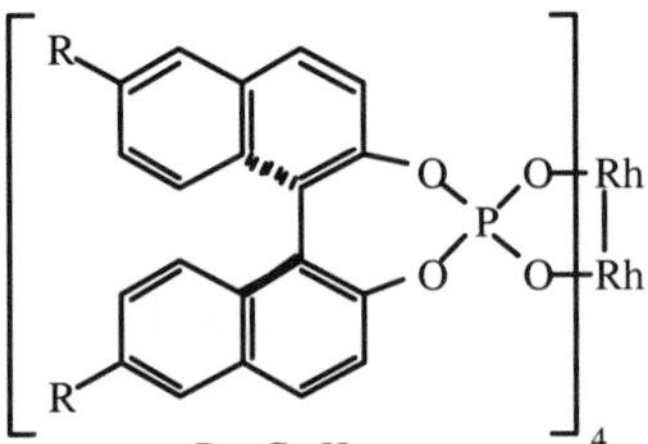

R = $C_{12}H_{25}$
(90)

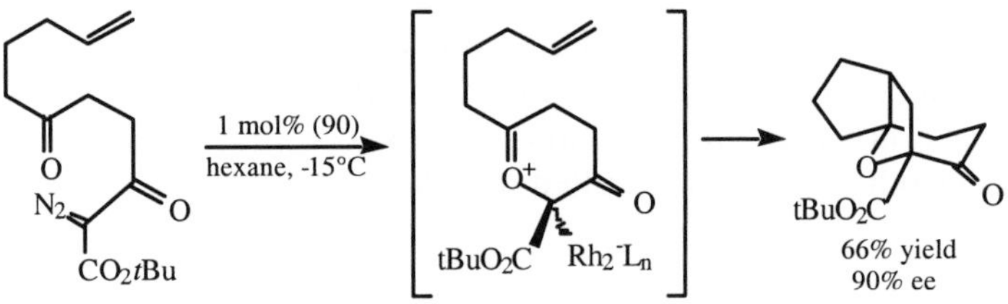

1 mol% (90)
hexane, -15°C

$tBuO_2C$ Rh_2L_n
(91)

66% yield
90% ee
(92)

5. [2+2] Cycloaddition reactions

Examples of [2+2] cycloadditions in synthesis are fewer than the reactions featured in the previous sections of this chapter, however some impressive enantioselective variants have been developed (for a review, see; Y. Hayashi and K. Narasaka. in "Comprehensive Asymmetric Catalysis", Vol. 3, Ed. E. N. Jacobsen, A. Pfaltz and H. Yamamoto, Springer, Berlin, 1999) There are essentially two types of [2+2] reactions that have been adapted to asymmetric catalysis; the reaction between a ketene and an aldehyde to generate a β-lactone product and the reaction between an electron poor alkene, usually an enone derivative, and an electron rich alkene, usually an enol ether or derivative, to generate a cyclobutane.

Imide derivatives such as (93) feature prominently as electron-poor dienophiles in enantioselective Diels-Alder reactions and have been adopted as the electron-poor alkenes of choice in a number of asymmetric [2+2] cycloadditions. The Narasaka group have developed effective methodology to combine imide (93) and dithioacetal (94) in the presence of Ti(IV)-TADDOL (35) (K. Narasaka *et al.*, *J. Am. Chem. Soc.*, 1992, **114**, 8869). Reaction with 10 mol% (35) delivers cyclobutane (95) in 83% yield with 98% ee. The range of functionality present in (95) has been exploited in a synthesis of the nucleoside antibiotic oxetanocin and derivatives thereof (Y.-I., Ichikawa *et al.*, *Chem. Commun.*, 1989, 1919). Several alternatives to dithioacetals have also been investigated as electron-rich alkenes, including alkenyl sulfides, alkynyl sulfides and allenyl sulfides (Y. Hayashi and K. Narasaka, *Chem. Lett.*, 1990, 1295).

(35)

(93) + (94) → 10 mol% (35), 4Å MS, PhMe/Hexane, 0°C → (95) 83% yield, 98% ee

(95) → steps → oxetanocin

Ti-TADDOL complexes have also been used to catalyse [2+2] cycloadditions between 1,4-benzoquinones and styrenes with the cyclobutane products being obtained with enantioselectivities of up to 83% (T. A. Engler, M. A. Letavic and J. P. Reddy, *J. Am. Chem. Soc.*, 1991, **113**, 5068).

In 1996 Kocienski and co-workers reported an enantioselective [2+2] cycloaddition between aldehydes and (trimethylsilyl)ketene (B. W. Dymock, P. J. Kocienski and J.-M. Pons, *Chem. Commun.*, 1996, 1053). The β-lactone products obtained from such reactions are valuable synthetic intermediates as they contain a masked aldol unit. (Trimethylsilyl)ketene is often employed in [2+2] cycloadditions due to its far greater stability compared to simple alkyl ketenes or ketene itself. In the Kocienski chemistry the cycloadditions are catalysed by chiral aluminium complex (96) to produce the corresponding lactones with ee's up to 83%.

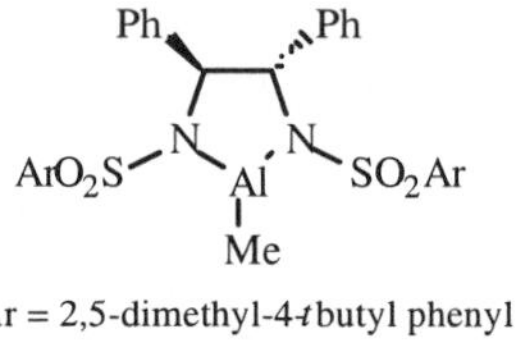

Ph Ph ArO2S N Al N SO2Ar Me

Ar = 2,5-dimethyl-4-*t*butyl phenyl

(96)

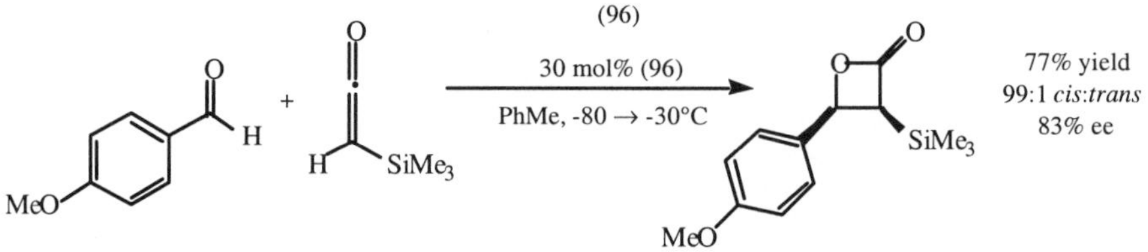

30 mol% (96)

PhMe, -80 → -30°C

77% yield
99:1 *cis*:*trans*
83% ee

(Trimethylsilyl)ketene cycloadditions have also been investigated using Ti-TADDOL complexes as catalysts (H. W. Yang and D. Romo, *Tetrahedron Lett.*, 1998, **39**, 2877). The enantioselectivities obtained using this catalyst family were only moderate. The same group have developed a novel aluminium based complex for similar cycloadditions (D. Romo *et al.*, *Bioorg. Med. Chem.*, 1998, **6**, 1255). As can be seen below, complex (97) catalyses the reaction between (trimethylsilyl)ketene and cyclohexane carboxaldehyde (98) delivering the product (99) in good yield with 84% ee. The chiral β-lactones produced in this study were investigated as potential inhibitors of HMG-CoA synthase.

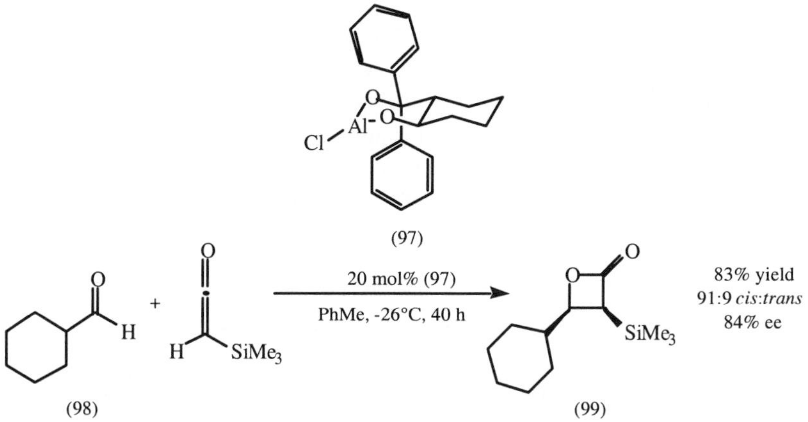

(97)

20 mol% (97)

PhMe, -26°C, 40 h

83% yield
91:9 *cis:trans*
84% ee

(98) (99)

Both of the previous examples are limited in that they employ (trimethylsilyl)ketene as a stable ketene equivalent; Nelson has recently developed methodology to generate ketenes *in-situ* and thus override this limitation (S. G. Nelson, T. J. Peelen and Z. Wan, *J. Am. Chem. Soc.*, 1999, **121**, 9742). Treatment of acetyl bromide (100) with aluminium complex (101) and diisopropylamine generates ketene (102) *in-situ*; catalysed reaction between (102) and aldehyde (103) then provides the β-lactone product (104) with excellent selectivity. Treatment of the adducts with $La(OtBu)_3$ and benzyl alcohol delivers the corresponding protected linear acetate aldol adducts (105) in good yield.

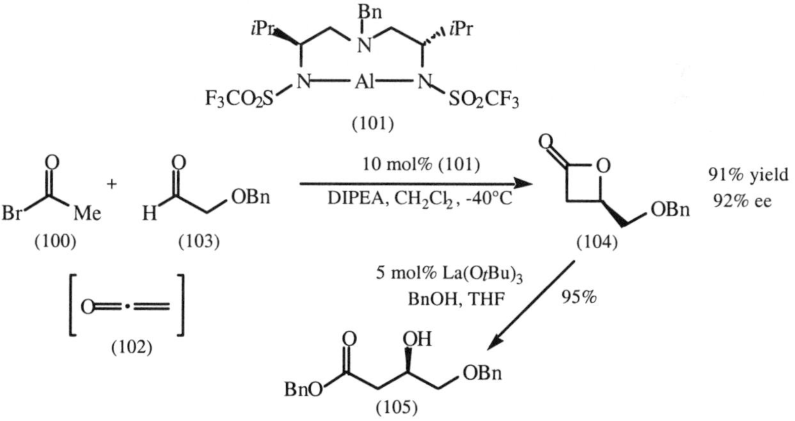

(101)

10 mol% (101)

DIPEA, CH_2Cl_2, -40°C

91% yield
92% ee

(100) (103) (104)

(102)

5 mol% $La(OtBu)_3$
BnOH, THF

95%

(105)

This methodology has been extended to include substituted ketenes and thus provide access to propionate-derived aldol adducts (S. G. Nelson and Z. Wan, *Organic Lett.*, 2000, **2**, 1883). Alkynyl substituted aldehydes such as (106) can also be successfully employed as substrates; treatment of the resultant β-lactones with Grignard reagents produces optically active allenes such as (107). This methodology has been utilised in a synthesis of (-)-malyngolide (108) (Z. Wan and S. G. Nelson, *J. Am. Chem. Soc.*, 2000, **122**, 10470).

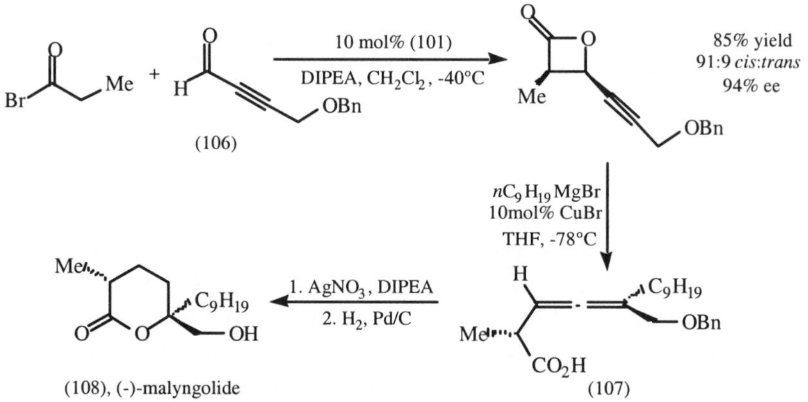

10 mol% (101)
DIPEA, CH_2Cl_2, -40°C

(106)

85% yield
91:9 *cis:trans*
94% ee

$nC_9H_{19}MgBr$
10mol% CuBr
THF, -78°C

1. $AgNO_3$, DIPEA
2. H_2, Pd/C

(108), (-)-malyngolide

(107)

The usefulness of the available chiral β-lactones has been further demonstrated in a concise route to protected β-amino acids (S. G. Nelson and K. L. Spencer, *Angew. Chem., Int. Ed. Engl.*, 2000, **39**, 1323). Treatment of the relevant β-lactone (109) with sodium sulfonamide (110) followed by treatment with diazomethane provided the corresponding protected amino acids (111) in excellent yield.

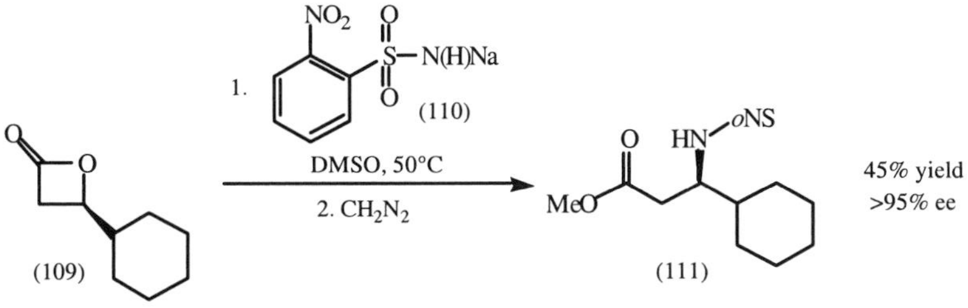

NO_2 O S N(H)Na O (110)

1. DMSO, 50°C

2. CH_2N_2

(109) → MeO O HN oNS (111) 45% yield >95% ee

Treatment of α-bromoacyl bromides with elemental zinc has also been used as a method for the *in-situ* generation of ketenes. Calter has developed such a system in combination with the use of chiral amine catalysts to produce chiral β-lactones *via* ketene dimerisations (M. A. Calter, *J. Org. Chem.*, 1996, **61**, 8006). The final system involves treatment of acylbromide (112) with zinc metal, distillation of ketene (113) and immediate treatment with 1 mol% of quinidine to deliver the dimer (114) in 55% yield with 98% ee. The use of chiral amines as catalysts for [2+2] cycloadditions has also been reported in studies on the reaction of ketene and chloral (H. Wynberg and E. G. Staring, *J. Am. Chem. Soc.*, 1982, **104**, 166). Lactone (114) has been employed in the asymmetric synthesis of the C_1-C_{10} segment of the natural product papamycin 621A (M. A. Calter and F. C. Bi, *Organic Lett.*, 2000, **2**, 1529).

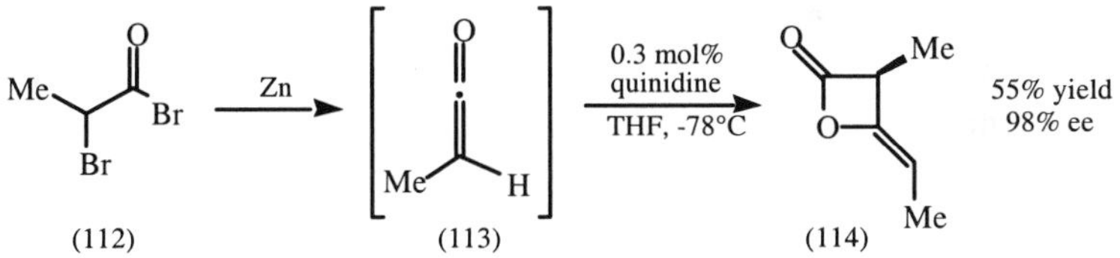

Me O Br Br (112) —Zn→ [O Me H] (113) —0.3 mol% quinidine, THF, -78°C→ O Me O Me (114) 55% yield 98% ee

Second Edition of Rodd's Chemistry of Carbon Compounds,
Volume V, Topical Volumes
Asymmetric Catalysis, edited by M.Sainsbury

Chapter 9

PALLADIUM CATALYSED COUPLING REACTIONS

C. G. FROST

This chapter describes leading advances in the specific areas of palladium catalysed enantioselective allylic substitution, cross-coupling and Heck reactions. The author has attempted to highlight the most important historic and contemporary contributions but as with all reviews the content is subject to preference and limitations of space.

1. Allylic substitution

The palladium catalysed nucleophilic substitution of allylic compounds is an established, efficient and reliable process, and has become an important tool for the synthetic organic chemist. The first η^3-allylpalladium complexes were isolated and characterised over 30 years ago, synthesised by the reaction of dienes with palladium(II) salts (B. L. Shaw, *Chem. Ind. (London)*, 1962, 1190; B. L. Shaw and N. Sheppard, *Chem. Ind. (London)*, 1961, 517). In 1965, the stoichiometric reaction of π-allylpalladium complexes with nucleophiles was reported, effecting an overall allylic substitution (J. Tsuji, H. Takashashi and M. Morikawa, *Tetrahedron Lett.*, 1965, **6**, 4387). Later, in the early 1970's it was noted that the allylic displacement of acetate with a variety of nucleophiles required only a catalytic amount of palladium (K. E. Atkins, W. E. Walker and R. M. Manyik, *Tetrahedron Lett.*, 1970, **11**, 3821; H. Hata, K. Takahashi and A. Miyake, *J. Chem. Soc., Chem. Commun.*, 1970, 1392). These findings opened the door to a vast area of further studies and applications. Since the mid–1970's the palladium catalysed allylic substitution reaction has evolved into a very mild, efficient process illustrated by the reaction of allyl acetate (1) with the sodium salt of dimethyl malonate in the presence of catalytic amounts of triphenylphosphine and palladium(0). Such reactions are typically conducted in a polar solvent such as THF to afford the substitution product (2) in good yield and with a high number of turnovers (B. M. Trost and T. R. Verhoeven, *J. Am. Chem. Soc.*, 1978, **100**, 3435).

$$\underset{(1)}{\text{CH}_2\text{=CHCH}_2\text{OAc}} \xrightarrow[\text{Pd cat./PPh}_3 \text{ THF}]{\text{NaCH(CO}_2\text{Me)}_2} \underset{(2)}{\text{CH}_2\text{=CHCH}_2\text{CH(CO}_2\text{Me)}_2}$$

The generally accepted mechanism of palladium catalysed substitution involves the initial co-ordination of palladium(0) to the alkene followed by oxidative addition to afford an intermediate η^3–allyl complex (T. Yamamoto, M. Akimoto, O. Saito and A. Yamamoto, *Organometallics*, 1986, **5**, 1559; H. Kurosawa, *J. Organomet. Chem.*, 1987, **334**, 243).

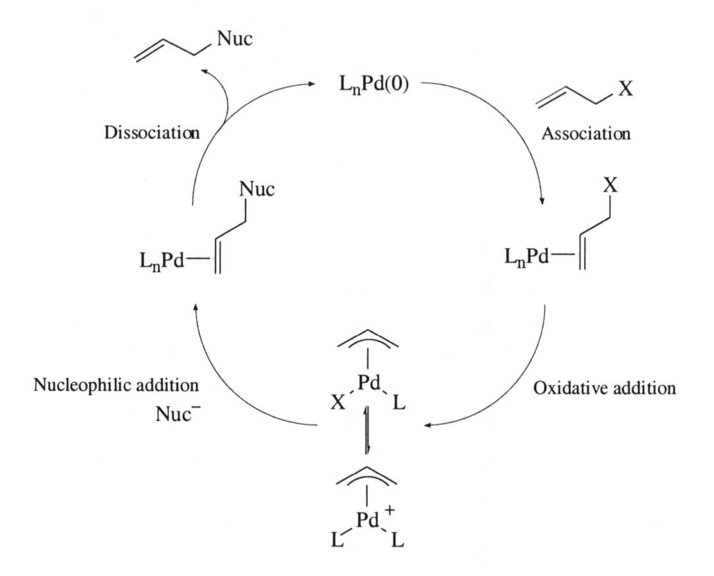

In the presence of triphenylphosphine, or other π-accepting ligands, an equilibrium between a neutral and cationic complex results. The use of bidentate ligands favours formation of the cationic complex. Nucleophilic addition to the cationic complex is favoured, and occurs at one of the allylic termini to furnish the palladium(0) complex of the product. Dissociation of the palladium(0) catalyst liberates the product, and regenerates the active palladium catalyst.

In 1973, Trost and Dietsche showed that the stoichiometric reaction of palladium chloride dimer (3) with $NaCH(CO_2Me)_2$ and enantiopure phosphine ligands, such as (+)-DIOP afforded the substitution product (4), achieving an enantiomeric excess of up to 23% (B. M. Trost and T. J. Dietsche, *J. Am. Chem. Soc.*, 1973, **95**, 8200).

Me Me Pd Cl 2 (3) $NaCH(CO_2Me)_2$ (+)-DIOP $CH(CO_2Me)_2$ Me Me (4) ee (%) 23

The first reported example of a catalytic, asymmetric palladium catalysed allylic substitution reaction was the conversion of racemic allyl acetate (5) to the enantiomerically enriched product (6); however only a modest enantioselectivity of up to 46% ee was achieved.

OAc (5) $NaCH(CO_2Me)_2$ Pd cat./(+)-DIOP $CH(CO_2Me)_2$ (6) ee (%) 46

Most of the reported asymmetric palladium catalysed allylic substitution processes start from a racemic allylic component (7), which in the absence of enantiopure ligands, forms an intermediate *meso* complex (8) with palladium(0). Since a nucleophile may attack at either terminus of the allylic component, the enantiomers (9) and (10) are formed. The degree of the enantioselectivity of a reaction depends on the ability of the enantiopure ligand to promote attack of the nucleophile to one terminus of the allylic component in preference to the other. The problem of achieving high levels of asymmetric induction results from the fact that a soft nucleophile attacks complex (8) from the opposite side to the ligand, and as a result, the distance between the reaction centre and the chirality-controlling centre is large.

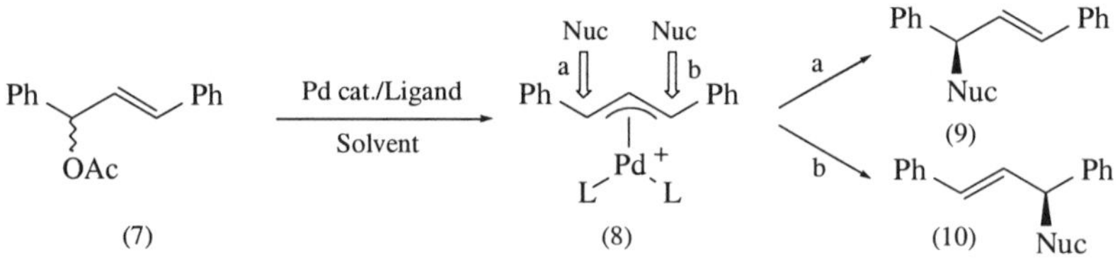

There have been many enantiopure ligands tested in the palladium catalysed enantioselective allylic substitution process (see below). This review will focus only on the important historical developments in terms of ligand design and the best ligands for specific processes (For previous reviews on the control of enantioselectivity in allylic substitution reactions including comprehensive lists of ligands tested, see; C. G. Frost, J. Howarth and J. M. J. Williams, *Tetrahedron: Asymmetry,* 1992, **3**, 1089; O. Reiser, *Angew. Chem. Int. Ed. Engl.*, 1993, **32**, 4, 547; B. M. Trost and D. L. Van Vranken, *Chem. Rev.*, 1996, **96**, 395; A. Pfaltz and M. Lautens, *Comprehensive Asymmetric Catalysis*, ed. E. N. Jacobsen, A. Pfaltz, H. Yamamoto, Springer Berlin, London, 1999, Chapter 24).
One of the most interesting early developments was the design of a family of enantiopure ligands with side-chain modifications, prepared from (R)-1-[(S)-1, 2-bis(diphenylphosphino)-ferrocenyl]ethyl acetate (T. Hayashi, K. Kanehira, T. Hagihara and M. Kumada, *J. Org. Chem.*, 1988, **53**, 113). The ligands achieved extremely high enantioselectivities in the palladium catalysed allylic alkylation of 1,3-disubstituted allylic acetates with sodium acetylacetonate and related stabilised carbon nucleophiles (T. Hayashi, A. Yamamoto, T. Hagihara and Y. Ito, *Tetrahedron Lett.*, 1986, **27**, 191). The

enantioselectivity in the reaction of 1,3-diphenyl-3-acetoxy-1-propene (7) with sodium acetylacetonate increases as the number of hydroxyl groups on the pendant side of the enantiopure ligand increases (Up to 96% ee with ligand (11)). It is noteworthy that the catalysts that show the highest levels of enantioselectivity are more catalytically active in general. The X-ray crystal structure of the π-allylpalladium complex bearing a hydroxylated ligand has been published (T. Hayashi, A. Yamamoto, Y. Ito, E. Nishioka, H. Miura and K. Yanagi, *J. Am. Chem. Soc.*, 1989, **111**, 6301).

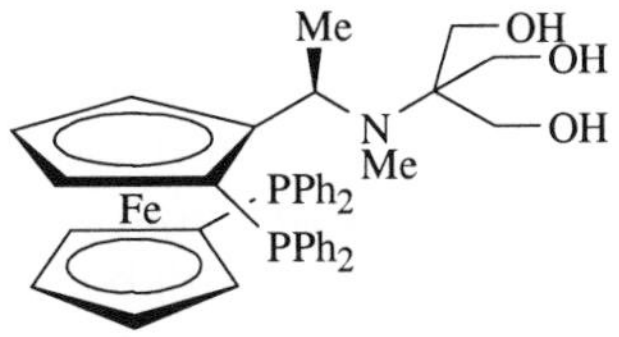

(11) 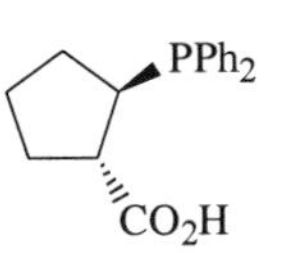 (12) 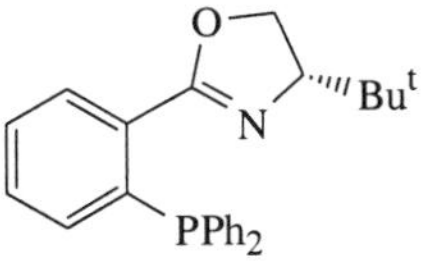 (13)

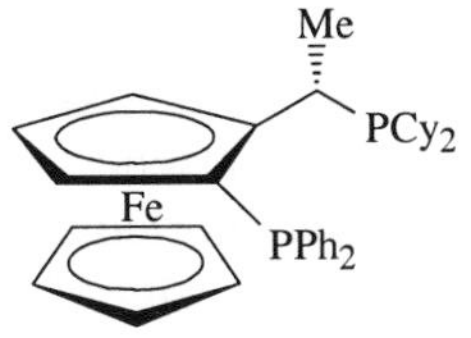

(14) 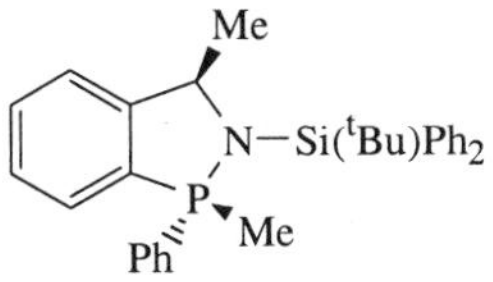 (15) 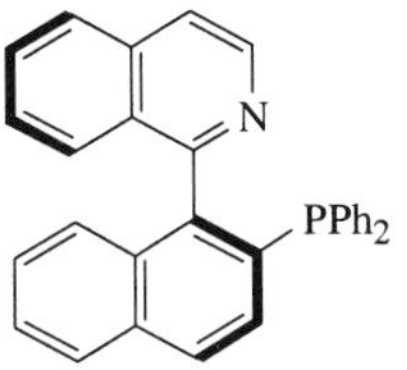 (16)

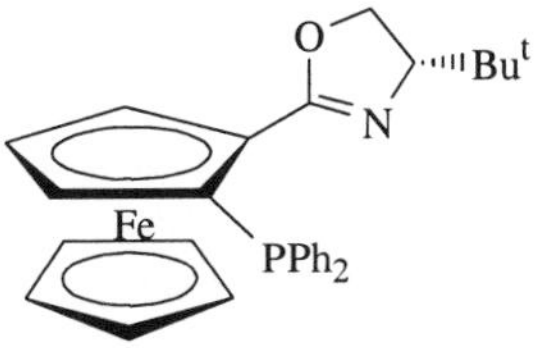

(17) 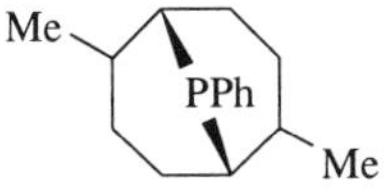 (18) 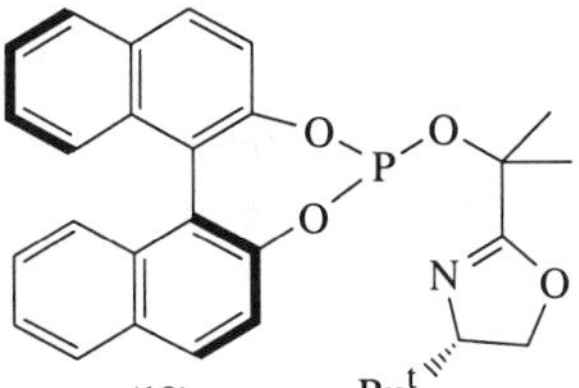 (19)

The pendant hydroxyl group is shown to reach over to the *exo* face of the π-allyl group and is located in close proximity to one of the π-allyl carbon atoms. Hayashi proposes a hypothetical transition state model where hydrogen bonding between the hydroxyl group and enolate anion causes

preferential attack of the enolate to one terminus of the π-allyl moiety. Other ferrocenylphosphine ligands such as (14) have been employed that achieve high enantioselectivity (up to 93% ee) yet are not capable of interacting with the incoming nucleophile through secondary interactions (A. Togni, *et al*, *J. Am. Chem. Soc.*, 1994, **116**, 4062).
An interesting series of enantiopure monodentate phosphine ligands, which have a carboxylic acid functionality and a cyclobutane or cyclopentane (12) backbone have been prepared. The ligands have been applied successfully to the asymmetric alkylation of 1,3-diphenyl-3-acetoxy-1-propene (7) with triethyl sodiophosphonoacetate and dimethyl malonate (Y. Okada, T. Minami, Y. Sasaki, Y. Umezu and M. Yamaguchi, *Tetrahedron Lett.*, 1990, **31**, 3905; Y. Okada, T. Minami, Y. Umezu, S. Nishikawa, R. Mori and Y. Nakayama, *Tetrahedron: Asymmetry*, 1991, **2**, 667). Although a high enantioselectivity of around 80% ee is obtained in the reaction with both nucleophiles in the presence of enantiopure ligand (12) and its cyclobutane analogue, a drastic decrease in the stereoselectivity is caused by the use of enantiopure ligands the carboxyl group of which is connected to the cycloalkane backbone *via* a methylene group. This indicates that the position of the carboxyl substituent is important for the enantioselective allylic alkylation. The importance of the carboxyl group is further supported by the decrease in enantioselectivity observed for the reaction using an ester of (12). It has been proposed that the high enantioselectivities observed with ligand (12) is caused by an electronic repulsion between the carboxylate anion on the ligands and the negative charge of the incoming nucleophiles, which directs the nucleophilic attack onto one of the π-allyl carbons.
Other valuable contributions include the monodentate phosphorus-based ligand (15), devised by Wills (M. Wills *et al*, *Tetrahedron Lett.*, 1994, **35**, 2791) and the QUINAP ligand (16) reported by Brown (J. M. Brown, D. I. Hulmes and P. J. Guiry, *Tetrahedron*, 1994, **50**, 4493).
One class of ligands that deserve a special mention is the phosphine-oxazolines, for example (13). The ligands were independently conceived and prepared by the three research groups of Helmchen (G. Helmchen and J. Sprinz, *Tetrahedron Lett.*, 1993, **34**, 1769), Pfaltz (A. Pfaltz and P. Von Matt, *Angew. Chem. Int. Ed. Engl.*, 1993, **32**, 4, 566) and Williams (J. M. J. Williams *et al*, *Tetrahedron Lett.*, 1993, **34**, 3149). The attraction of these

compounds lies in their ease of preparation, the ability to tailor the steric and electronic properties of the ligand and versatility across a range of applications. The introduction of phosphine-oxazoline ligands allowed for the first time, a predictable and highly enantioselective allylic substitution of symmetrically substituted allyl systems. This is demonstrated by the enantioselective substitution of (7) with malonate nucleophile to afford the product (20) with excellent enantioselectivity.

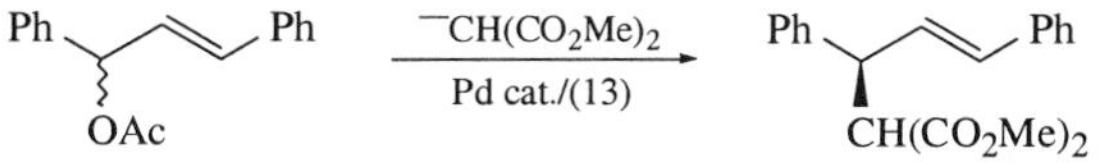

Ph Ph OAc (7) $\xrightarrow[\text{Pd cat./(13)}]{^{-}CH(CO_2Me)_2}$ Ph Ph $CH(CO_2Me)_2$ (20)

ee (%)

99

The phosphine-oxazolines are now regarded as privileged ligand structures that have found widespread application in a range of catalytic processes and are now commercially available. The use of phosphine-oxazoline ligands in enantioselective allylic substitution and other reactions have been reviewed by their inventors (J. M. J. Williams, *Synlett*, 1996, 705; G. Helmchen and A. Pfaltz, *Acc. Chem. Res.*, 2000, **33**, 336).

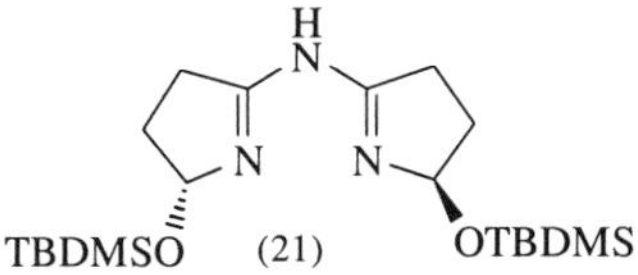

H N N N TBDMSO (21) OTBDMS

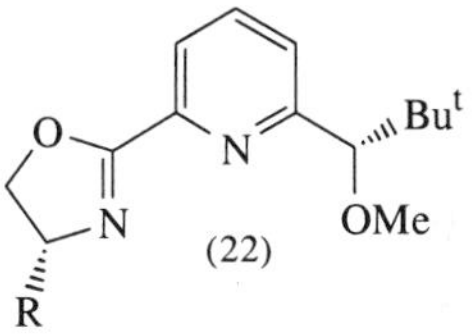

O N N R (22) Bu^t OMe

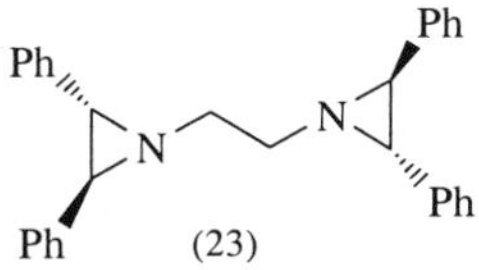

Ph N N Ph Ph Ph (23)

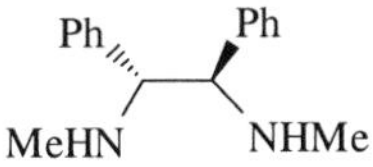

Ph Ph MeHN NHMe (24)

Many nitrogen and sulfur ligand systems are able to impart impressive levels of asymmetric induction. Ligands which may be considered in this category include the 5-azasemicorrins (A. Pfaltz *et al*, *Tetrahedron*, 1992, **48**, 2143). The selectivities and reaction rates were found to be solvent-dependent. The best results were obtained in polar media using a mixture of dimethyl malonate and N,O-bis(trimethylsilyl)acetamide, according to a procedure described by Trost (B. M. Trost and D. J. Murphy, *Organometallics*, 1985, **4**, 1143). The catalytic procedure is smoothly initiated by the addition of a catalytic amount of potassium acetate. Under these conditions, in the presence of 1-2 mol% of catalyst the reaction proceeded smoothly at room temperature to give the desired alkylation product in very high enantiomeric purity and essentially quantitative yield. The most effective ligands were found to be the azasemicorrin (21) which carries silyloxymethyl groups at the stereogenic centre adjacent to the co-ordination site. In a related approach, the use of enantiopure aziridines has been explored for palladium catalysed allylic substitution (D. Tanner *et al*, *Tetrahedron Lett.*, 1994, **35**, 4631). The readily available *bis*(aziridines) (23), effect an enantioselective nucleophilic substitution of allylic acetate (7) furnishing the addition product (20) with >99% ee.

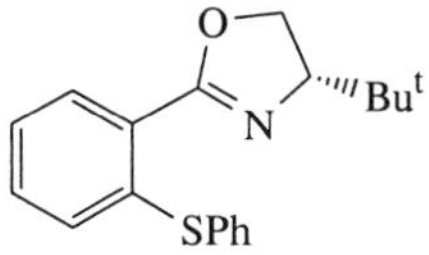

(25) 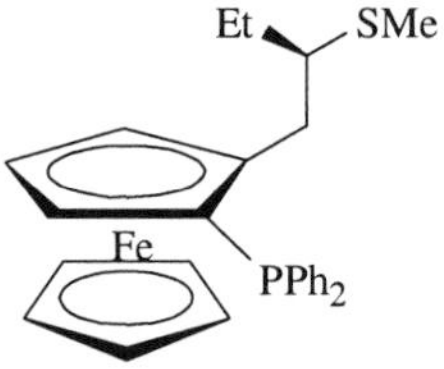 (26) 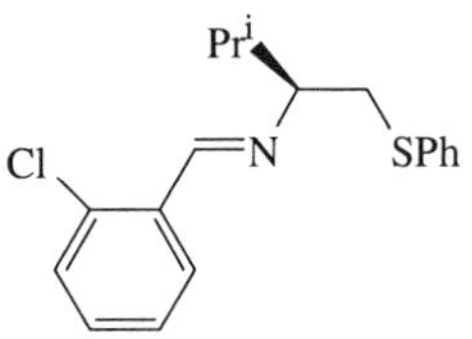 (27)

Other nitrogen containing ligands worthy of note are the pyridine-oxazoline (22) (C. Moberg *et al*, *J. Org. Chem.*, 1997, **62**, 1604) and the diamine ligand (24) (M. Lemaire *et al*, *Chem Commun.*, 1994, 1417).There are several useful sulfur-based ligands that are capable of providing high enantioselectivity in allylic substitution reactions. A selection of these ligands are represented by the structure (25) (J. M. J. Williams *et al*, *Tetrahedron Lett.*, **1993**, 34, 7793; G. Helmchen *et al*, *Tetrahedron Lett.*, 1994, **35**, 1523), (26) (D. Enders *et al*, Org. Lett., 1999, **1**, 1863) and (27) (J. C. Anderson, D. S. James and J. P. Mathias, *Tetrahedron:Asymmetry*, 1998, **9**, 753).

The π-allyl complex generated from cyclic alkenyl acetate starting materials has a different configuration from that generated by an acyclic precursor. It follows that the ligands which afford high enantioselectivity with acyclic precursors do not achieve similar results with cyclic substrates. The ligand of choice for cyclic substrates (e.g. (28) and (30)) is (32), which produces the products (29) and (31) with high enantioselectivity using a typical soft carbon nucleophile such as the sodium salt of dimethylmalonate (B. M. Trost and D. L. Van Vranken, *Chem. Rev.*, 1996, **96**, 395).

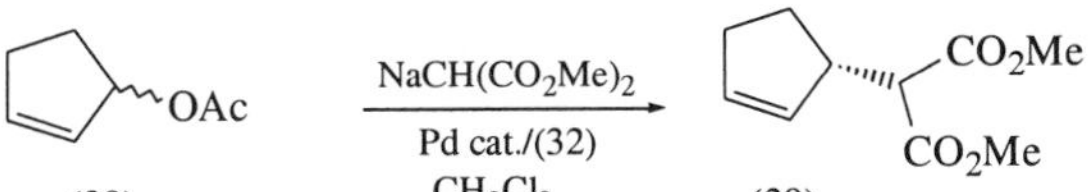

OAc (28) → $NaCH(CO_2Me)_2$, Pd cat./(32), CH_2Cl_2 → (29) CO_2Me, CO_2Me

Yield (%) 81; ee (%) 98

OAc (30) → $NaCH(CO_2Me)_2$, Pd cat./(32), CH_2Cl_2 → (31) CO_2Me, CO_2Me

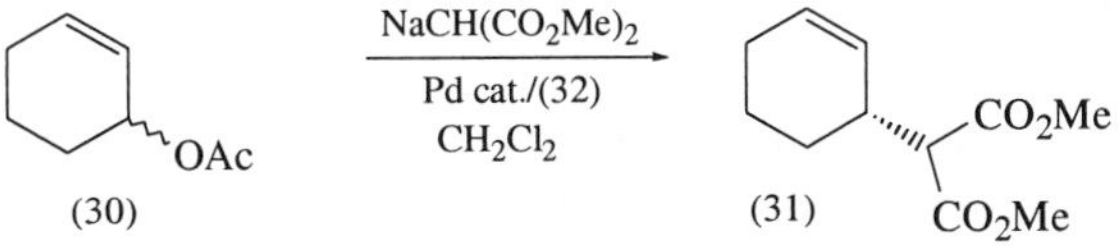

Yield (%) 86; ee (%) 96

O, NH, HN, O, PPh_2, Ph_2P

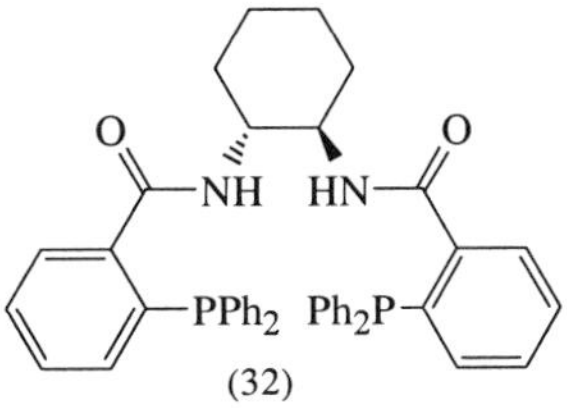

(32)

A large number of nucleophiles have been employed in allylic substitution reactions, and there is much interest from both a synthetic and mechanistic stand-point in the development of enantioselective variants of these reactions. The most commonly employed nucleophiles are the 'soft' stabilised carbanions such as dimethyl malonate, but under suitable conditions, nitrogen based nucleophiles have afforded highly enantio-enriched products.

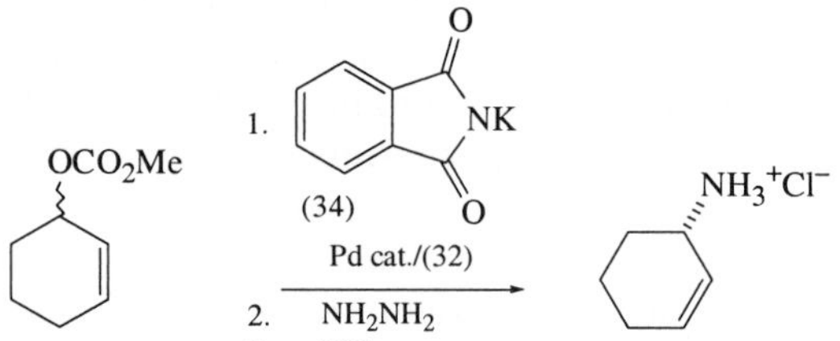

(33) (35)

	Yield (%)	ee (%)
	95	97

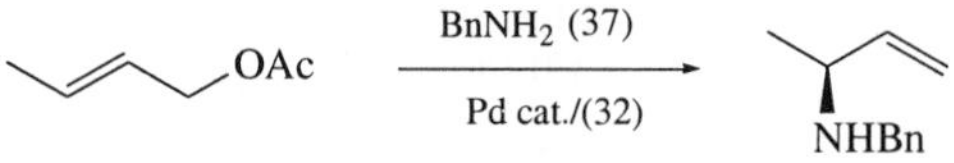

(36) (38)

32:1 (branch:linear) ee (%) 84

In the case of simple cyclic substrates (such as (33)), the use of (34) as nucleophile provides a convenient means of obtaining the allylic amine product (35), after deprotection (B. M. Trost *et al*, *J. Am Chem. Soc.*, 1994, **116**, 4089). The use of (32) as ligand enables an efficient regioselective and enantioselective allylic amination of crotyl acetate (36) with benzylamine (37) to afford the product (38) in high ee (T. Hayashi *et al*, *Tetrahedron Lett.*, 1990, **31**, 1743).

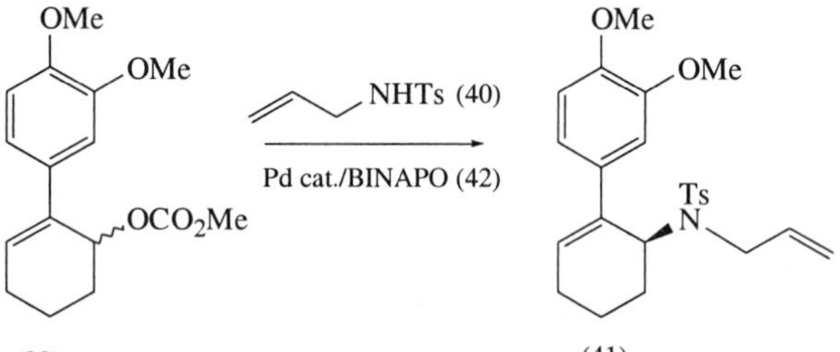

(39) (41)

ee (%)
86

An approach to the mesembrine and mesembrane alkaloids relies on the enantioslective allylic amination of (39) with tosyl amine (40) to produce the intermediate (41). High enantioselectivity was observed with BINAPO (42) as ligand (M. Mori *et al*, *J. Org. Chem.*, 1997, **62**, 3263).

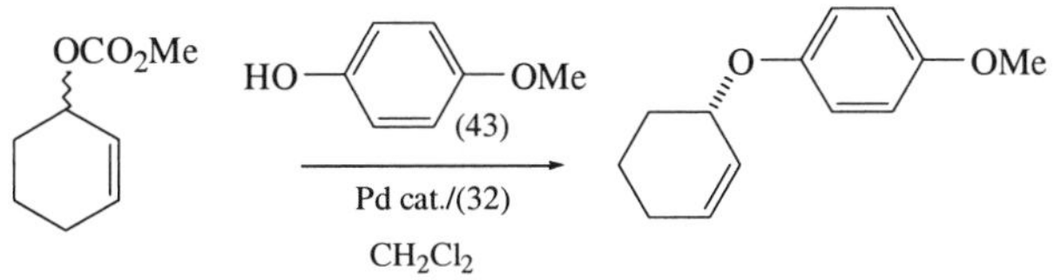

OCO_2Me HO—C₆H₄—OMe (43) O—C₆H₄—OMe

Pd cat./(32)

CH_2Cl_2

(33) (44)

Yield (%) ee (%)

88 97

OCO_2Me $OCOCH_2CH_3$

$CH_3CH_2CO_2^-$

Pd cat./(32)

CO_2Me CH_2Cl_2 CO_2Me

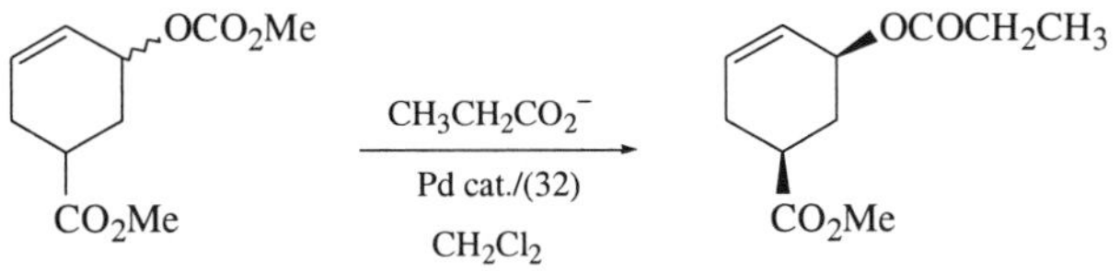

(45) (46)

Yield (%) ee (%)

95 98

The use of the Trost ligand (32) allows for the incorporation of oxygen nucleophiles with very high enantioselectivity. This is effective for the reaction of allylic carbonate (33) with phenol (43) to afford the allylic phenyl ether product (44). Some of the products from this process have been shown to undergo Claisen rearrangements to furnish highly functionalised products (B. M. Trost and F. D. Toste, *J. Am. Chem. Soc.*, 1998, **120**, 815). Similar conditions allow the effective deracemisation of (45) under mild conditions to afford (46) (B. M. Trost and M. G. Organ, *J. Am. Chem. Soc.*, 1994, **116**, 10320).

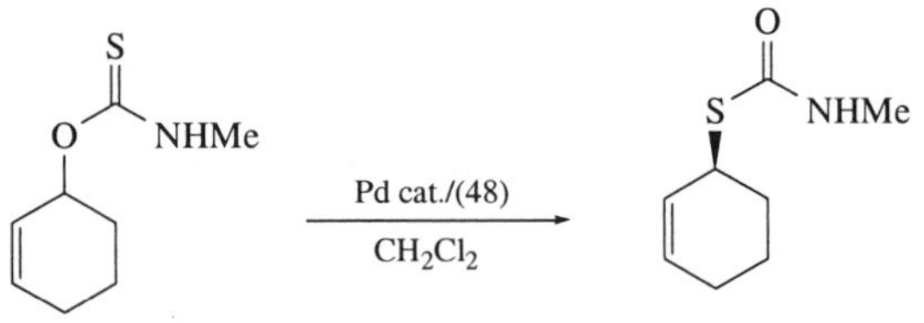

S
O NHMe
Pd cat./(48)
CH_2Cl_2
O
S NHMe
(47) (48)

Yield (%) ee (%)
94 97

A concise route to enantiopure thiols relies on the palladium catalysed rearrangement reaction of *O*-allylic thiocarbonates (A. Bohme and H. Gais, *Tetrahedron:Asymmetry*. 1999, **10**, 2511). In the presence of ligand (32) the substrate (47) rearranges in high yield and enantioselectivity to product (48). This provides an ingenious solution to the common problem concerning low reactivity of thiols in palladium(0) catalysed allylic substitution reactions. Using a modified version of the Trost ligand (54) promotes the highly enantioselective addition of phthalimide (50) to vinyl epoxide (49).
The usual deprotection affords the allylic amine (51) as product (B. M. Trost *et al*, *J. Am. Chem. Soc.*, 1996, **118**, 6520).

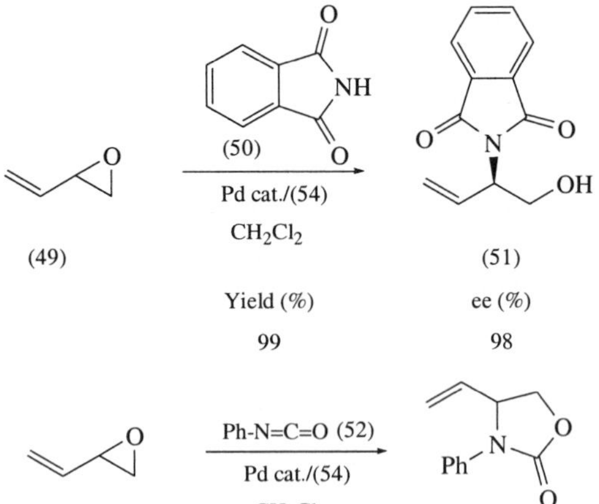

O
NH
(50) O
O
Pd cat./(54)
CH_2Cl_2
O N O
OH
(49) (51)

Yield (%) ee (%)
99 98

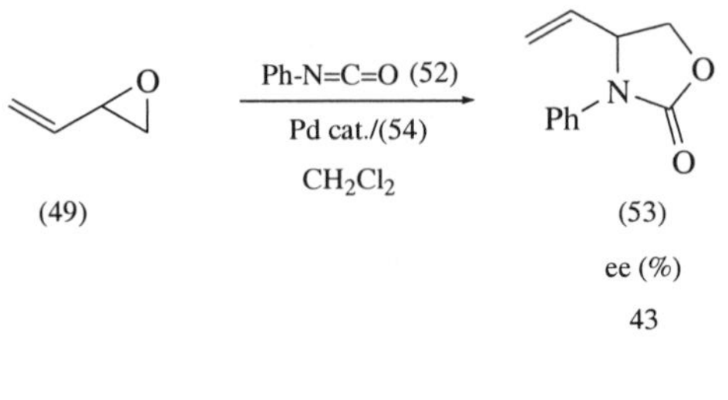

O
Ph-N=C=O (52)
Pd cat./(54)
CH_2Cl_2
O
Ph N
O
(49) (53)

ee (%)
43

(54)

In the presence of phenyl isocyanate (52) treatment of vinyl epoxide (49) with palladium catalyst sets up an intramolecular addition of nitrogen nucleophile to give product (53) with modest selectivity (T. Hayashi, A. Yamamoto and Y. Ito, *Tetrahedron Lett.*, 1988, **29**, 669).

(55) → (57): (56), Pd cat./BINAP (58), base, toluene

Yield (%) 87; ee (%) 94

(59) → (61): (60), Pd cat./(32), LiTMP $(CH_3)_3SnCl$ DME

Yield (%) 99; ee (%) 86

The enantioselective electrophilic attack of a π-allyl palladium(II) to a stabilised prochiral nucleophile offers a succinct route to highly functionalised molecules. This has been achieved in the allylation of nucleophile (56) with cinnamyl acetate (55) catalysed by enantiopure palladium-BINAP (58) complexes to afford product (57) having a quaternary stereogenic centre at the α-carbon (R. Kuwano and Y. Ito, *J. Am. Chem. Soc.* 1999, **121**, 3236). Trost and Schroeder have taken the ligand (32), developed in their laboratories and report that it is effective in the palladium catalysed

allylation of non-stabilised ketone enolates. Thus, ketone (60) is treated with (59) to afford product (61) in excellent yield and good enantioselectivity (B. M. Trost and G. M. Schroeder, *J. Am. Chem. Soc.* 1999, **121**, 6759). The choice of base and Lewis acid was crucial to high enantioselectivity, trimethyltin chloride affords the best results under the reported conditions. The allylated products obtained are suitable for further elaboration.
The azalactone (63) has also been exploited as a prochiral nucleophile by Trost and Ariza to give exceptionally high enantioselectivity in the palladium catalysed prenylation of allylic acetate (62) to afford product (64) (B. M. Trost and X. Ariza, *J. Am. Chem. Soc.*, 1999, **121**, 10727).

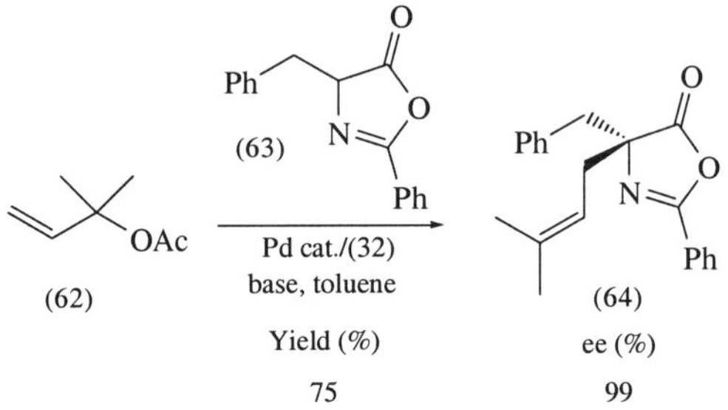

(62) OAc + (63) → (64)
Pd cat./(32)
base, toluene

Yield (%) 75
ee (%) 99

The intramolecular cyclisation of *meso*-substrates such as (65) allows the formation of product (66) with good enantioselectivity using ligand (32). The initial ionisation step is important for high enantioselectivity and is induced by approach of the palladium complex to the face opposite the carbamate (B. M. Trost and D. E. Patterson, *J. Org. Chem.*, 1998, **63**, 1339).

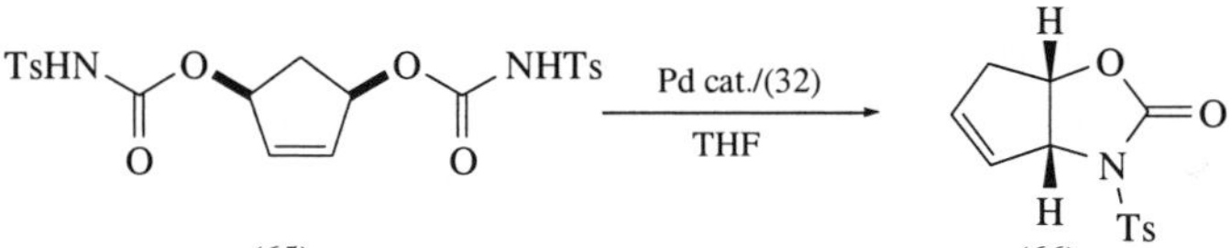

TsHN O O NHTs — Pd cat./(32), THF →

(65) (66)

ee (%)

80

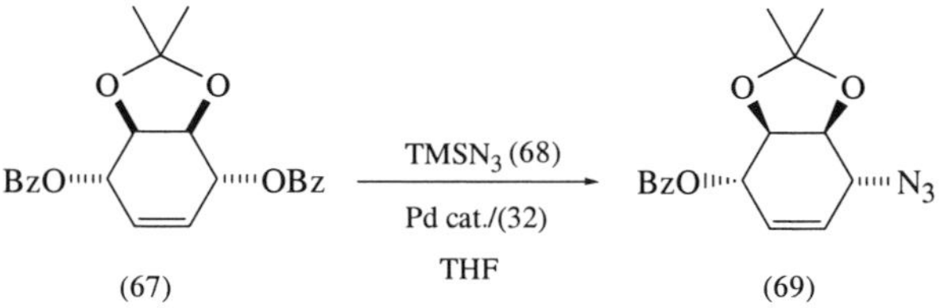

BzO OBz — $TMSN_3$ (68), Pd cat./(32), THF → BzO N_3

(67) (69)

ee (%)

>98

The related desymmetrisation of highly functionalised *meso*-substrates such as (67) has been achieved using an intermolecular substitution process. The product (69) is obtained by addition of the azide (68) (B. M. Trost, L. Li and S. D. Guile, *J. Am. Chem. Soc.*, 1992, **114**, 8745). A similar strategy has been used in the synthesis of conduramines and pancratistatin (B. M. Trost and S. R. Pulley, *J. Am. Chem. Soc.*, 1995, **117**, 10143). Enantioselective intramolecular alkylations have been investigated by a number of research groups with limited success.

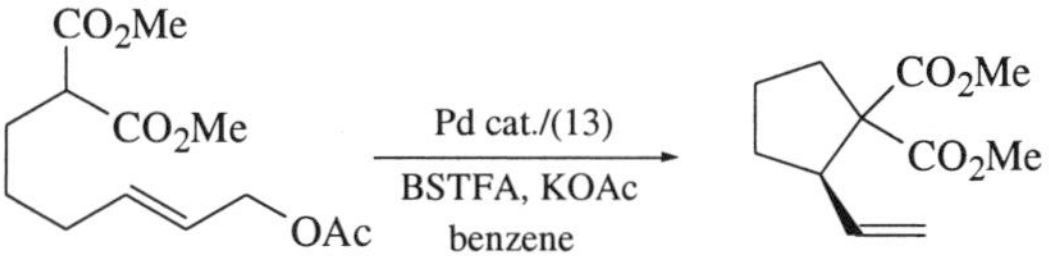

CO_2Me CO_2Me OAc — Pd cat./(13), BSTFA, KOAc, benzene → CO_2Me CO_2Me

(70) (71)

Yield (%)	ee (%)
60	87

However, there are several examples where good levels of asymmetric induction have been revealed. In the case of allylic acetate (70), the product (71) can be obtained in reasonable yield and good enantioselectivity by using

the archetypal phosphine-oxazoline ligand (13) (A. Pfaltz and G. Koch, *Tetrahedron:Asymmetry*, 1996, **7**, 2213).
In a similar fashion, heterocyclic compounds can be prepared by the cyclisation of amine (72) to afford the product (73) with excellent enantioselectivity (B. M. Trost *et al*, *J. Am. Chem. Soc.*, 1996, **118**, 6297). Other examples of different ring sizes and the effect on rate and enantioselectivity were reported.

NH OAc MeO (72) — Pd cat./(32), THF → N MeO (73)

ee (%)
91

2. Cross-coupling reactions

The palladium catalysed cross-coupling process is established as a practical and efficient method of forming carbon-carbon and carbon-heteroatom bonds utilising a wide range of coupling partners. The generally accepted mechanism involves the oxidative addition of an aryl or alkenyl halide (or triflate) R^1-X producing an intermediate $L_nPd(II)(R^1)X$. Transfer of a group R^2 by transmetallation generates $L_nPd(II)(R^1)(R^2)$ which affords the product by reductive elimination (Palladium Reagents and Catalysts: Innovations in Organic Synthesis, Jiro Tsuji 1996, Chapter 4, Wiley).
To achieve an enantioselective cross-coupling reaction enantiopure palladium complexes are employed where the ligand (L) is often an enantiopure phosphine. Historically, the most extensively studied enantioselective cross-coupling process is the coupling of alkenyl halides with secondary alkyl organometallics (Mg, Zn).

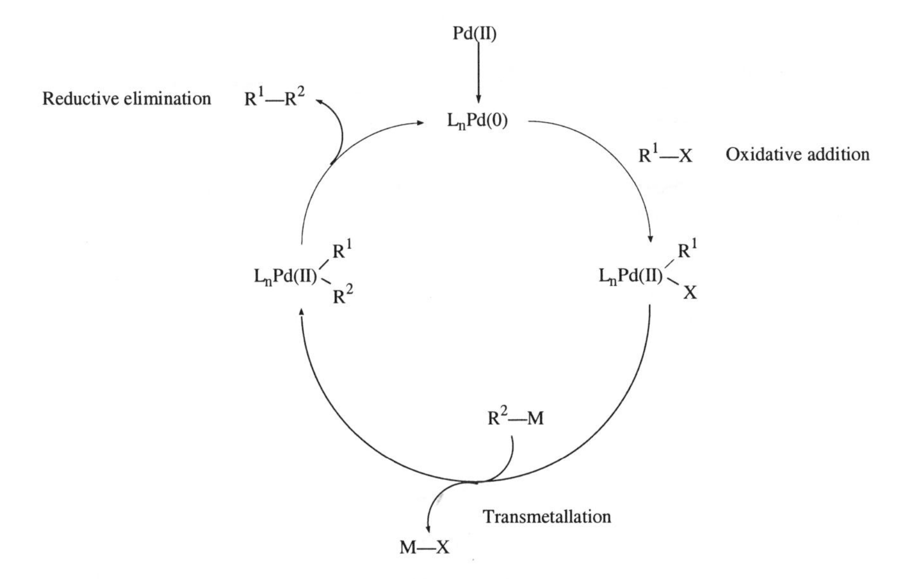

In this process a kinetic resolution of the racemic organometallic affords enantio-enriched products and since the rate of racemisation (of remaining organometallic) is comparable to the rate of cross-coupling, no loss of enantiopurity occurs when conversions are high. Interestingly, the use of enantiopure diphosphines as ligand affords products with low enantioselectivity.

The presence of a dimethylamino moiety is essential for high enantioselectivity. This is attributed to the selective coordination of one of the enantiomers of the organometallic at the transmetallation stage (J. M. Brown *et al*, *J. Organomet. Chem.*, 1989, **370**, 397).

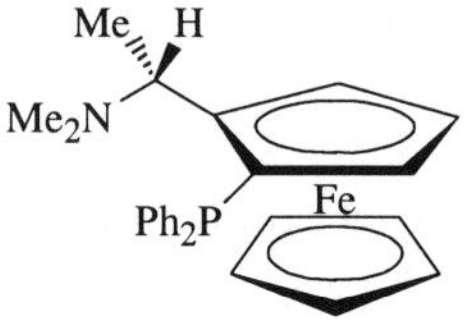

Me H Me2N Ph2P Fe (74) 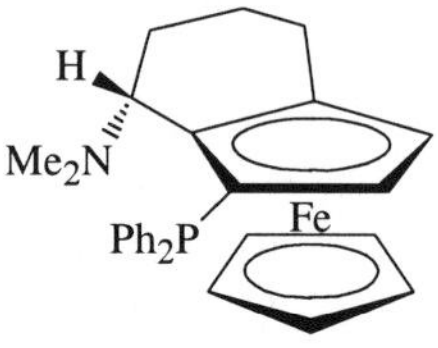
H Me2N Ph2P Fe (75) 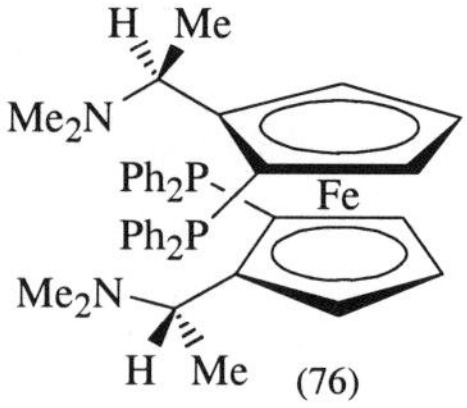
H Me Me2N Ph2P Fe Ph2P Me2N H Me (76)

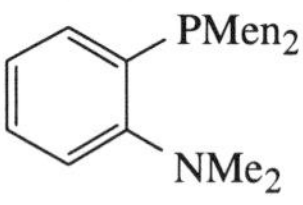

PMen2 NMe2 (77) 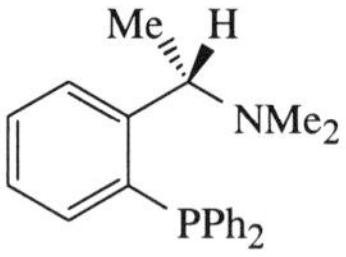
Me H NMe2 PPh2 (78) 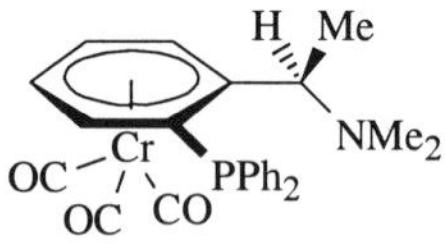
H Me NMe2 Cr OC OC CO PPh2 (79)

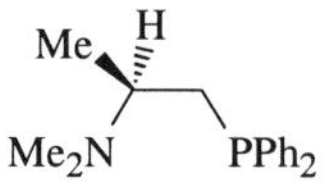

Me H Me2N PPh2 (80) 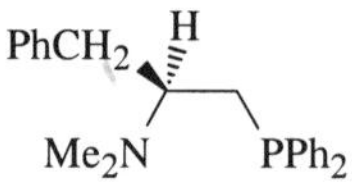
PhCH2 H Me2N PPh2 (81) 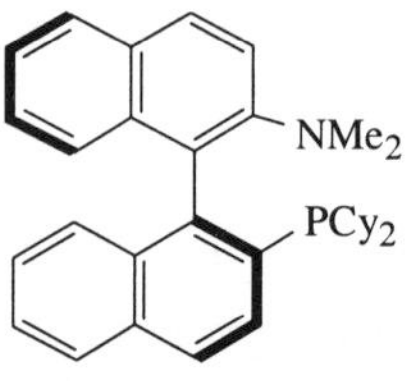
NMe2 PCy2 (82)

The use of ferrocenylphosphine ligands such as (74) and (75) have proved effective in this transformation. This is illustrated in the coupling of Grignard reagent (84) with vinyl bromide (83) to afford the product (85) (Ligand (74): T. Hayashi *et al*, *J. Am. Chem. Soc.*, 1982, **104**, 180; Ligand (75): B. Jedlicka *et al*, *J. Chem. Soc., Chem. Commun.*, 1993, 1329).

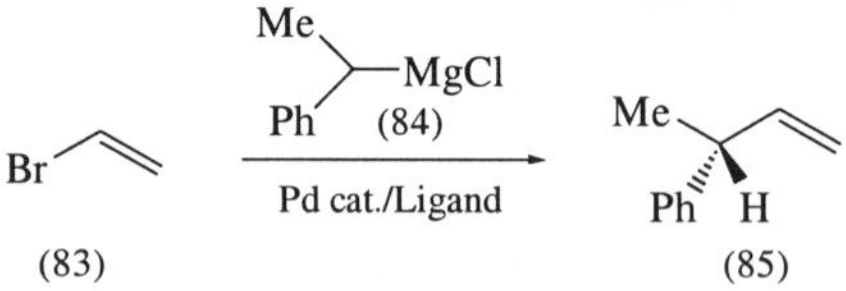

Br (83) — Me, Ph, MgCl (84); Pd cat./Ligand → Me, Ph, H (85)

Ligand	Yield (%)	ee (%)
(74)	82	61 (R)
(75)	95	79 (R)

Many other (aminoalkyl)phosphines ligands have been tested in this reaction including very simple compounds such as (77) and (78). However, as demonstrated by the cross-coupling of (84) with (86) to afford (87), the enantioselectivity is modest (Ligand (77): C. Döbler and A. Kinting, , *J. Organomet. Chem.*, 1991, **401**, C23; Ligand (78): H. –J. Kreuzfeld, C. Döbler and H. –P. Abicht, *J. Organomet. Chem.*, 1987, **336**, 287).

Br–CH=CH–Ph (86) + Me(Ph)CH–MgCl (84) —[Pd cat./Ligand]→ Me(Ph)(H)C–CH=CH–Ph (87)

Ligand	Yield (%)	ee (%)
(77)	65	11 (R)
(78)	95	40 (R)

The methodology has been employed to prepare enantiomerically-enriched allylsilanes. Using a palladium complex in combination with ligand (74) the allyl silanes (89) and (90) could be prepared from vinyl bromides (83) and (86) with high enantioselectivity using (88). The coupling of dienyl bromide (91) with Grignard reagent (92) afforded the novel dienylsilane (93) with modest enantioselectivity (T. Hayashi *et al*, *J. Am. Chem. Soc.*, 1982, **104**, 4962; T. Hayashi *et al*, *J. Org. Chem.*, 1986, **51**, 3772).

Me$_3$Si
MgCl
Ph (88)
Br
Pd cat./(74)
Me$_3$Si
Ph H
(83)
(89)
ee (%)
95

Me$_3$Si
MgCl
Ph
Ph (88)
Br
Pd cat./(74)
Me$_3$Si
Ph
Ph H
(86)
(90)
ee (%)
95

Me$_2$PhSi
MgCl
Me (92)
Me
Br
Pd cat./(74)
Me$_2$PhSi
Me
Me H
(91)
(93)
ee (%)
45

The replacement of Grignard reagent with the corresponding zinc reagent has been shown in some cases to lead to higher enantioselectivity. For example, the coupling of 1-phenylethylzinc reagent (95) with vinyl chloride (94) or vinyl bromide (86). The use of ligand (74) affords product (85) with high yield and improved enantioselectivity. The highest enantioselectivty of product (85) was observed with the C_2-symmetric ferrocenylphosphine (76). The chromium complex (79) possess the required (aminoalkyl)phosphine groups to achieve a respectable enantioselectivity in theformation of product (85) (Ligand (74): T. Hayashi *et al*, *Bull. Chem. Soc. Jpn.*, 1983, **56**, 363; Ligand (76): T. Hayashi *et al*, *J. Chem. Soc., Chem. Commun.*, 1989, 495; Ligand (79): M. Uemura *et al*, *Tetrahedron: Asymmetry*, 1992, **3**, 213).

Me
ZnCl
Ph (95)
Pd cat./Ligand

Cl (94) → Me, H, Ph (85)

Ligand	Yield (%)	ee (%)
(74)	95	85 (S)
(76)	95	93 (R)
(79)	67	61 (S)

Me
ZnCl
Ph (95)
Pd cat./(74)

Br, Ph (86) → Me, Ph, H, Ph (87)

Yield (%)	ee (%)
88	60 (S)

The asymmetric synthesis of axially chiral molecules such as biaryls is an important area of investigation. The products may form an important constituent of natural products or the key structural element of novel enantiopure ligands. Axially chiral biaryl compounds such as (98) can be prepared by an enantioselective cross-coupling protocol employing (aminoalkyl)phosphine ligands such as (80) and (81) derived from amino acids. The achiral ditriflate (96) is treated with two equivalents of phenylmagnesium bromide (97) resulting in the formation of (98) and the product from diphenylation. The high enantioselectivity of (98) arises from a kinetic resolution in the formation of the diphenylation product. Thus, the enantiopurity of (98) increases as the yield of diphenylation product increases (T. Hayashi *et al*, *J. Am. Chem. Soc.*, 1995, **117**, 9101). The same strategy can be found in an enantioselective alkynylation process to prepare the product (101) from (99) and (100) in very high enantiopurity (T. Hayashi *et al*, *Tetrahedron Lett.*, 1996, **37**, 3161).

TfO OTf Me $\xrightarrow[\text{Pd cat./(81)}]{\text{PhMgBr (97)}}$ TfO Ph Me

(96) (98)

ee (%)

84

TfO OTf Ph $\xrightarrow[\text{Pd cat./(80)}]{Ph_3SiC{\equiv}CMgBr\ (100)}$ TfO Ph $SiPh_3$

(99) (101)

ee (%)

99

There are very few examples of asymmetric Suzuki reactions despite the potential applications of this powerful and popular transformation. Recently, two groups have reported the use of (alkylamino)phosphines as effective ligands in the intermolecular asymmetric Suzuki reaction to prepare axially chiral biaryls. In the first example of such a coupling reaction, the naphthylboronate (103) is united with naphthyl iodide (102) to afford binaphthalene product (104) with excellent enantioselectivity. A selection of ligands were originally tested with the ferrocenylphosphine (74) being the most effective in controlling asymmetric induction (A. N. Cammidge and K. V. L. Crépy, *Chem. Commun.*, 2000, 1723). An important point relating to this work is that the methodology permits the use of functional groups that are incompatible with magnesium or zinc derived organometallic reagents. The potential scope of this transformation is further highlighted in an independent report relating to the preparation of enantiomerically-enriched biaryl phosphonates, for example (107). In the presence of the binaphthyl-(N,N-dimethylamino)phoshine ligand (82), the iodide (105) is coupled with the boronic acid (106) with impressive efficiency and unmatched enantioselectivity. The products could be easily converted into novel enantiopure phosphine ligands (J. Yin and S. L. Buchwald, *J. Am. Chem. Soc.*, 2000, **122**, 12051).

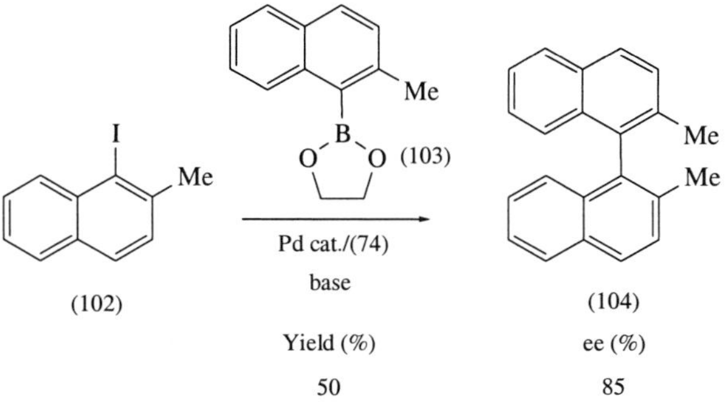

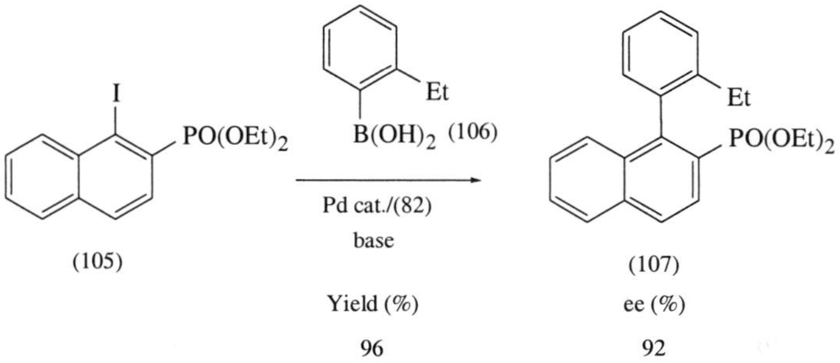

I Me (102)

Me B O O (103)

Pd cat./(74) base

Me Me (104)

Yield (%) 50

ee (%) 85

I $PO(OEt)_2$ (105)

Et $B(OH)_2$ (106)

Pd cat./(82) base

Et $PO(OEt)_2$ (107)

Yield (%) 96

ee (%) 92

The preparation of enantio-enriched planar chiral tricarbonyl(η^6-arene) chromium complexes can be prepared by the cross-coupling of (108) with alkenyl boronic acid (109). Using catalysts derived from (74) the product (110) can be obtained with modest enantioselectivity. The reaction was sensitive to the organometallic and the use of alkenyl tributyltin derivatives resulted in racemic products (M. Uemura *et al*, *J. Organomet. Chem.*, 1994, **473**, 129).

(HO)$_2$B (109)
Pd cat./(74)

(108) → (110)

ee (%)
42 (1S, 2R)

An intramolecular asymmetric Suzuki reaction allows for the preparation of enantio-enriched cylopentane derivatives (113) after oxidation. The strategy involves the enantiotopic group selective ring closure of the intermediate obtained from the hydroboration of (111) with (112). Although the enantioselectivities obtained are low, the desymmetrisation approach offers much potential (S. Young Cho and M. Shibasaki, *Tetrahedron: Asymmetry*, 1998, **9**, 3751).

OTf

1. 9-BBN (112)
2. Pd cat./(74)
3. NaOH, H_2O_2

(111) → (113)

Yield (%)	ee (%)
58	28

One of the most exciting developments in catalysis over the last few years is undoubtedly the advent of a practical, mild and efficient protocol for catalytic carbon-heteroatom coupling reactions. In independent investigations, Buchwald and Hartwig reported the first catalytic aminations of aryl bromides with free amines (A. S. Guram, R. A. Rennels and S. L. Buchwald, *Angew. Chem. Int. Ed. Engl.*, 1995, **34**, 1348; J. Louie and J. F. Hartwig, *Tetrahedron Lett.*, 1995, **36,** 3609). The group of Pye and Rossen have reported a fascinating kinetic resolution of racemic dibromide (114) employing an enantiomerically pure ligand in a palladium catalysed amination with benzylamine (K. Rossen, P. J. Pye, A. Maliakal and R. P. Volante, *J. Org. Chem.*, 1997, **62**, 6462). The combination of $Pd_2(dba)_3$/(S)-[2.2]PHANEPHOS (115) is effective in providing practical enantiomeric discrimination when halide is removed from the reaction mixture using the halide scavenger $TlPF_6$. The products from the reaction are the monoaddition compound, the bis(benzylamine) and the dehalogenated compound. After 90% conversion the remaining 10% of (114) is enantiomerically pure at >99.9% ee!

Br
Br
rac-(114)
$BnNH_2$
Pd cat./(115)
NaO^tBu
toluene
$TlPF_6$
Br
NHBn
+
H
NHBn
PPh_2
PPh_2
(S)-(115)
NHBn
NHBn
+
Br
Br
(R)-(114)
ee (%)
>99.9

The Buchwald group have described the first examples of catalytic, asymmetric arylation of ketone enolates to produce all carbon quaternary centres using BINAP (58) as enantiopure ligand. A typical example is illustrated by the coupling of (116) with (117) to afford the product (118) in good enantioselectivity. In an impressive example, (119) is coupled to aryl bromide (117) under the optimised conditions to afford the product (120) in 98% ee (S. L. Buchwald *et al*, *J. Am. Chem. Soc.*, 1998, **120**, 1918). Although the scope of the reaction is not yet wide and the origin of the enantioselectivity in this exciting process has not fully been ascertained, this represents a promising solution to one of the remaining challenges in asymmetric catalysis.

O
Me
(116)
Bu^t — — Br (117)
Pd cat./(58)
NaO^tBu
O
Me
^tBu
(118)

Yield (%)	ee (%)
73	88

O
Ph
Me
(119)
Bu^t — — Br (117)
Pd cat./(58)
NaO^tBu
O
Ph
Me
^tBu
(120)

Yield (%)	ee (%)
75	98

3. Heck Reactions

The Heck reaction is the common description of a palladium catalysed coupling of alkenes with aryl or vinyl halides or triflates (R. F. J. Heck, *J. Am. Chem. Soc.*, 1968, **90**, 5518). The reaction is of enormous appeal to synthetic chemists as a wide range of coupling partners and functional groups are tolerated. For some references on mechanistic aspects of the Heck reaction, see: (A. de Meijere and F. E. Meyer, *Angew. Chem. Int. Ed. Engl.*, 1994, **33**, 2379; W. Cabri and I. Candiani, *Acc. Chem. Res.*, 1995, **28**, 2; B. L. Shaw, S. D. Perera and E. A. Staley, *Chem. Commun.*, 1998, 1361; B. L. Shaw and S. D. Perera, *Chem. Commun.*, 1998, 1863).
The control of asymmetric induction is complicated by both regioselectivity and facial selectivity and there have been a number of stimulating discussions

on the topic (J. M. Brown et al, Organometallics, 1995, **14**, 207; J. M. Brown and K. K. Hii, *Angew. Chem. Int. Ed. Engl.*, 1996, **35**, 657; M. Shibasaki and E. M. Vogel, *Comprehensive Asymmetric Catalysis*, ed. E. N. Jacobsen, A. Pfaltz, H. Yamamoto, Springer Berlin, London, 1999, Chapter 14). The advent of a successful enantioselective variant of the process stems from the use of enantiopure phosphine ligands. As with any catalytic asymmetric process there is large body of work that serves to optimise the enantioselectivity through ligand design. However, often the ligands retain the core structure of priviliged ligands such as those illustrated below. A phosphine-oxazoline ligand system based on hydroxyproline is used as a catalyst for the asymmetric Heck reaction (S. R. Gilbertson, Z. Fu and D. Zie, *Tetrahedron Lett.*, 2001, **42**, 365). New enantiopure P,N-ligands have been prepared from (1S)-(+)-ketopinic acid using a palladium-catalysed coupling reaction of a vinyl triflate and either a diarylphosphine or a dialkylphosphine as the key step. Palladium complexes of these ligands are reported to be efficient catalysts for asymmetric Heck reaction between aryl or alkenyl triflates and cyclic alkenes (S. R. Gilbertson and Z. Fu, *Org. Lett.*, 2001, **3**, 161).

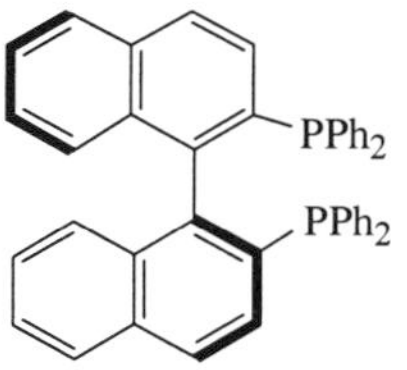

(58) 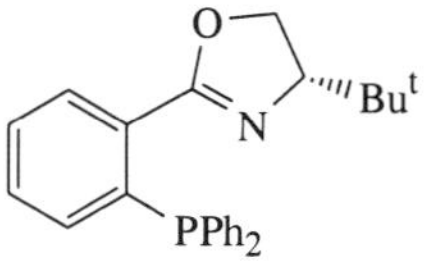 (13) 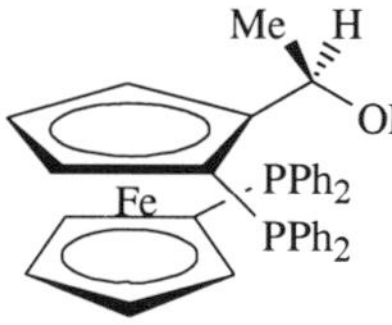 (121) 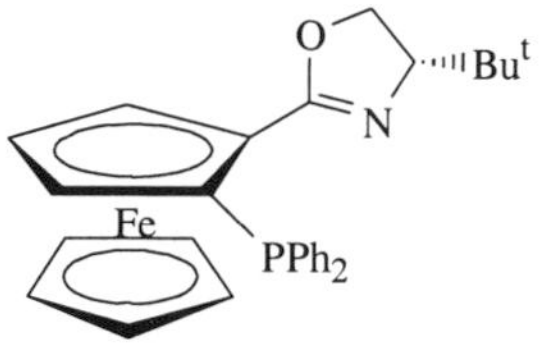 (17)

The enantioselective intermolecular Heck reaction is desirable and many researchers have explored this process. However, success in this area has been limited in the majority of cases to fairly reactive substrates containing heteroatoms such as dihydrofurans and dihydropyrroles. In general the use of triflates is necessary as iodides often results in racemic products. This is illustrated by the arylation of dihydrofuran (122) with phenyl triflate (123) in the presence of a palladium catalyst with BINAP (58) as ligand (T. Hayashi *et al*, *J. Am. Chem. Soc.*, 1991, **113**, 1417; T. Hayashi *et al*, *Tetrahedron*

Lett., 1993, **34**, 2505). The enantioselectivity obtained was significantly enhanced (up to 97% ee) when the phosphine-oxazoline ligand (13) was employed (A. Pfaltz *et al*, *Angew. Chem. Int. Ed. Engl.*, 1996, **35**, 200). It was important to have the bulky ^{t}Bu group on the oxazoline as moving to smaller groups reduced the activity of the catalyst. The reactions are very slow (2-9 days) which is an area for possible improvement. Other phosphine-oxazoline ligands have been prepared including one derived from *cis*-2-amino-3,3-dimethyl-1-indanol, which proved to be an efficient ligand for the palladium-catalysed asymmetric Heck reaction of phenyl triflate (123) with 2,3-dihydrofuran (122) in the presence of ethyldiisopropylamine in benzene affording the product (125) in 96% ee (Y. Hashimoto *et al*, *Tetrahedron: Asymmetry*, 2000, **11**, 2205).

(122) → PhOTf (123), Pd cat./(58), base, C_6H_6 → (124)

Yield (%) 70; ee (%) 87 (R)

(122) → PhOTf (123), Pd cat./(13), base, THF → (125)

Yield (%) 87; ee (%) 97 (R)

The two different ligands afford the two regioisomeric products (124) and (125). In the example using BINAP (58) as ligand, both products are encountered with (124) being the major product. In the first instance the catalyst can bind to either face of the substrate and the usual insertion and elimination steps occur. It is postulated that on one of the diastereomeric complexes unfavourable steric intereractions involving the ligand and associated product cause a dissociation to produce product (125). In the other diastereomeric complex the absence of significant steric interactions results in the reinsertion of the alkene to the Pd-H bond followed by a second elimination to furnish the thermodynamic product (124). The overall result is a kinetic resolution through a double-bond isomerisation process. In the example using phosphine-oxazoline (13) as ligand, the double bond isomerisation process does not occur. The product (125) is obtained in high enantioselectivity with none of the 2,3-isomer (124) being detected.

PhOTf (123), Pd cat.; Double bond isomerisation

(122) (124) Thermodynamic product

(125)

A short five-step synthesis of (±)-2,2'-bis(diphenylphosphino)-3,3'-binaphtho[2,1-b]furan (BINAPFu,) starting from 2-naphthoxyacetic acid has been revealed. The resolution of BINAPFu is possible using an interesting new procedure for phosphines where (1S)-camphorsulfonyl azide is employed to prepare the bis(phosphinimine) of BINAPFu via the Staundinger reaction (N. G. Anderson *et al*, *Org. Lett.*, 2000, **2**, 2817). BINAPFu consistently outperformed BINAP (58) in an asymmetric Heck reaction between 2,3-dihydrofuran (122) and phenyl triflate (123).
The use of 2,2-dimethyl-2,3-dihydrofuran (126) as a new test substrate for the intermolecular Heck reaction has been reported. It provides an elegant solution to the issue of double bond isomerisation as only one regioisomer can be formed. This allows for easy and direct comparison of a wide range of ligands in this useful process. The initial application of this substrate was in the asymmetric phenylation of (126), which proceeded in high yield and enantioselectivities of up to 98% of product (127) with the ferrocenyl phosphine-oxazoline (17) as ligand (A. J. Hennessy, Y. M. and P. J. Guiry, *Tetrahedron Lett.* 1999, **40**, 9163). Catalysts derived from a range of ligands

have been directly compared in the related asymmetric cyclohexenylation of (126). The diphosphine ligand BINAP (58) gave poor results in comparison to the use of 2,3-dihydrofuran (122) as substrate. More reactive and more enantioselective catalysts were derived from phosphine-oxazoline ligands and the highest enantioselectivity of 97% was obtained using complexes derived from ligand (13). The choice of amine base was reported to be crucial to success with proton sponge giving better results than trialkylamines (A. J. Hennessy, Y. M. and P. J. Guiry, *Tetrahedron Lett.* 2000, **41**, 2261).

(126) → PhOTf (123), Pd cat./(17), base, C_6H_6 → (127)

Yield (%)	ee (%)
90	98 (R)

(126) → TfO-cyclohexenyl (128), Pd cat./(13), base, C_6H_6 → (129)

Yield (%)	ee (%)
68	97 (R)

A similar kinetic resolution sequence has also been observed in the reactions of dihydropyrroles. For example, dihydropyrrole (130) reacts with (131) to afford the product (132) with good enantioselectivity in the presence of a palladium catalyst with BINAP (58) as ligand (T. Hayashi *et al*, *Tetrahedron Lett.*, 1993, **34**, 2505). In this example no regioisomeric products were observed. The palladium catalysed arylations of dihydropyrrole (133) with aryl triflates afforded the products (134) with moderate enantioselectivity again using BINAP (58). In these examples it was necessary to add excess acetate in the form of thallium acetate to surpress initial double-bond isomerisation which would afford the alternative 2-substituted arylation products (C. Sonnesson *et al, J. Org. Chem.*, 1996, **61**, 4756).

(130) → (131), Pd cat./(58), base, C_6H_6 → (132)

Yield (%)	ee (%)
95	91 (R)

(133) → ArOTf, Pd cat./(58), base, DMF, TlOAc → (134)

Yield (%)	ee (%)
34	58 (R)

The arylation of 4,7-dihydro-1,3-dioxepin (135) provides a facile route to some useful enantiopure intermediates. The coupling of phenyl triflate (123) proceeds in good yields using catalysts derived from ligands (58) and (13) to provide the enol ether product (136) in good to excellent enantioselectivity (S. Takano, K. Samizu and K. Ogasawara, *Synlett*, 1993, 393; A. Pfaltz *et al*, *Angew. Chem. Int. Ed. Engl.*, 1996, **35**, 200). The use of molecular sieves was found to enhance both yield and enantioselectivity.

(135) → PhOTf (123), Pd cat./Ligand, base, C_6H_6 → (136)

Ligand	Yield (%)	ee (%)
(58)	84	72 (S)
(13)	70	92 (R)

The enantioselective reductive Heck reaction of norbornene and heterocyclic analogues such as (137) have been performed using BINAP (58) as ligand. The coupling of alkenyl bromide (86) leads to the product (138) in good yield and high enantioselectivity (T. Hayashi *et al*, *Chem. Commun.*, 1994, 1323). A further example allows the formation of N-protected epibatidine (141) with good enantioselectivity. In the presence of palladium acetate and BINAP (58), (139) is coupled with pyridine (140) to afford the product in reasonable yield (J. C. Namyslo and D. E. Kaufmann, *Synlett*, 1999, 804).

MeO_2C MeO_2C O (137) —[Br, Ph (86); Pd cat./(58), base, HCO_2H]→ MeO_2C MeO_2C O Ph (138)

Yield (%)	ee (%)
63	96

MeO_2C N (139) —[I, N, Cl (140); Pd cat./(58), base, HCO_2H]→ MeO_2C N N Cl (141)

Yield (%)	ee (%)
66	81

The wide utility of the intramolecular Heck reaction in the synthesis of natural products has stimulated the search for efficient enantioselective catalysts. Some of the earliest success in this process were reported for the asymmetric synthesis of decalins and hydrindans (Y. Sato, M. Sodeoka and M. Shibasaki, *J. Org. Chem.*, 1989, **54**, 4738; Y. Sato, T. Honda and M. Shibasaki, *Tetrahedron Lett.*, 1992, **33**, 2593). The use of palladium catalysts in combination with BINAP ligand (58) leads to good yields and enantioselectivities in the cyclisation of vinyl iodides (142) and (144) to the respective products (143) and (145). Enantiopure 2-diphenylarsino-2'-diphenylphosphino-1,1'-binaphthyl (BINAPAs) has been prepared and is an effective ligand in similar asymmetric Heck cyclisations (S. Y. Cho and M. Shibasaki, *Tetrahedron Lett.*, 1998, **39**, 1773).

OTBS I (142) → Pd cat./(58), base, NMP, Ag_3PO_4 → OTBS H (143)

Yield (%)	ee (%)
71	87

OTBS I (144) → Pd cat./(58), base, NMP, Ag_3PO_4 → OTBS H (145)

Yield (%)	ee (%)
78	82

The effect of ligand structure, method of catalyst generation, reaction solvent, and HI scavenger on the formation of an enantioenriched 3,3-disubstituted 2-oxindole (147) from asymmetric Heck cyclisation of (146) has been reported.

O NMe I O O (146) → Pd cat./(58), Ag_3PO_4, DMA → Me N O O O (147)

ee (%)

71 (S)

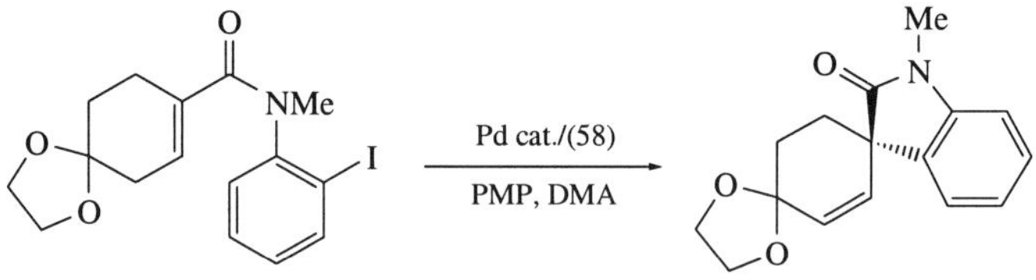

(146) → Pd cat./(58), PMP, DMA → ent-(147)

ee (%)

66 (R)

Depending upon whether the HI scavenger was a silver salt or a basic tertiary amine, *either* enantiomer of (147) could be formed with good selectivity using the same enantiomer of BINAP ligand (58). Using Pd-BINAP as catalyst, a variety of enantioenriched 3,3-disubstituted oxindoles, indolines, and dihydrobenzofurans have been prepared. from (E)-α,β-unsaturated-2-haloaniline substrates. In the great majority of cases, cyclisations conducted in the presence of Ag_3PO_4 or 1,2,2,6,6-pentamethylpiperidine (PMP) afforded opposite enantiomers of the spirocyclic product (147). Which HI acceptor results in highest enantioselection is substrate dependent. These studies illustrate, for the first time, that asymmetric Heck reactions of halide substrates can proceed with good levels of enantioselectivity in the absence of silver or thallium salts (L. E. Overman *et al*, *J. Am. Chem. Soc.*, 1998, **120**, 6477).

The intramolecular asymmetric Heck reaction provides a suitable method for preparing enantiomerically enriched indolizidines.

(148) → Pd cat./(58), base, DMSO-DMF, Ag-zeolite → (149)

Yield (%)	ee (%)
94	86

MeO I O OTIPS N Me Me (150) → Pd cat./(58), PMP, DMA → MeO Me CHO O N Me (151)

Yield (%)	ee (%)
84	95

The use of a silver-exchanged zeolite appears to give better results than the usual silver salts (M. Shibasaki *et al*, *Tetrahedron*, 1994, **50**, 371). A further example of the highly enantioselective synthesis of oxindoles has been reported which proceeds under neutral conditions. The palladium catalysed cyclisation of (150) in the presence of ligand (58) occurs with high enantioselectivity to afford product (151) which can be converted to the natural product physostigmine (L. E. Overman *et al, J. Org. Chem.*, 1993, **58**, 6949). Historically the synthesis of enantio-enriched spirocyclic tetrahydropyridines using an intramolecular Heck cyclisation has met with limited success. The migration of the double bond accounts for a leakage in asymmetric induction which variation of the reaction parameters could not improve (K. Karabelas, C. Westerlund and A. Hallberg, *J. Org. Chem.*, 1985, **50**, 3896). The use of phosphine-oxazoline ligand (13) improves the situation considerably. In the cyclisation of tetrahydropyridine (152) the two products (153) and (154) are observed with good regioselectivity and excellent enantioselectivity (L. Ripa and A. Hallberg, *J. Org. Chem.*, 1997, **62**, 595).

TfO N CHO (152) → Pd cat./(13), base, Toluene → N CHO (153) + N CHO (154) (6 : 1)

Yield (%)	ee (%) (153)	ee (%) (154)
71	87	99

Asymmetric Heck cyclisations can be incorporated into palladium catalysed cascade processes. In an elegant example the exposure of triflate (155) to typical Heck conditions affords the pentacyclic product (156) with reasonable enantioselectivity using BINAP (58) as ligand (B. A. Keay *et al*, *J. Am. Chem. Soc.*, 1996, **118**, 10766). The product (156) is a precursor in the enantioselective total synthesis of (+)-xestoquinone.

OMe OTf OMe O O (155)

Pd cat./(58)
PMP, toluene

OMe OMe O O (156)

Yield (%)	ee (%)
82	68

The same product has been assembled using the aryl bromide in the presence of a silver salt to form xestoquinone precursor (156) in up to 63% ee. The use of a larger amount of silver exchanged zeolite resulted in a decrease in ee and yield. This is the first example of an effect of the amount of a silver salt on the ee and yield of a product.

Guide to the Index

This index is constructed in a similar manner to the volume indexes of the first edition of the Chemistry of Carbon Compounds. However, to make the index easier to use, more descriptive entries have been made for the commonly occurring individual, and groups of chemicals.

The indexes cover primarily the chemical compounds mentioned in the text, and also include reactions and techniques, where named, and some sources of chemical compounds such as plant and animal species, oils, etc.

Chemical compounds have been indexed alphabetically under the names used by authors, editing being restricted to ensuring uniformity of entries under the same heading. In view of the alternative nomenclature that can often be used, a limited amount of cross-referencing has been done where it is considered to be helpful, but attention is particularly drawn to Convention 2 below.

For this and the succeeding volumes, the indexing conventions listed below have been adopted.

1. Alphabetisation

(a) A letter by letter alphabetical sequence is followed for entries, firstly for the main entry, followed by the descriptive entry.

(b) The following prefixes have not been counted for alphabetising:

n-	*o-*	*as-*	*meso-*	*C-*	*E-*
	m-	*sym-*	*cis-*	*O-*	*Z-*
	p-	*gem-*	*trans-*	*N-*	
	vic-			*S-*	
		lin-		*Bz-*	
				Py-	

Some prefixes and numbering have been omitted in the index, where they do not usefully contribute to the reference.

(c) The following prefixes have been alphabetised:

Allo	Epi	Neo
Anti	Hetero	Nor
Bis	Homo	Pseudo
Cyclo	Iso	

2. Cross references

In view of the many alternative trivial and systematic names for chemical compounds, the indexes should be searched under any alternative names which may be indicated in the main body of the text. Only a limited amount of cross-referencing has been carried out, where it is considered that it would be helpful to the user.

3. Derivatives

Simple derivatives are not normally indexed if they follow in the same short section of the text.

4. Collective and plural entries

In place of "– derivatives" the plural entry has normally been used. Plural entries have occasionally been used where compounds of the same name but differing numbering appear in the same section of the text.

5. Main entries

The main entry of the more common individual compounds is indicated by heavy type. Multiple entries, such as headings and sub-headings over several pages are shown by "–", e.g., 67–74, 137–139, etc.

Index